Fast Fourier Transforms

Second Edition

Studies in Advanced Mathematics

Series Editor

STEVEN G. KRANTZ
Washington University in St. Louis

Editorial Board

R. Michael Beals
Rutgers University

Dennis de Turck
University of Pennsylvania

Ronald DeVore
University of South Carolina

Lawrence C. Evans
University of California at Berkeley

Gerald B. Folland
University of Washington

William Helton
University of California at San Diego

Norberto Salinas
University of Kansas

Michael E. Taylor
University of North Carolina

Titles Included in the Series

Fast Fourier Transforms

Second Edition

James S. Walker
University of Wisconsin
Eau Claire

CRC Press
Boca Raton   New York   London   Tokyo

Acquiring Editor:	Tim Pletscher
Editorial Assistant:	Nora Konopka
Project Editor:	Gail Renard
Marketing Manager:	Susie Carlisle
Direct Marketing Manager:	Becky McEldowney
Cover Designer:	Denise Craig
PrePress:	Kevin Luong
Manufacturing:	Sheri Schwartz

LIMITED WARRANTY

CRC Press warrants the physical diskette(s) enclosed herein to be free of defects in materials and workmanship for a period of thirty days from the date of purchase. If within the warranty period CRC Press receives written notification of defects in materials or workmanship, and such notification is determined by CRC Press to be correct, CRC Press will replace the defective diskette(s).

The entire and exclusive liability and remedy for breach of this Limited Warranty shall be limited to replacement of defective diskette(s) and shall not include or extend to any claim for or right to cover any other damages, including but not limited to, loss of profit, data, or use of the software, or special, incidental, or consequential damages or other similar claims, even if CRC Press has been specifically advised of the possibility of such damages. In no event will the liability of CRC Press for any damages to you or any other person ever exceed the lower suggested list price or actual price paid for the software, regardless of any form of the claim.

CRC Press specifically disclaims all other warranties, express or implied, including but not limited to, any implied warranty of merchantability or fitness for a particular purpose. Specifically, CRC Press makes no representation or warranty that the software is fit for any particular purpose and any implied warranty of merchantability is limited to the thirty-day duration of the Limited Warranty covering the physical diskette(s) only (and not the software) and is otherwise expressly and specifically disclaimed.

Since some states do not allow the exclusion of incidental or consequential damages, or the limitation on how long an implied warranty lasts, some of the above may not apply to you.

DISCLAIMER OF WARRANTY AND LIMITS OF LIABILITY: The author(s) of this book have used their best efforts in preparing this material. These efforts include the development, research, and testing of the theories and programs to determine their effectiveness. Neither the author(s) nor the publisher make warranties of any kind, express or implied, with regard to these programs or the documentation contained in this book, including without limitation warranties of merchantability or fitness for a particular purpose. No liability is accepted in any event for any damages, including incidental or consequential damages, lost profits, costs of lost data or program material, or otherwise in connection with or arising out of the furnishing, performance, or use of the programs in this book.

Library of Congress Cataloging-in-Publication Data

Walker, James S.
    Fast Fourier transforms/James S. Walker, — 2nd ed.
      p. cm. — (Studies in advanced mathematics)
    Includes bibliographical references (p.433–436) and index.
    ISBN 0-8493-7163-5 (alk. paper)
    1. Fourier transformations. I. Title. II. Series.
QA403.W33 1996
515´.723—dc20

                  96–16852
                  CIP

© 1996 by CRC Press, Inc.

No claim to original U.S. Government works
International Standard Book Number 0-8493-7163-5
Library of Congress Card Number 96-16852
Printed in the United States of America 1 2 3 4 5 6 7 8 9 0
Printed on acid-free paper

To my wife,

and my mother,

and the memory of my father.

Acknowledgments

I would like to take this opportunity to thank Steve Krantz for first suggesting this project to me; it has turned out to be a most rewarding endeavor. I would also like to thank Bob Stern at CRC Press for encouraging me to write a second edition. My students have played a major role in helping me to write this book (especially in spurring me on to write better code for the software). If I were to thank all of them by name the list would be too long; let me just say a warm thank you to all of them for their continuing interest.

Finally, I want to express my deep appreciation for my dear wife Ching. She has helped me with the design of *FAS* and shown me how to eliminate some bugs in the program. She also has been a source of inspiration to me. Without her love and support (including her delicious lunch boxes) I would not have been able to complete this work.

About the Author

James S. Walker received his doctorate from the University of Illinois at Chicago in 1982. Since then he has been teaching in the Department of Mathematics at the University of Wisconsin – Eau Claire. In addition to his book *Fast Fourier Transforms,* he is also the author of *Fourier Analysis* (published by Oxford University Press). He has also published papers in the fields of Fourier analysis and complex variables.

Dr. Walker is a member of the American Mathematical Society.

Contents

Preface

This book is an introduction to the *Fast Fourier Transform (FFT)* and some of its applications. Since its creation in the mid-1960s, the field of computerized Fourier analysis, based on the FFT, has been near the heart of a revolution in scientific understanding made possible through the aid of the digital computer. No single book can describe all the features of this new science. I have tried to show how the FFT arises from classical Fourier analysis and describe how the FFT is used to do Fourier analysis.

This book is now in its second edition. As with the first edition, I have tried to write for as wide an audience as I could, given the mathematical requirements needed to appreciate the basic definitions of Fourier analysis. I hope the text will continue to be helpful to students of electrical engineering, optical engineering, physics, physical chemistry, and mathematics. As an additional aid to study, in this second edition I have included solutions for all of the odd-numbered exercises. The mathematical prerequisites for reading the book are a solid understanding of calculus and, for some of the applications, differential equations. To understand the proofs in Sections 4.9 and 5.9 to 5.12 a first course in advanced calculus would be ideal, although a determined reader could gain the necessary knowledge by pursuing a few references (for example, [Kr] or [Ru,2]). I have generally tried to keep the mathematical proofs to a bare minimum. Readers who wish to pursue the pure mathematical theory might begin with the following reference:

James S. Walker, *Fourier Analysis,* Oxford University Press, Oxford, 1988.

This text is referred to as [Wa] in this book. Since I wrote [Wa] for the purpose of explaining the mathematical theory of classical Fourier analysis, I saw no need to reproduce its pure mathematical arguments in this book.

What makes this book unique among the vast literature on the FFT is that it comes with computer software (*FAS*) for doing Fourier analysis on any PC (using MS–DOS version 5.0 and up). Using *FAS,* readers will be able to generate immediately on their computer screens images of Fourier series, sine and cosine series, Fourier transforms, convolutions, and all the other aspects of Fourier analysis described in the text. This second edition includes a new version of this software. This new version provides extensive online help, mouse support, and vastly improved function compiling and screen printing. The online help for creating functions should be particularly helpful for students, since *it contains every formula used in the text* (except the exercises), *and these formulas can be entered into FAS by simply clicking on them with the mouse.*

Here is a summary of the main topics covered. In Chapter 1, I discuss the principal features of Fourier series. The emphasis here is on using *FAS* to gain a firm grasp of the main aspects of Fourier series, as well as Fourier sine and cosine series. In Chapter 2, the digital (discrete) version of Fourier analysis is introduced, showing how the Fourier series can be put into a discrete form intended for computer computation. An efficient means for carrying out this computer computation is called a *Fast Fourier Transform* or FFT. Chapter 3 is an introduction to one type of FFT. Although there are a great many FFT algorithms, I discuss just one. Any readers who are familiar with Bracewell's book on the Hartley transform, [Br], will recognize the debt I owe to him and his former colleague at Stanford, the late Oscar Buneman. The work done by Buneman significantly improved the original FFT algorithms. Besides the basic FFT algorithm, I also describe the standard method for computing a real FFT and for computing discrete sine and discrete cosine transforms efficiently.

Since the purpose of the FFT is to do Fourier analysis, Chapter 4 consists of applications of Fourier series. The classic Fourier series solutions to the heat and wave equations are described with the principal emphasis being on using *FAS* to study the time evolution of these solutions. Using *FAS* we can create *computer animations* of vibrating strings and of the changing temperature of a thin wire. These animations enliven the purely mathematical solutions of the wave and heat equations. Another area where the computer allows one to do much more than the typical textbook discussion is the famous *particle in a box problem,* in one dimension, from quantum mechanics. Section 4.3 shows how *FAS* allows one to study the time evolution of a potential-free quantum mechanical particle. This problem, which has important applications to electron diffraction, is too often treated in a sterile fashion with too much emphasis on stationary states. Many students are left with the impression that the stationary states are the *only* states. The rest of Chapter 4 describes aspects of Fourier series that are important in signal processing, in particular, emphasizing the concept of *filtering* of Fourier series. *FAS* makes it possible for students to see with their own eyes why hanning and Hamming filters, as well as other types of filters, are used. The subtle notion of *point spread functions (kernels)* should be easier to learn with the aid of *FAS,* too.

In Chapter 5, I take up the other main area of Fourier analysis, *Fourier transforms*. Here, *FAS* is used to aid in understanding the fundamentals of Fourier transforms, Fourier inversion, and convolution. In particular, the difficult concept of convolution integrals should be more accessible with the help of *FAS*. A student will be able to see very quickly what the convolution of two functions looks like. The examples I have treated (heat equation, Laplace's equation, potential-free Schrödinger equation) will show the practical importance of convolutions. I have made a point of showing that *FAS* can be used to implement the classic convolution solutions to these problems in a practical way. The second half of Chapter 5 (Sections 5.9 to 5.13) covers some very important, but sometimes difficult, material on Poisson summation and sampling theory, stressing the close connection between these two topics. It would be hard to overestimate the importance of sampling theory in modern electrical engineering and communication systems design. *FAS* should help make some of the principles of sampling theory easier to understand.

The book concludes with a chapter on Fourier optics. I think that it is important to describe in as much detail as possible one major area of application of computerized Fourier analysis. In Chapter 6, the topics of Fresnel diffraction, diffraction from circular apertures, interference, diffraction gratings, Fourier transforming properties of a lens, and imaging with a single lens, are all discussed. Fourier optics is an essential element of modern optics. It plays a vital role in understanding diffraction and imaging. And these two areas, through their roles in crystallography, chemical analysis, and optical and electron microscopy, underlie a great deal of what we have learned about the physical world during the last century.

James S. Walker
Eau Claire, Wisconsin
February 1996

1

Basic Aspects of Fourier Series

This chapter is a summary of the basic theory of Fourier series. Some of the deeper theorems are only quoted; their proofs can be found in [Wa], listed in this book's bibliography. Besides their importance in applications, Fourier series provide a foundation for understanding the FFT.

1.1 Definition of Fourier Series

To understand the definition of Fourier series we will begin with the essential idea: representing a wave form in terms of frequency as opposed to time. We will denote time by x rather than t.

Suppose our wave form is described by $4 \cos 2\pi \nu x$, which has frequency ν. Using *Euler's identity*

$$e^{i\phi} = \cos \phi + i \sin \phi \tag{1.1}$$

we can write $4 \cos 2\pi \nu x$ in complex exponential form

$$4 \cos 2\pi \nu x = 2e^{i2\pi \nu x} + 2e^{-i2\pi \nu x}$$

where the complex exponentials have amplitudes of 2 and frequencies of ν and $-\nu$. See Figure 1.1. (To see how Figure 1.1(b) was produced, consult Exercise 1.3.)

Or, suppose our wave form is described by $6 \sin 2\pi \nu x$, which also has frequency ν. Using Euler's identity (1.1) again, we obtain

$$6 \sin 2\pi \nu x = 3i e^{-i2\pi \nu x} - 3i e^{i2\pi \nu x}$$

where the complex exponentials have complex amplitudes of $3i$ and $-3i$ and frequencies of $-\nu$ and ν. See Figure 1.2. (To see how Figure 1.2(b) was produced, consult Exercise 1.4.)

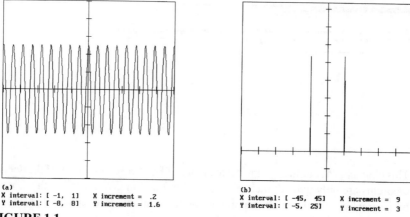

(a)
X interval: [-1, 1]   X increment = .2
Y interval: [-8, 8]   Y increment = 1.6

(b)
X interval: [-45, 45]   X increment = 9
Y interval: [-5, 25]   Y increment = 3

FIGURE 1.1
Frequency representation of $4\cos(2\pi\nu x)$, $\nu = 9$. **(a) Graph in x domain (time or space). (b) Graph in frequency domain.**

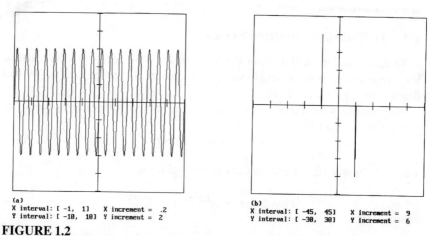

(a)
X interval: [-1, 1]   X increment = .2
Y interval: [-10, 10]   Y increment = 2

(b)
X interval: [-45, 45]   X increment = 9
Y interval: [-30, 30]   Y increment = 6

FIGURE 1.2
Frequency representation of $6\sin(2\pi\nu x)$, $\nu = 9$. **(a) Graph in x domain (time or space). (b) Graph in frequency domain.**

These two examples show how the waves $\cos 2\pi\nu x$ and $\sin 2\pi\nu x$ can be expressed in frequency terms *and distinguished from each other* using complex exponentials. Also, Figures 1.1(b) and 1.2(b) show how, in a certain sense, the frequency representation of these waves is simpler.

The basic idea in Fourier series is to express a periodic wave as a sum of complex exponentials all of which have the same period. This is made feasible by the following property of complex exponentials having the same period.

THEOREM 1.1: ORTHOGONALITY OF COMPLEX EXPONENTIALS

*For all integers m and n, the complex exponentials $\{e^{i2\pi nx/P}\}$ with period P satisfy the following **orthogonality relation***

$$\frac{1}{P} \int_0^P e^{i2\pi mx/P} e^{-i2\pi nx/P} \, dx = \begin{cases} 0 & \text{for } m \neq n \\ 1 & \text{for } m = n. \end{cases}$$

PROOF   We will prove the theorem for the case of $P = 2\pi$. For this case, we have

$$\frac{1}{2\pi} \int_0^{2\pi} e^{imx} e^{-inx} \, dx = \frac{1}{2\pi} \int_0^{2\pi} e^{i(m-n)x} \, dx$$

$$= \frac{1}{2\pi} \int_0^{2\pi} \cos(m-n)x \, dx$$

$$+ \frac{i}{2\pi} \int_0^{2\pi} \sin(m-n)x \, dx. \qquad (1.2)$$

If $m \neq n$, then by calculus we get

$$\frac{1}{2\pi} \int_0^{2\pi} \cos(m-n)x \, dx = \frac{\sin(m-n)x}{2\pi(m-n)} \Bigg|_0^{2\pi} = 0$$

and similarly

$$\frac{i}{2\pi} \int_0^{2\pi} \sin(m-n)x \, dx = 0.$$

That takes care of the case when $m \neq n$. If $m = n$, then $\sin(m-n)x = \sin 0 = 0$ and $\cos(m-n)x = \cos 0 = 1$. Hence (1.2) becomes

$$\frac{1}{2\pi} \int_0^{2\pi} e^{imx} e^{-inx} \, dx = \frac{1}{2\pi} \int_0^{2\pi} 1 \, dx = 1.$$

This proves the theorem.   ∎

Now, suppose g is a periodic function, period P, and g is expanded in a series of complex exponentials having the same period

$$g(x) = \sum_{n=-\infty}^{\infty} c_n e^{i2\pi nx/P}. \tag{1.3}$$

We will show that a reasonable formula for the general coefficient c_n can be obtained using the orthogonality theorem just proved. If we multiply both sides of (1.3) by $(1/P)e^{-i2\pi nx/P}$ (being careful to change the name of the index in the series) and integrate term by term, then we get

$$\frac{1}{P}\int_0^P g(x)e^{-i2\pi nx/P}\,dx = \sum_{m=-\infty}^{\infty} c_m \frac{1}{P}\int_0^P e^{i2\pi mx/P} e^{-i2\pi nx/P}\,dx$$

$$= c_n$$

because of Theorem 1.1. Based on this result, we make the following definition.

DEFINITION 1.1: *If the function g has period P, then the **Fourier coefficients** $\{c_n\}$ for g are defined by*

$$c_n = \frac{1}{P}\int_0^P g(x)e^{-i2\pi nx/P}\,dx$$

*for all integers n. Using these coefficients $\{c_n\}$, the **Fourier series** for g is defined by the right side of the following correspondence*

$$g \sim \sum_{n=-\infty}^{\infty} c_n e^{i2\pi nx/P}.$$

REMARK 1.1:   (a) We have used the correspondence symbol $\sim$ since we have not yet discussed the conditions of validity of Equation (1.3); we will do that in Section 1.4. (b) Note that when g is a real valued function, then $c_{-n} = c_n^*$ where c_n^* is the *complex conjugate* of c_n. (c) The Fourier coefficients of g can also be found from the formula

$$c_n = \frac{1}{P}\int_{-\frac{1}{2}P}^{\frac{1}{2}P} g(x)e^{-i2\pi nx/P}\,dx.$$

■

1.2 Examples of Fourier Series

In this section we will cover some basic examples of Fourier series expansions.

Example 1.1:
Expand the function $g(x) = x$ in a Fourier series, period 2π, using the interval $[-\pi, \pi]$.

SOLUTION For $n = 0$, we have

$$c_0 = \frac{1}{2\pi} \int_{-\pi}^{\pi} x \, dx = 0.$$

For $n \neq 0$, we split the integral for c_n into real and imaginary terms and then integrate by parts, obtaining

$$c_n = \frac{1}{2\pi} \int_{-\pi}^{\pi} x e^{-inx} \, dx$$

$$= \frac{1}{2\pi} \int_{-\pi}^{\pi} x \cos nx \, dx - \frac{i}{2\pi} \int_{-\pi}^{\pi} x \sin nx \, dx = \frac{i(-1)^n}{n}.$$

Hence,

$$g \sim \sum_{n=-\infty}^{\infty}{}' \frac{i(-1)^n}{n} e^{inx} \tag{1.4}$$

where the prime on the sum indicates that the $n = 0$ term is omitted.

If we group the terms for n and $-n$, then we can rewrite Formula (1.4) in the *real form*

$$g \sim \sum_{n=1}^{\infty} \frac{2(-1)^{n+1}}{n} \sin nx.$$

In Figure 1.3, we show the graphs of the *partial sums*

$$S_M(x) = \sum_{n=1}^{M} \frac{2(-1)^{n+1}}{n} \sin nx$$

for $M = 1, 2, 4$, and 8. The number M denotes the *number of harmonics* in the partial sum.

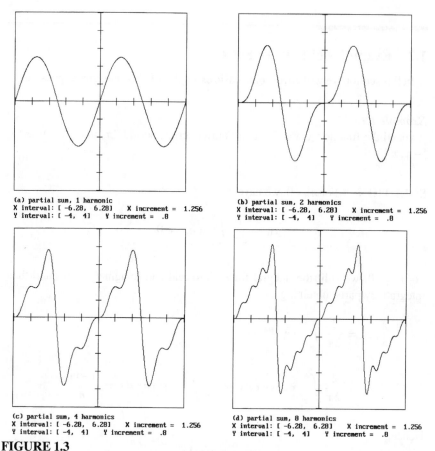

(a) partial sum, 1 harmonic
X interval: [-6.28, 6.28]   X increment = 1.256
Y interval: [-4, 4]   Y increment = .8

(b) partial sum, 2 harmonics
X interval: [-6.28, 6.28]   X increment = 1.256
Y interval: [-4, 4]   Y increment = .8

(c) partial sum, 4 harmonics
X interval: [-6.28, 6.28]   X increment = 1.256
Y interval: [-4, 4]   Y increment = .8

(d) partial sum, 8 harmonics
X interval: [-6.28, 6.28]   X increment = 1.256
Y interval: [-4, 4]   Y increment = .8

FIGURE 1.3
Graphs of some partial sums of the Fourier series for $g(x) = x$, period 2π, using the interval $[-\pi, \pi]$ for computing Fourier series coefficients.

Notice that in Figure 1.3(d) the graph of S_8, on the interval $[-\pi, \pi]$ that we used to calculate the Fourier series, is a wavy (or wiggly) approximation to the original function $g(x) = x$. Outside the original interval, the graph of S_8 is approximating the *periodic extension* of $g(x) = x$. See Figure 1.4.   ∎

Example 1.2:

Expand the function $g(x) = x^2 - 2x$ in a Fourier series, period 2, using the interval $[0, 2]$.

SOLUTION   For c_0 we obtain $c_0 = (1/2) \int_0^2 x^2 - 2x \, dx = -2/3$. For $n \neq 0$,

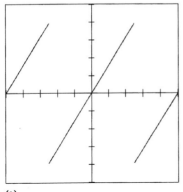

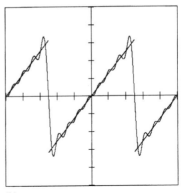

(a)
X interval: [-6.28, 6.28]   X increment = 1.256
Y interval: [-4, 4]   Y increment = .8

(b)
X interval: [-6.28, 6.28]   X increment = 1.256
Y interval: [-5, 5]   Y increment = 1

FIGURE 1.4
(a) The periodic extension, period 2π, of the function $g(x) = x$. (b) The graph of the partial sum S_8 together with the graph from (a).

we obtain

$$c_n = \frac{1}{2} \int_0^2 (x^2 - 2x) e^{-i\pi nx} \, dx$$

$$= \frac{1}{2} \int_0^2 (x^2 - 2x) \cos n\pi x \, dx - \frac{i}{2} \int_0^2 (x^2 - 2x) \sin n\pi x \, dx = \frac{2}{n^2 \pi^2}$$

after integrating by parts twice. Thus,

$$g \sim -\frac{2}{3} + \sideset{}{'}\sum_{n=-\infty}^{\infty} \frac{2}{n^2 \pi^2} e^{i n\pi x}.$$

Grouping terms for n and $-n$, we obtain the real form of the Fourier series for g

$$g \sim -\frac{2}{3} + \sum_{n=1}^{\infty} \frac{4}{n^2 \pi^2} \cos n\pi x.$$

In Figure 1.5, we have graphed the partial sums

$$S_M(x) = -\frac{2}{3} + \sum_{n=1}^{M} \frac{4}{n^2 \pi^2} \cos n\pi x$$

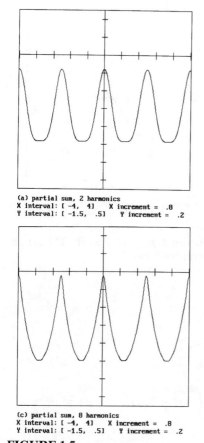

(a) partial sum, 2 harmonics
X interval: [-4, 4]   X increment = .8
Y interval: [-1.5, .5]   Y increment = .2

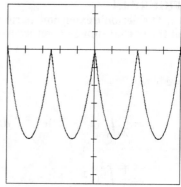

(b) partial sum, 4 harmonics
X interval: [-4, 4]   X increment = .8
Y interval: [-1.5, .5]   Y increment = .2

(c) partial sum, 8 harmonics
X interval: [-4, 4]   X increment = .8
Y interval: [-1.5, .5]   Y increment = .2

(d) periodic extension and 8 harmonic sum
X interval: [-4, 4]   X increment = .8
Y interval: [-1.5, .5]   Y increment = .2

FIGURE 1.5
Graphs illustrating Example 1.2.

for $M = 2, 4$, and 8 harmonics, and the graph of the periodic extension of g with the graph of S_8 superimposed on it. ∎

Example 1.3:
Expand the function $g(x) = x^2 + x$ in a Fourier series, period 1, over the interval $[-1/2, 1/2]$.

SOLUTION   For c_0 we obtain $c_0 = \int_{-1/2}^{1/2} x^2 + x \, dx = 1/12$. For $n \neq 0$, we obtain

$$c_n = \int_{-\frac{1}{2}}^{\frac{1}{2}} (x^2 + x) \cos 2\pi nx \, dx - i \int_{-\frac{1}{2}}^{\frac{1}{2}} (x^2 + x) \sin 2\pi nx \, dx$$

$$= \frac{(-1)^n}{2} \left[\frac{1}{n^2\pi^2} + \frac{i}{n\pi} \right]$$

after integrating by parts twice. Thus,

$$g \sim \frac{1}{12} + \sum_{n=-\infty}^{\infty} {}' \frac{(-1)^n}{2} \left[\frac{1}{n^2\pi^2} + \frac{i}{n\pi} \right] e^{i2\pi nx}.$$

Grouping terms for n and $-n$, we obtain the real form of the Fourier series for g

$$g \sim \frac{1}{12} + \sum_{n=1}^{\infty} (-1)^n \left[\frac{\cos 2\pi nx}{n^2\pi^2} - \frac{\sin 2\pi nx}{n\pi} \right].$$

In Figure 1.6, we show the graphs of the partial sums

$$S_M(x) = \frac{1}{12} + \sum_{n=1}^{M} (-1)^n \left[\frac{\cos 2\pi nx}{n^2\pi^2} - \frac{\sin 2\pi nx}{n\pi} \right]$$

for $M = 3, 12$, and 36 harmonics, and the graph of the periodic extension of $x^2 + x$ superimposed on S_{36}. ∎

1.3 Fourier Series of Real Functions

In each of the examples in the previous section we computed Fourier series for real valued functions. We were able in each case to express those Fourier series in real forms, series involving real valued sines and cosines. We will now show that this is possible in general.

Suppose that g is a *real valued* function. Then the Fourier coefficients of g over the interval $[0, P]$ are given by

$$c_n = \frac{1}{P} \int_0^P g(x) e^{-i2\pi nx/P} \, dx. \tag{1.5}$$

If we take complex conjugates of both sides of (1.5), then by keeping in mind that $g(x)$ and x are real we can write

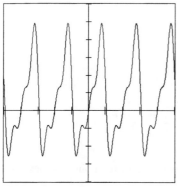

(a) partial sum, 3 harmonics
X interval: [-2.5, 2.5]   X increment = .5
Y interval: [-.5, .75]   Y increment = .125

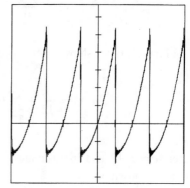

(b) partial sum, 12 harmonics
X interval: [-2.5, 2.5]   X increment = .5
Y interval: [-.5, 1]   Y increment = .15

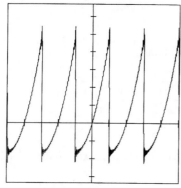

(c) partial sum, 36 harmonics
X interval: [-2.5, 2.5]   X increment = .5
Y interval: [-.5, 1]   Y increment = .15

(d) periodic extension and 36 harmonic sum
X interval: [-2.5, 2.5]   X increment = .5
Y interval: [-.5, 1]   Y increment = .15

FIGURE 1.6
Graphs illustrating Example 1.3.

$$c_n^* = \frac{1}{P} \left[\int_0^P g(x) e^{-i2\pi nx/P} \, dx \right]^*$$

$$= \frac{1}{P} \int_0^P g(x)^* \left[e^{-i2\pi nx/P} \right]^* \, dx$$

$$= \frac{1}{P} \int_0^P g(x) e^{i2\pi nx/P} \, dx = c_{-n} \, .$$

Thus, whenever g is real valued, we have

$$c_n^* = c_{-n} \, . \tag{1.6}$$

Using (1.6) we can cast the Fourier series for g into a real form by grouping the nth and $-n^{th}$ terms as follows

$$g \sim \sum_{n=-\infty}^{\infty} c_n e^{i2\pi nx/P} = c_0 + \sum_{n=1}^{\infty} \left\{ c_n e^{i2\pi nx/P} + c_{-n} e^{-i2\pi nx/P} \right\}$$

$$= c_0 + \sum_{n=1}^{\infty} \left\{ c_n e^{i2\pi nx/P} + (c_n e^{i2\pi nx/P})^* \right\}$$

$$= c_0 + \sum_{n=1}^{\infty} 2\mathrm{Re} \left\{ c_n e^{i2\pi nx/P} \right\}$$

where Re means taking the real part of a complex number.

Thus, we have

$$g \sim c_0 + \sum_{n=1}^{\infty} 2\mathrm{Re} \left\{ c_n e^{i2\pi nx/P} \right\}. \tag{1.7}$$

Now, we write c_n more explicitly as a complex number

$$c_n = \frac{1}{P} \int_0^P g(x) e^{-i2\pi nx/P} \, dx$$

$$= \frac{1}{P} \int_0^P g(x) \cos \frac{2\pi nx}{P} \, dx - \frac{i}{P} \int_0^P g(x) \sin \frac{2\pi nx}{P} \, dx$$

$$= \frac{1}{2} A_n - \frac{i}{2} B_n$$

where we *define* A_n and B_n by

$$A_n = \frac{2}{P} \int_0^P g(x) \cos \frac{2\pi nx}{P} \, dx \,, \qquad B_n = \frac{2}{P} \int_0^P g(x) \sin \frac{2\pi nx}{P} \, dx. \tag{1.8}$$

Using $c_n = (1/2)A_n - (i/2)B_n$ and $e^{i2\pi nx/P} = \cos(2\pi nx/P) + i \sin(2\pi nx/P)$ in (1.7) we obtain

$$g \sim c_0 + \sum_{n=1}^{\infty} \left\{ A_n \cos \frac{2\pi nx}{P} + B_n \sin \frac{2\pi nx}{P} \right\}.$$

For notational uniformity, we define $(1/2)A_0$ to be equal to c_0. We then have the following formulas:

$$g \sim \frac{1}{2}A_0 + \sum_{n=1}^{\infty} \left\{ A_n \cos \frac{2\pi nx}{P} + B_n \sin \frac{2\pi nx}{P} \right\}$$

$$A_n = \frac{2}{P} \int_0^P g(x) \cos \frac{2\pi nx}{P} \, dx, \qquad (n = 0, 1, 2, \ldots)$$

$$B_n = \frac{2}{P} \int_0^P g(x) \sin \frac{2\pi nx}{P} \, dx, \qquad (n = 1, 2, 3, \ldots). \qquad (1.9)$$

The formulas in (1.9) define the real form of the Fourier series for g, using the period interval $[0, P]$. Similar formulas hold if the interval is $[-P/2, P/2]$. In particular, the coefficients A_n and B_n would then be given by

$$A_n = \frac{2}{P} \int_{-\frac{1}{2}P}^{\frac{1}{2}P} g(x) \cos \frac{2\pi nx}{P} \, dx, \qquad (n = 0, 1, 2, \ldots)$$

$$B_n = \frac{2}{P} \int_{-\frac{1}{2}P}^{\frac{1}{2}P} g(x) \sin \frac{2\pi nx}{P} \, dx, \qquad (n = 1, 2, 3, \ldots). \qquad (1.10)$$

If we define the partial sum S_M having M harmonics by

$$S_M(x) = \sum_{n=-M}^{M} c_n e^{i2\pi nx/P} \qquad (1.11)$$

then, by the same reasoning as above, it follows that S_M has a real form given by

$$S_M(x) = \frac{1}{2}A_0 + \sum_{n=1}^{M} \left\{ A_n \cos \frac{2\pi nx}{P} + B_n \sin \frac{2\pi nx}{P} \right\}. \qquad (1.12)$$

REMARK 1.2:   The convention of using $1/2$ as a factor on A_0 in the constant term in the real form of Fourier series is troublesome. But, it is a common convention since it simplifies the form of (1.9). It is also the convention adopted in [Wa], to which we make frequent reference.   ∎

1.4 Pointwise Convergence of Fourier Series

In this section we will examine the validity of equating a periodic function with its Fourier series. Our approach to this problem will be to examine the limit of partial sums of the Fourier series at any given point.

To be more specific, we examine under what conditions we can have

$$\sum_{n=-\infty}^{\infty} c_n e^{i2\pi nx/P} = g(x) \tag{1.13}$$

where the left side of (1.13) is the Fourier series, period P, for g. A formal definition of (1.13) is

$$\lim_{M \to \infty} S_M(x) = g(x)$$

where S_M is the partial sum, containing M harmonics, of the Fourier series for g [see Formula (1.11)].

The types of functions that we will be interested in are mostly covered by the following definition.

DEFINITION 1.2: *A function g on a finite interval* $[a, b]$ *is **piecewise continuous on*** $[a, b]$ *if the interval* $[a, b]$ *can be divided into a finite number of subintervals on each of which g is continuous. If g is piecewise continuous on every finite interval, then g is called **piecewise continuous**.*

Example 1.4:
(a) The function

$$g(x) = \begin{cases} x+1 & \text{for } -1 \le x \le 0 \\ 0 & \text{for } 0 < x \le 2 \end{cases}$$

is piecewise continuous on the interval $[-1, 2]$. (b) The function g defined by $g(x) = x$ for $-\pi \le x < \pi$ and having period 2π is piecewise continuous. (c) The function $g(x) = x^2$ is piecewise continuous. In fact, every continuous function is piecewise continuous.

For a piecewise continuous function, the following limits always make sense and are finite:

$$g(x+) = \lim_{h \to 0+} g(x+h), \qquad g(x-) = \lim_{h \to 0-} g(x+h).$$

The first limit above is called the *right-hand limit of g at x*, since $x+h$ always lies to the right of x when $h > 0$. The second limit above is called the *left-hand limit of g at x*, since $x + h$ always lies to the left of x when $h < 0$. We can also define some useful generalizations of derivative. Whenever the following limit exists and is finite,

$$\lim_{h \to 0+} \frac{g(x + h) - g(x+)}{h}$$

then g is said to have a *right-hand derivative at x*, denoted by $g'(x+)$ (which equals the value of this limit). Similarly, whenever the limit

$$\lim_{h \to 0-} \frac{g(x + h) - g(x-)}{h}$$

exists and is finite, then g is said to have a *left-hand derivative at x*, denoted by $g'(x-)$ (which equals the value of this limit). Of course, if the left-hand and right-hand derivatives are equal and g is continuous at x, then g has a derivative, $g'(x) = g'(x+) = g'(x-)$, at x.

We can now state a pointwise convergence theorem for Fourier series. This theorem covers many of the functions dealt with in an analytic setting.

THEOREM 1.2:

Let g be a piecewise continuous function, period P. At each point x where g has a right- and left-hand derivative, the Fourier series for g converges to $[g(x+) + g(x-)]/2$. Thus, we can write

$$\sum_{n=-\infty}^{\infty} c_n e^{i2\pi nx/P} = \frac{1}{2}[g(x+) + g(x-)].$$

If x is also a point of continuity for g, then this result simplifies to

$$\sum_{n=-\infty}^{\infty} c_n e^{i2\pi nx/P} = g(x).$$

PROOF   For a proof of this theorem, see [Wa, Chapter 2.3].   ∎

Suppose we are given a function g on the interval $[0, P]$ and we compute its Fourier series, period P. To apply Theorem 1.2 we look at the *periodic extension* $g\mathbf{p}$ of our function g defined by

$$g\mathbf{p}(x + kP) = g(x), \qquad (k = 0, \pm 1, \pm 2, \ldots).$$

Since $g_\mathbf{p} = g$ on $[0, P]$, $g_\mathbf{p}$ has the same Fourier series as g and we may apply Theorem 1.2 to the periodic extension $g_\mathbf{p}$. Similar remarks apply if we are given a function on $[-P/2, P/2]$.

Example 1.5:
Let g be defined by

$$g(x) = \begin{cases} 1 & \text{for } 0 < x < 1 \\ 0 & \text{for } -1 < x < 0. \end{cases}$$

Graphs of g and its partial sums S_M for $M = 12, 36$, and 108 harmonics are shown in Figure 1.7. Fixing $x \neq 0, \pm 1$, we see $S_M(x)$ converging to $g(x)$ in these graphs. For $x = 0$, the graphs show $S_M(0)$ converging to $[g(0+) + g(0-)]/2 = 1/2$. While, for $x = \pm 1$, the graphs show $S_M(\pm 1)$ converging to half the sum of the right- and left-hand limits of the periodic extension of g at $x = \pm 1$, which is $1/2$ in both cases.

Example 1.6:
Let $g(x) = |x - 2| + 1$ on the interval $[0, 4]$. Graphs of the function g and S_M for $M = 12, 36$, and 108 harmonics are shown in Figure 1.8. The tendency of S_M to converge to g at all points in the interval $[0, 4]$ is clearly shown in these graphs.

1.5 Further Aspects of Convergence of Fourier Series

In this section we will discuss the following aspects of convergence of Fourier series: Gibbs' phenomenon, uniform convergence, and a more profound convergence theorem which applies to a wider class of functions that are encountered in measuring signals.

We begin by mentioning Gibbs' phenomenon. Consider Example 1.5 again. In Figure 1.7 notice the sharp peaks in S_{36} and S_{108} near 0, the discontinuity point of g. The height of these peaks is greater than $g(0+)$ by about 9%. Figure 1.9 shows that this phenomenon, known as *Gibbs' phenomenon,* does not go away as the number of harmonics is increased. Further treatment of Gibbs' phenomenon will be given in Chapter 4. See also [Wa, Chapter 2.5].

Some functions do not exhibit Gibbs' phenomenon in their Fourier series. For example, look again at Figure 1.8. That Fourier series exhibits a property known as *uniform convergence.* Uniform convergence occurs when the partial sums tend uniformly over some interval (or over the whole real line) to a function g. To be more precise, we have the following theorem. (The notation *sup* in this theorem means the same thing as *maximum.*)

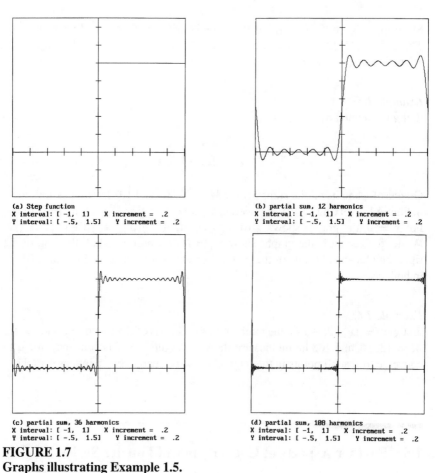

(a) Step function
X interval: [-1, 1]   X increment = .2
Y interval: [-.5, 1.5]   Y increment = .2

(b) partial sum, 12 harmonics
X interval: [-1, 1]   X increment = .2
Y interval: [-.5, 1.5]   Y increment = .2

(c) partial sum, 36 harmonics
X interval: [-1, 1]   X increment = .2
Y interval: [-.5, 1.5]   Y increment = .2

(d) partial sum, 108 harmonics
X interval: [-1, 1]   X increment = .2
Y interval: [-.5, 1.5]   Y increment = .2

FIGURE 1.7
Graphs illustrating Example 1.5.

THEOREM 1.3:

If g is a continuous function with period P and its derivative g′ is piecewise continuous, then the partial sums of the Fourier series for g converge uniformly to g over the whole real line. In particular, we have

$$\lim_{M \to \infty} \left\{ \sup_{x \in \mathbf{R}} |g(x) - S_M(x)| \right\} = 0.$$

REMARK 1.3:   As we mentioned above, the notation *sup* in this theorem can be viewed as just shorthand for *maximum*. Hence, the theorem says that as $M \to \infty$ the maximum difference between g and S_M tends to 0. ∎

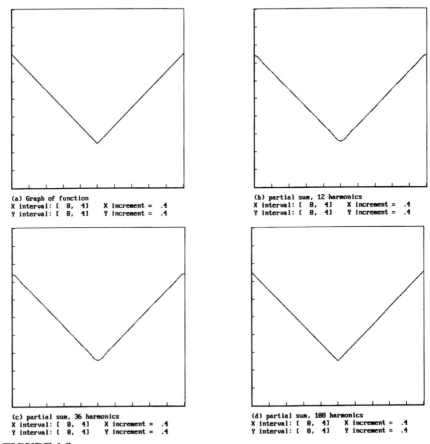

(a) Graph of function
X interval: [0, 4]   X increment = .4
Y interval: [0, 4]   Y increment = .4

(b) partial sum, 12 harmonics
X interval: [0, 4]   X increment = .4
Y interval: [0, 4]   Y increment = .4

(c) partial sum, 36 harmonics
X interval: [0, 4]   X increment = .4
Y interval: [0, 4]   Y increment = .4

(d) partial sum, 108 harmonics
X interval: [0, 4]   X increment = .4
Y interval: [0, 4]   Y increment = .4

FIGURE 1.8
Graphs illustrating Example 1.6.

PROOF OF THEOREM 1.3.   See [Wa, Chapter 2.4].   ∎

Theorem 1.3 applies to the periodic extension of the function g in Example 1.6. We also saw in Figure 1.8 the excellent convergence properties of the sequence of partial sums $\{S_M\}$ to g. It is interesting to note, however, that g is being thought of here as a limit of the differentiable functions S_M (in fact, S_M can be differentiated as many times as you please). Yet, g itself is not differentiable. In fact, $g'(2)$ does not exist. Figure 1.10 shows the Fourier series partial sums S_M near the point $(2, 1)$ and how they compare to g. These graphs show the failure of the Fourier series partial sums to reproduce the sharp corner in the graph of g at the point $(2, 1)$, a failure that is only eliminated in the limit as $M \to \infty$.

In an analytic setting we sometimes have formulas for the Fourier coefficients of a specific function, such as the examples given in Section 1.2, for instance. When

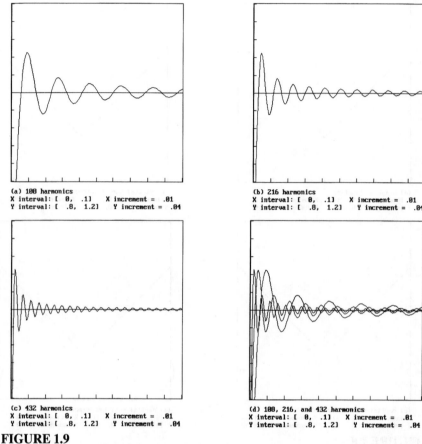

(a) 100 harmonics
X interval: [0,   .1]   X increment = .01
Y interval: [.8, 1.2]   Y increment = .04

(b) 216 harmonics
X interval: [0,   .1]   X increment = .01
Y interval: [.8, 1.2]   Y increment = .04

(c) 432 harmonics
X interval: [0,   .1]   X increment = .01
Y interval: [.8, 1.2]   Y increment = .04

(d) 100, 216, and 432 harmonics
X interval: [0,   .1]   X increment = .01
Y interval: [.8, 1.2]   Y increment = .04

FIGURE 1.9
Graphs illustrating Gibbs' phenomenon.

this happens there is a very useful theorem which allows us to obtain estimates on the closeness of S_M to our function.

THEOREM 1.4:

If $\sum_{n=-\infty}^{\infty} |c_n|$ converges, then the Fourier series $\sum_{n=-\infty}^{\infty} c_n e^{i2\pi nx/P}$ converges uniformly to a continuous function g with period P.

PROOF   See [Wa, Chapter 1.5]. ∎

Here is an interesting result that is related to Theorem 1.4. Consider the real

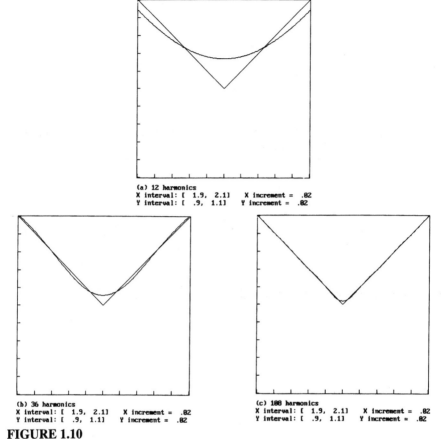

(a) 12 harmonics
X interval: [1.9, 2.1]   X increment = .02
Y interval: [.9, 1.1]   Y increment = .02

(b) 36 harmonics
X interval: [1.9, 2.1]   X increment = .02
Y interval: [.9, 1.1]   Y increment = .02

(c) 108 harmonics
X interval: [1.9, 2.1]   X increment = .02
Y interval: [.9, 1.1]   Y increment = .02

FIGURE 1.10
Graphs of $g(x) = |x - 2| + 1$ **and** S_M **for** $M = 12, 36,$ **and** 108, **near the point** $(2, 1)$.

Fourier series

$$1 + \sum_{n=1}^{\infty} \frac{\cos 2^n x}{2^n}.$$

Since

$$\frac{|\cos 2^n x|}{2^n} \leq \frac{1}{2^n}$$

the series converges to a function g for all x-values (by the Comparison Test). And

$$|g(x) - S_{2^N}(x)| = \left| \sum_{n > N}^{\infty} \frac{\cos 2^n x}{2^n} \right|$$

$$\leq \sum_{n > N}^{\infty} \frac{|\cos 2^n x|}{2^n} .$$

The inequality holds because of possible cancellation of positive and negative values. Since $|\cos 2^n x| \leq 1$ and $\sum_{n > N}^{\infty} 1/2^n = 1/2^N$ we get $|g(x) - S_{2^N}(x)| \leq 1/2^N$. Hence,

$$\sup_{x \in \mathbf{R}} |g(x) - S_{2^N}(x)| \leq \frac{1}{2^N} .$$

For example, if $N = 9$, then g and $S_{512}(x) = 1 + \sum_{n=1}^{9} (\cos 2^n x)/2^n$ differ by no more than $\pm 1/512$ over all of $\mathbf{R}$. In Figure 1.11, we have shown the graph of S_{512}. The function g, to which S_M converges as $M \to \infty$, is called the *Weierstrass nowhere differentiable function.* It is a continuous function for all x-values, but it has no derivative at any x-value.

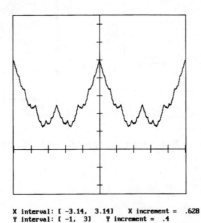

X interval: [-3.14, 3.14]   X increment =   .628
Y interval: [-1, 3]   Y increment =   .4

FIGURE 1.11
Ten-term approximation of Weierstrass nowhere differentiable function.

We close this section by describing a theorem that is broad enough to cover most data measured from periodic signals.

DEFINITION 1.3:   *A function g is said to be **Lipschitz from the right at** x if for some positive constants A, α, and δ, we have*

$$|g(x + h) - g(x+)| \leq A h^{\alpha} , \qquad \text{for } 0 < h < \delta.$$

*Similarly, g is said to be **Lipschitz from the left at** x if for some positive constants B, β, and ε, we have*

$$|g(x - h) - g(x-)| \leq Bh^{\beta}, \qquad \text{for } 0 < h < \epsilon.$$

The following theorem covers most of the function data (signals) that arise from measurements.

THEOREM 1.5:
Let g be a piecewise continuous function with period P. At each point x where g is Lipschitz from the left and right,

$$\sum_{n=-\infty}^{\infty} c_n e^{i2\pi nx/P} = \frac{1}{2}[g(x+) + g(x-)].$$

If x is also a point of continuity for g, then this result simplifies to

$$\sum_{n=-\infty}^{\infty} c_n e^{i2\pi nx/P} = g(x).$$

PROOF The techniques needed to prove this theorem are discussed in [Wa, Chapter 2.3]. ∎

1.6 Fourier Sine Series and Cosine Series

Fourier sine series and cosine series are special forms of Fourier series for functions possessing either odd or even symmetry.

DEFINITION 1.4: *A function g is called **odd** on the interval $[-L, L]$ if $g(-x) = -g(x)$ for each x-value. A function g is called **even** on the interval $[-L, L]$ if $g(-x) = g(x)$ for each x-value.*

It is easy to see that the product of two odd functions, or the product of two even functions, is an even function, while the product of an odd and an even function is an odd function. We leave as an exercise for the reader the proofs of the following facts: whenever g is odd on the interval $[-L, L]$, then

$$\int_{-L}^{L} g(x)\,dx = 0 \qquad (1.14a)$$

and, whenever g is even on the interval $[-L, L]$, then

$$\int_{-L}^{L} g(x)\,dx = 2\int_{0}^{L} g(x)\,dx. \qquad (1.14b)$$

Now, suppose we compute the Fourier series for an odd function g over the interval $[-L, L]$. Here the period of our complex exponentials will be $2L$ and we get for c_n

$$c_n = \frac{1}{2L}\int_{-L}^{L} g(x)e^{-i2\pi nx/(2L)}\,dx$$

$$= \frac{1}{2L}\int_{-L}^{L} g(x)\cos\frac{n\pi x}{L}\,dx - \frac{i}{2L}\int_{-L}^{L} g(x)\sin\frac{n\pi x}{L}\,dx.$$

Since g is odd, so is $g(x)\cos(n\pi x/L)$, and we have

$$\int_{-L}^{L} g(x)\cos\frac{n\pi x}{L}\,dx = 0.$$

And, $g(x)\sin(n\pi x/L)$ is even, so

$$\int_{-L}^{L} g(x)\sin\frac{n\pi x}{L}\,dx = 2\int_{0}^{L} g(x)\sin\frac{n\pi x}{L}\,dx.$$

Returning to our calculation of c_n, these last two results allow us to simplify c_n to the following form:

$$c_n = \frac{-i}{L}\int_{0}^{L} g(x)\sin\frac{n\pi x}{L}\,dx. \qquad (1.15)$$

From (1.15) we obtain, using the oddness of the sine function,

$$c_{-n} = -c_n, \qquad \text{(and } c_0 = 0\text{).} \qquad (1.16)$$

If we now examine the Fourier series for g, then (1.16) allows us to express this series as a series of sines

$$g \sim \sum_{n=-\infty}^{\infty} c_n e^{i2\pi nx/(2L)} = \sum_{n=1}^{\infty} c_n e^{i\pi nx/L} + \sum_{n=1}^{\infty} c_{-n} e^{-i\pi nx/L}$$

$$= \sum_{n=1}^{\infty} c_n \left\{ e^{i\pi nx/L} - e^{-i\pi nx/L} \right\}$$

$$= \sum_{n=1}^{\infty} (2ic_n) \sin \frac{n\pi x}{L} .$$

If we *define B_n* to be $2ic_n$, then we can write

$$g \sim \sum_{n=1}^{\infty} B_n \sin \frac{n\pi x}{L}, \qquad \left(B_n = \frac{2}{L} \int_0^L g(x) \sin \frac{n\pi x}{L} \, dx \right). \qquad (1.17)$$

Based on our results, we now make a formal definition of Fourier sine series.

DEFINITION 1.5: *If g is a function defined on the interval $[0, L]$, then the Fourier sine series for g is defined in (1.17).*

Example 1.7:
Expand $g(x) = 1$ on the interval $[0, 1]$ in a Fourier sine series.

SOLUTION Using (1.17) we have

$$B_n = \frac{2}{1} \int_0^1 1 \sin n\pi x \, dx = \frac{-2}{n\pi} \cos n\pi x \Big|_0^1 = \frac{2}{n\pi} [1 - (-1)^n]$$

$$= \begin{cases} 4/(n\pi) & \text{for } n \text{ odd} \\ 0 & \text{for } n \text{ even.} \end{cases}$$

Thus, *on the interval* $[0, 1]$

$$1 \sim \sum_{k=0}^{\infty} \frac{4 \sin (2k + 1)\pi x}{(2k + 1)\pi} .$$

∎

Suppose we have an even function g on the interval $[-L, L]$. The Fourier series for g can be cast into a series of cosines. In fact, we get for the nth Fourier series coefficient c_n

$$c_n = \frac{1}{2L} \int_{-L}^{L} g(x) \cos \frac{n\pi x}{L} \, dx - \frac{i}{2L} \int_{-L}^{L} g(x) \sin \frac{n\pi x}{L} \, dx$$

$$= \frac{1}{L} \int_{0}^{L} g(x) \cos \frac{n\pi x}{L} \, dx . \tag{1.18}$$

From (1.18) we get for each n,

$$c_{-n} = c_n \tag{1.19}$$

and then

$$g \sim \sum_{n=-\infty}^{\infty} c_n e^{i2\pi nx/(2L)} = c_0 + \sum_{n=1}^{\infty} c_n \left[e^{i\pi nx/L} + e^{-i\pi nx/L} \right]$$

$$= c_0 + \sum_{n=1}^{\infty} 2c_n \cos \frac{n\pi x}{L} . \tag{1.20}$$

Therefore, if we *define* A_n by $2c_n$ for $n = 0, 1, 2, \ldots$, then

$$g \sim \frac{1}{2} A_0 + \sum_{n=1}^{\infty} A_n \cos \frac{n\pi x}{L}, \qquad \left(A_n = \frac{2}{L} \int_{0}^{L} g(x) \cos \frac{n\pi x}{L} \, dx \right). \tag{1.21}$$

REMARK 1.4:   Note that the constant term in (1.21) is $A_0/2$. Although this may be troublesome to remember, it follows from (1.20).   ∎

Based on our results we can make a formal definition of Fourier cosine series.

DEFINITION 1.6:   *If g is a function defined on the interval $[0, L]$, then the* **Fourier cosine series** *for g is defined in (1.21).*

Example 1.8:

Expand $\sin x$ on the interval $[0, \pi/2]$ in a Fourier cosine series.

SOLUTION   Using (1.21) we get for A_n

$$A_n = \frac{4}{\pi} \int_0^{\frac{1}{2}\pi} \sin x \cos 2nx \, dx$$

$$= \frac{2}{\pi} \int_0^{\frac{1}{2}\pi} \sin(2n+1)x - \sin(2n-1)x \, dx = \frac{-4}{\pi} \frac{1}{4n^2 - 1}.$$

Thus, *on the interval* $[0, \pi/2]$

$$\sin x \sim \frac{2}{\pi} + \sum_{n=1}^{\infty} \frac{-4}{\pi} \frac{\cos 2nx}{4n^2 - 1}.$$

■

1.7   Convergence of Fourier Sine and Cosine Series

The pointwise convergence of Fourier sine and cosine series can be deduced from the pointwise convergence of Fourier series. The key to this is to properly extend a function from its original domain of $[0, L]$.

Let's begin with Fourier sine series. Here the proper extension to make is an *odd, periodic extension*. Given a function g on the interval $[0, L]$ one makes an *odd extension* to $[-L, L]$. See Figure 1.12. This odd extension, if we also call it g, would have to satisfy $g(-x) = -g(x)$. Then the function can be extended to a periodic function, period $2L$. See Figure 1.13.   Our discussion in Section 1.6 shows that the Fourier series for this odd, periodic extension is just the Fourier sine series for our original function on $[0, L]$. Therefore, the convergence theorems described in Section 1.5 can be applied.   In particular, we have the following convergence theorem.

THEOREM 1.6:   *POINTWISE CONVERGENCE OF FOURIER SINE SERIES*

Let g be a piecewise continuous function on $[0, L]$. At the points 0 and L the sine series for g equals 0. For $0 < x < L$ we have

$$\sum_{n=1}^{\infty} B_n \sin \frac{n\pi x}{L} = \frac{1}{2}[g(x+) + g(x-)]$$

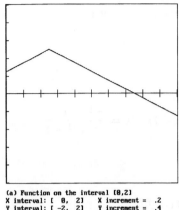

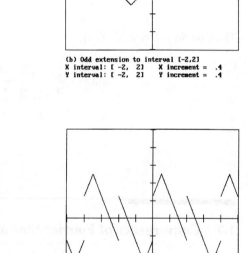

(a) Function on the interval [0,2]
X interval: [0, 2]   X increment = .2
Y interval: [-2, 2]   Y increment = .4

(b) Odd extension to interval [-2,2]
X interval: [-2, 2]   X increment = .4
Y interval: [-2, 2]   Y increment = .4

FIGURE 1.12
Odd extension of a function.

(a) Function on the interval [0,2]
X interval: [0, 2]   X increment = .2
Y interval: [-2, 2]   Y increment = .4

(b) Odd, periodic extension
X interval: [-5, 5]   X increment = 1
Y interval: [-2, 2]   Y increment = .4

FIGURE 1.13
Odd, periodic extension of a function.

provided x is a point where g is Lipschitz from the right and left (or where g has right- and left-hand derivatives).

Of course, when x is also a point of continuity for g, then

$$\sum_{n=1}^{\infty} B_n \sin \frac{n\pi x}{L} = g(x)$$

provided g is Lipschitz from the left and the right at x (or g has right- and left-hand derivatives at x).

The rigorous definition of convergence in Theorem 1.6 is that

$$\lim_{M\to\infty} S_M(x) = \frac{1}{2}[\,g(x+) + g(x-)\,]$$

where

$$S_M(x) = \sum_{n=1}^{M} B_n \sin\frac{n\pi x}{L}.$$

The function S_M is called the *(sine series) partial sum containing M harmonics.*

Example 1.9:
Let $g(x) = 1$ on the interval $[0, 1]$. In Figure 1.14 we have graphed S_M for $M = 10$, 20, and 40 harmonics. Notice the Gibbs' phenomenon near $x = 0$ and $x = 1$. This is a consequence of the discontinuity at these points of the odd, periodic extension of g.

For Fourier cosine series the proper extension to make is an *even, periodic extension*. Given a function g on the interval $[0, L]$ one first makes an even extension to $[-L, L]$. See Figure 1.15. This even extension, if we also call it g, would have to satisfy $g(-x) = g(x)$. Then one extends this even extension to a periodic function, period $2L$. See Figure 1.16. Our discussion in Section 1.6 shows that the Fourier series for this even, periodic extension is just the Fourier cosine series for our original function on $[0, L]$. Therefore, the convergence theorems from Section 1.5 lead to the following convergence theorem for Fourier cosine series.

THEOREM 1.7: *POINTWISE CONVERGENCE OF FOURIER COSINE SERIES*

Let g be a piecewise continuous function on $[0, L]$. At the point 0 the Fourier cosine series converges to $g(0+)$ provided g is Lipschitz from the right at 0 (or has a right-hand derivative at 0). At the point L the Fourier cosine series converges to $g(L-)$ provided g is Lipschitz from the left at L (or has a left-hand derivative at L). For $0 < x < L$, we have

$$\frac{1}{2}A_0 + \sum_{n=1}^{\infty} A_n \cos\frac{n\pi x}{L} = \frac{1}{2}[\,g(x+) + g(x-)\,]$$

provided g is Lipschitz from the left and right at x (or has left- and right-hand derivatives at x).

We remind the reader again, that in Theorem 1.7, $(1/2)[\,g(x+)+g(x-)\,]$ equals $g(x)$ when x is a point of continuity for g. The rigorous definition of convergence

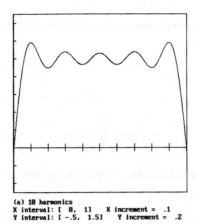

(a) 10 harmonics
X interval: [0, 1]   X increment = .1
Y interval: [-.5, 1.5]   Y increment = .2

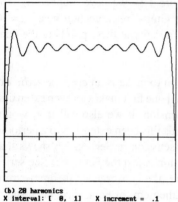

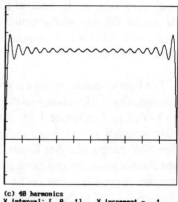

(b) 20 harmonics
X interval: [0, 1]   X increment = .1
Y interval: [-.5, 1.5]   Y increment = .2

(c) 40 harmonics
X interval: [0, 1]   X increment = .1
Y interval: [-.5, 1.5]   Y increment = .2

FIGURE 1.14
Graphs of Fourier sine series partial sums for the function in Example 1.9.

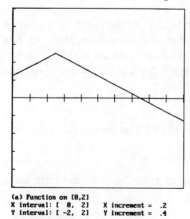

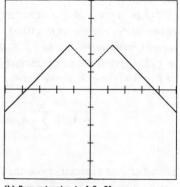

(a) Function on [0,2]
X interval: [0, 2]   X increment = .2
Y interval: [-2, 2]   Y increment = .4

(b) Even extension to [-2, 2]
X interval: [-2, 2]   X increment = .4
Y interval: [-2, 2]   Y increment = .4

FIGURE 1.15
Even extension of a function.

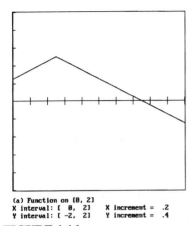

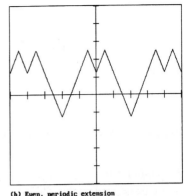

(a) Function on [0, 2]
X interval: [0, 2]   X increment = .2
Y interval: [-2, 2]   Y increment = .4

(b) Even, periodic extension
X interval: [-5, 5]   X increment = 1
Y interval: [-2, 2]   Y increment = .4

FIGURE 1.16
Even, periodic extension of a function.

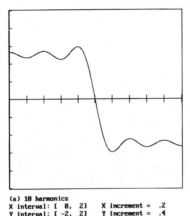

(a) 10 harmonics
X interval: [0, 2]   X increment = .2
Y interval: [-2, 2]   Y increment = .4

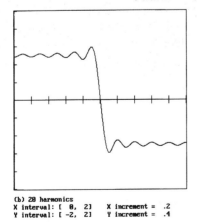

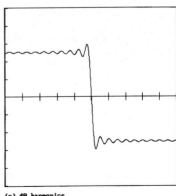

(b) 20 harmonics
X interval: [0, 2]   X increment = .2
Y interval: [-2, 2]   Y increment = .4

(c) 40 harmonics
X interval: [0, 2]   X increment = .2
Y interval: [-2, 2]   Y increment = .4

FIGURE 1.17
Graphs of Fourier cosine series partial sums for the function in Example 1.10.

is

$$\lim_{M \to \infty} S_M(x) = \frac{1}{2}[g(x+) + g(x-)]$$

where

$$S_M(x) = \frac{1}{2}A_0 + \sum_{n=1}^{M} A_n \cos \frac{n\pi x}{L}.$$

The function S_M is called the *(cosine series) partial sum containing M harmonics.*

Example 1.10:

Let g be defined on the interval $[0, 2]$ by

$$g(x) = \begin{cases} 1 & \text{for } 0 < x < 1 \\ -1 & \text{for } 1 < x < 2. \end{cases}$$

In Figure 1.17 we have graphed S_M for $M = 10, 20,$ and 40 harmonics. Notice the Gibbs' phenomenon near the jump discontinuity of g at $x = 1$.

References

For further discussion of the theory of Fourier series, consult the following: [Ch-B], [Da], [Ka], [To], [Wa], and [Zy]. A good history of Fourier analysis can be found in [Da-H].

Exercises

Section 1

1.1 Using power series expansions, justify Euler's identity (1.1).

1.2 Draw figures similar to Figures 1.1 and 1.2 for the following functions.

   (a) $3 \cos 10\pi x - 6 \cos 20\pi x$

   (b) $7 \sin 8\pi x + 3 \sin 16\pi x - \sin 32\pi x$

   (c) $6 \cos 4\pi x + 3 \sin 12\pi x$

1.3 Using *FAS* (*Fourier Analysis Software,* contained on the disk accompanying this volume), graph the Fourier transform of

$$4 \cos 2\pi v x \setminus v=9$$

using 2048 points over each of the intervals $[-2, 2], [-5, 5],$ and $[-10, 10]$. (*Note:* press n [= neg. exponent] when you are asked for which type of transform.)

Notice the similarity of your transform's real part to Figure 1.1(b). In particular, by changing the x-interval to $[-45, 45]$, check that the frequency 9 is located correctly on the horizontal axis.

1.4 Repeat Exercise 1.3, but use

$$6 \sin 2\pi vx \ \backslash v{=}9$$

and compare your transform's imaginary part to Figure 1.2(b).

1.5 Repeat Exercise 1.3, but use the functions given in Exercise 1.2. Check that the frequencies involved for each function are accurately determined.

1.6 Prove the identity $c_{-n} = c_n^*$ described in Remark 1.1(b).

1.7 Prove the identity described in Remark 1.1(c), *which amounts to proving the identity*

$$\int_{-\frac{1}{2}P}^{\frac{1}{2}P} h(x)\, dx = \int_0^P h(x)\, dx$$

for a periodic function h of period P.

1.8 Explain why, for a function h having period P, the following equation is true for *all* constants c and d:

$$\int_c^{c+P} h(x)\, dx = \int_d^{d+P} h(x)\, dx.$$

Note: this shows that the integral of a function of period P is always the same over all intervals of length P.

Section 2

1.9 Expand the following functions in Fourier series using the periods specified and the given intervals:

(a) $\cos x$, period π, interval $[-\pi/2, \pi/2]$

(b) x^2, period 3, interval $[0, 3]$

(c) $\exp(x)$, period 2, interval $[-1, 1]$. [*Note:* $\exp(x)$ is the same as e^x.]

(d) $g(x) = \begin{cases} -1 & \text{for } -1/2 < x < 0 \\ 1 & \text{for } 0 < x < 1/2, \end{cases}$ period 1, interval $[-1/2, 1/2]$

(e) $h(x) = \begin{cases} 1 & \text{for } 0 < x < 1 \\ 0 & \text{for } 1 < x < 2, \end{cases}$ period 2, interval $[0, 2]$

1.10 Obtain the real forms of the Fourier series of the functions in Exercise 1.9.

1.11 Using *FAS*, graph the partial sums S_M for the Fourier series of the functions in Exercise 1.9. Use 1024 points and $M = 1, 4, 16$, and 64 harmonics.

Section 3

1.12 Using the formulas in either (1.9) or (1.10), calculate the real form of the Fourier series for the following functions over the given intervals.

(a) $f(x) = x^2 - 3x$, over $[-\pi, \pi]$

(b) $f(x) = \cos x$, over $[-\pi/2, \pi/2]$

(c) $f(x) = |x|$, over $[-1, 1]$

(d) $f(x) = x^2 + 3$, over $[0, 4]$

(e) $f(x) = \sin x$, over $[0, \pi/2]$

(f) $f(x) = 3x$, over $[0, 4]$

1.13 Using the *Draw graphs* procedure of *FAS*, graph the sum of the first eight non-zero terms of each Fourier series that you found in Exercise 1.12. Also, graph each function as a check of your calculations.

Section 4

1.14 Using the *Fourier series* procedure of *FAS*, draw graphs of partial sums of the Fourier series, using 10, 30, and 90 harmonics, for the following functions.

(a) $\exp(x)$,   for $0 \le x \le 1$

(b) $x^2 + x$,   for $-2 \le x \le 2$

(c) $g(x) = \begin{cases} 0 & \text{for } 0 < x < 1 \\ 1 & \text{for } 1 < x < 2 \\ 2 & \text{for } 2 < x < 3, \end{cases}$ for $0 \le x \le 3$

(d) $g(x) = |x|$,   for $-1 \le x \le 1$

(e) $\sin x$,   for $-\pi/2 \le x \le \pi/2$

(f) $\exp(-x)$,   for $-1 \le x \le 1$

1.15 Discuss the convergence of the Fourier series of the functions in Exercise 1.14. In particular, draw graphs of the functions to which the Fourier series converge over the whole real line.

1.16 For the function given in Exercise 1.14(c), find out how many harmonics are needed in the partial sum of the Fourier series in order to approximate $g(x)$ to within ± 0.005 error on the interval $[1.4, 1.6]$. (*Note:* after initially graphing g change the x-interval to $[1.4, 1.6]$ and the y-interval to $[0.9, 1.1]$.) Repeat this exercise for the intervals $[0.1, 0.3]$ and $[0.3, 0.5]$.

1.17 Same problem as Exercise 1.16, but use the function in Exercise 1.14(d) and x-interval $[-0.1, 0.1]$. Repeat this exercise for the intervals $[0.1, 0.3]$ and $[0.3, 0.5]$.

Section 5

1.18 By changing x- and y-intervals, in order to magnify appropriate regions, display Gibbs' phenomenon for the functions in Exercise 1.14.

1.19 A more quantitative treatment of Gibbs' phenomenon (see [Wa, Chapter 2.5]) shows that as $M \to \infty$ the maxima (minima) of the Fourier series partial sum S_M near a jump discontinuity tend to overshoot (undershoot) by about 9% of the magnitude of the jump. Check this using the graphs displayed in solving Exercise 1.18.

1.20 If g is a piecewise continuous function with period P and its derivative g' is piecewise continuous, then it is known that on any closed interval where g is continuous, the Fourier series for g converges uniformly to g. (For a proof, see [Wa, Chapter 2.5].) Check this result for the function

$$g(x) = \begin{cases} -1 & \text{for } 0 < x < 1 \\ 0 & \text{for } 1 < x < 2 \\ 1 & \text{for } 2 < x < 3 \end{cases}$$

using the intervals $[0.1, 0.9]$, $[1.1, 1.9]$, and $[2.2, 2.8]$, and using 5, 10, 20, and 40 harmonics for the partial sums. Use 1024 points.

1.21 The function $g(x) = |x|$ has a uniformly convergent Fourier series, using the interval $[-1, 1]$. Show this using Theorem 1.3. Then compute the Fourier series for g and estimate how many harmonics are needed in the Fourier series partial sum S_M in order to ensure that $\sup_{x \in [-1,1]} |g(x) - S_M(x)| \le 0.01$. Using *FAS*, confirm this estimate. Why is $S_M(x)$ farthest from $g(x)$ at $x = 0$?

1.22 Which of the following functions has a uniformly convergent Fourier series? You are not required to compute the Fourier series, but you might check your results using *FAS*.

(a) $g(x) = x^2 - x$, interval $[0, 1]$

(b) $g(x) = \exp(x)$, interval $[-1, 1]$

(c) $g(x) = x^3$, interval $[-2, 2]$

(d) $g(x) = x^4$, interval $[-1, 1]$

Section 6

1.23 Expand the following functions in Fourier sine and cosine series over the given intervals.

(a) $\cos x$, interval $[0, 3]$

(b) x, interval $[0, 1]$

(c) $x^2 - x$, interval $[0, 2]$

(d) $g(x) = \begin{cases} 1 & \text{for } 0 < x < 1 \\ -1 & \text{for } 1 < x < 2, \end{cases}$ interval $[0, 2]$

(e) $g(x) = \begin{cases} 0 & \text{for } 0 < x < 1 \\ 1 & \text{for } 1 < x < 2 \\ 0 & \text{for } 2 < x < 3, \end{cases}$ interval $[0, 3]$

1.24 *Orthogonality of Sines.* Prove the following: For all positive integers m and n

$$\frac{2}{L} \int_0^L \sin \frac{m\pi x}{L} \sin \frac{n\pi x}{L} \, dx = \begin{cases} 0 & \text{for } m \neq n \\ 1 & \text{for } m = n. \end{cases}$$

1.25 Use the result of Exercise 1.24 to derive (1.17) for a function g defined on the interval $[0, L]$, using a similar argument to the one used in Section 1.1 to derive Fourier series.

1.26 *Orthogonality of Cosines.* Prove the following: For all positive integers m and n

$$\int_0^L \cos \frac{n\pi x}{L} \, dx = 0$$

$$\frac{2}{L} \int_0^L \cos \frac{m\pi x}{L} \cos \frac{n\pi x}{L} \, dx = \begin{cases} 0 & \text{for } m \neq n \\ 1 & \text{for } m = n. \end{cases}$$

1.27 Use the result of Exercise 1.26 to derive (1.21) for a function g defined on the interval $[0, L]$, using a similar argument to the one used in Section 1.1 to derive Fourier series.

1.28 Using the *Sine series* procedure of *FAS*, graph the partial sum containing M harmonics for the functions in Exercise 1.23. Use $M = 10, 20$, and 40 harmonics and 1024 points.

1.29 Using the *Cosine series* procedure of *FAS*, graph the partial sum containing M harmonics for the functions in Exercise 1.23. Use $M = 10, 20$, and 40 harmonics and 1024 points.

Section 7

1.30 Using *FAS*, graph partial sums of *Fourier series* for odd extensions of the functions in Exercise 1.23 for $M = 10, 20$, and 40 harmonics. Compare your results to the Fourier sine series partial sums graphed in Exercise 1.28.

1.31 Using *FAS*, graph partial sums of *Fourier series* for even extensions of the functions in Exercise 1.23 for $M = 10, 20$, and 40 harmonics. Compare your results to the Fourier cosine series partial sums graphed in Exercise 1.29.

2

The Discrete Fourier Transform (DFT)

In this chapter we describe the discrete version of Fourier analysis. We show how Fourier series can be discretized, resulting in the DFT. We also describe discretization of the Fourier sine and Fourier cosine series.

2.1 Derivation of the DFT

This section will derive the DFT by approximating Fourier coefficients.

Consider the kth Fourier series coefficient c_k for a function g, using period P complex exponentials

$$c_k = \frac{1}{P} \int_0^P g(x) e^{-i2\pi kx/P} \, dx. \tag{2.1}$$

We shall approximate the integral in (2.1) by a *left-endpoint, uniform Riemann sum*

$$c_k \approx \frac{1}{P} \sum_{j=0}^{N-1} g(x_j) e^{-i2\pi kx_j/P} \frac{P}{N} \tag{2.2}$$

where $x_j = jP/N$ for $j = 0, 1, \ldots, N-1$. For this choice of points $\{x_j\}$ we have

$$c_k \approx \frac{1}{N} \sum_{j=0}^{N-1} g\left(j\frac{P}{N}\right) e^{-i2\pi jk/N}. \tag{2.3}$$

Formula (2.3) motivates the following definition of the DFT.

DEFINITION 2.1:   *Given the N complex numbers, $\{h_j\}_{j=0}^{N-1}$, their N-point DFT is denoted by $\{H_k\}$ where H_k is defined by*

$$H_k = \sum_{j=0}^{N-1} h_j e^{-i2\pi jk/N}$$

for all integers $k = 0, \pm 1, \pm 2, \ldots$.

REMARK 2.1:   (a) We see that Formula (2.3) describes the kth Fourier coefficient of the function g as approximately G_k/N where $\{G_k\}$ is the N-point DFT of $\{g(jP/N)\}_{j=0}^{N-1}$. (b) It is a useful notational convention to use a small letter for a given finite sequence of numbers and a capital version of that letter for the DFT of the sequence. (c) In Definition 2.1 the integer j takes values from 0 to $N-1$, but the integer k takes values from all the integers. This apparent asymmetry will be removed in the next section when we discuss the *inverse* DFT. ∎

Example 2.1:
Let $h_j = e^{-jc}$ for $j = 0, 1, \ldots, N-1$ and c a constant. Show that the N-point DFT of $\{h_j\}$ is given by

$$H_k = \frac{1 - e^{-cN}}{1 - e^{-c-i2\pi k/N}}. \tag{2.4}$$

SOLUTION   By the definition of an N-point DFT, we have

$$H_k = \sum_{j=0}^{N-1} e^{-jc} e^{-i2\pi jk/N}. \tag{2.5}$$

Hence, because $e^x e^z = e^{x+z}$, we can write

$$H_k = \sum_{j=0}^{N-1} e^{j(-c-i2\pi k/N)}. \tag{2.6}$$

To simplify (2.6), we make use of the formula for the sum of a finite geometric series

$$1 + r + r^2 + \ldots + r^{N-1} = \frac{1 - r^N}{1 - r}. \tag{2.7}$$

Putting $r = e^{-c-i2\pi k/N}$ in (2.7) and substituting into (2.6), we get

$$H_k = \frac{1 - e^{N(-c-i2\pi k/N)}}{1 - e^{-c-i2\pi k/N}} . \qquad (2.8)$$

The right side of (2.8) can be further simplified by making use of the identities

$$e^{N(-c-i2\pi k/N)} = e^{-cN}e^{-i2\pi k}$$

$$= e^{-cN}. \qquad (2.9)$$

The last equality in (2.9) holds because $e^{-i2\pi k} = 1$. Using (2.9) we rewrite (2.8), obtaining

$$H_k = \frac{1 - e^{-cN}}{1 - e^{-c-i2\pi k/N}}$$

which shows that (2.4) is true. ∎

REMARK 2.2: In *FAS*, the method used for approximating Fourier coefficients is the one described in Exercise 2.5. To make our description simpler in the text, however, we stick to left-endpoint sums. ∎

2.2 Basic Properties of the DFT

In this section we shall discuss some basic properties of the DFT. These properties are *linearity, periodicity*, and *inversion*. This last property will allow us to define the *inverse DFT* and remove the asymmetry between the original sequence (of length N) versus the transformed sequence (of infinite length) mentioned in the previous section.

From now on we will use the letter W to stand for either $e^{-i2\pi/N}$ or $e^{i2\pi/N}$. Which of the two exponentials is meant should be clear from the context. In either case we have

$$W^N = 1 \qquad (2.10)$$

which is the only special property of W that we will need in this section.

THEOREM 2.1:
Suppose that the sequence $\{h_j\}_{j=0}^{N-1}$ has N-point DFT $\{H_k\}$ and the sequence $\{g_j\}_{j=0}^{N-1}$ has N-point DFT $\{G_k\}$, then the following properties hold:

(a) **Linearity.** *For all complex constants a and b, the sequence* $\{ah_j + bg_j\}_{j=0}^{N-1}$ *has N-point DFT* $\{aH_k + bG_k\}$.

(b) **Periodicity.** *For all integers k we have* $H_{k+N} = H_k$.

(c) **Inversion.** *For* $j = 0, 1, \ldots, N-1$,

$$h_j = \frac{1}{N} \sum_{k=0}^{N-1} H_k e^{i2\pi jk/N} \, .$$

PROOF   (a) is left to the reader as an exercise. To prove (b) and (c) we put W equal to $e^{-i2\pi/N}$ and make use of (2.10). The DFT $\{H_k\}$ is defined by

$$H_k = \sum_{j=0}^{N-1} h_j W^{jk}.$$

Hence

$$H_{k+N} = \sum_{j=0}^{N-1} h_j W^{j(k+N)} = \sum_{j=0}^{N-1} h_j W^{jk}(W^N)^j$$

$$= \sum_{j=0}^{N-1} h_j W^{jk}$$

which proves (b). To prove (c) we note that

$$W^{-1} = e^{i2\pi/N} \tag{2.11}$$

and then, by raising both sides of this last equation to the power jk, we get

$$W^{-jk} = e^{i2\pi jk/N} \, .$$

It then follows that

$$\frac{1}{N} \sum_{k=0}^{N-1} H_k e^{i2\pi jk/N} = \frac{1}{N} \sum_{k=0}^{N-1} H_k W^{-jk}$$

$$= \frac{1}{N} \sum_{k=0}^{N-1} \left[\sum_{m=0}^{N-1} h_m W^{mk} \right] W^{-jk}$$

$$= \frac{1}{N} \sum_{k=0}^{N-1} \left[\sum_{m=0}^{N-1} h_m W^{mk} W^{-jk} \right].$$

Now, since $W^{mk} W^{-jk} = W^{(m-j)k}$, we obtain upon switching the order of sums

$$\frac{1}{N} \sum_{k=0}^{N-1} H_k e^{i2\pi jk/N} = \frac{1}{N} \sum_{m=0}^{N-1} h_m \left[\sum_{k=0}^{N-1} W^{(m-j)k} \right]. \tag{2.12}$$

For *fixed* j, if $m \neq j$, then putting r equal to W^{m-j} in (2.7) yields

$$\sum_{k=0}^{N-1} W^{(m-j)k} = \frac{1 - (W^{m-j})^N}{1 - W^{m-j}} = \frac{0}{1 - W^{m-j}}$$

$$= 0$$

since $W^{m-j} \neq 1$. If, however, $m = j$, then $W^{(m-j)k} = W^0 = 1$ and (2.12) becomes

$$\frac{1}{N} \sum_{k=0}^{N-1} H_k e^{i2\pi jk/N} = \frac{1}{N} h_j \sum_{k=0}^{N-1} 1 = h_j$$

and (c) is proved. ∎

A consequence of inversion is that *no two distinct sequences can have the same DFT*. Based on the inversion property in Theorem 2.1(c) we make the following definition.

DEFINITION 2.2: If $\{G_k\}_{k=0}^{N-1}$ *is a sequence of N complex numbers, then its N-point* **inverse DFT** *is defined by*

$$g_j = \sum_{k=0}^{N-1} G_k e^{i2\pi jk/N}$$

for all integers $j = 0, \pm 1, \pm 2, \ldots.$

We can see from putting W equal to $e^{i2\pi/N}$ that (using essentially the same proof as above for Theorem 2.1) the inverse DFT has the properties of linearity, periodicity, and inversion. For the inverse DFT the inversion formula is

$$G_k = \frac{1}{N} \sum_{j=0}^{N-1} g_j e^{-i2\pi jk/N}.$$

Periodicity for the inverse DFT shows that $\{g_j\}_{j=0}^{N-1}$ is a finite subsequence of a larger periodic sequence $\{g_j\}$ defined for all integers j by the inverse DFT of $\{G_k\}_{k=0}^{N-1}$.

There are two other important properties of the DFT that are worth noting. First, by periodicity, it follows that every DFT $\{H_k\}$ satisfies

$$H_{N-k} = H_{-k}. \tag{2.13}$$

And, there is the property known as *Parseval's equality*

$$\sum_{j=0}^{N-1} |h_j|^2 = \frac{1}{N} \sum_{k=0}^{N-1} |H_k|^2 \tag{2.14}$$

where $\{H_k\}$ is the N-point DFT of $\{h_j\}$. We leave the proof of (2.14) to the reader as an exercise.

2.3   Relation of the DFT to Fourier Coefficients

This section will explain further the approximation involved in replacing Fourier coefficients by the DFT. The approximation described in (2.3) must be interpreted carefully. Since the right side of (2.3) is $1/N$ times the DFT of $\{g(jP/N)\}_{j=0}^{N-1}$ it has period N in the variable k. The left side of (2.3) is c_k, the kth Fourier coefficient of g, and will typically *not* have period N.

To make sure that (2.3) is valid we will make the assumption that $|k| \leq N/8$. To explain the origin of this *rule of thumb*, let's consider a specific example.

Example 2.2:

Let $g(x) = e^{-x}$ on the interval $[0, 10]$. Compare the two sides of (2.3).

SOLUTION   For c_k we have

$$c_k = \frac{1}{10} \int_0^{10} e^{-x} e^{-i2\pi kx/10} \, dx$$

$$= \frac{1}{10} \int_0^{10} e^{(-1-i2\pi k/10)x} \, dx$$

$$= \frac{1 - e^{-10}e^{-i2\pi k}}{10 + i2\pi k} \; .$$

Thus, we have found c_k,

$$c_k = \frac{1 - e^{-10}}{10 + i2\pi k} \; . \tag{2.15}$$

On the other hand, for G_k/N we have

$$\frac{1}{N}G_k = \frac{1}{N} \sum_{j=0}^{N-1} g\left(j\frac{P}{N}\right) e^{-i2\pi jk/N}$$

$$= \frac{1}{N} \sum_{j=0}^{N-1} e^{-j(10/N)} e^{-i2\pi jk/N} \; .$$

Applying the result of Example 2.1, by putting c equal to $10/N$, we have

$$\frac{1}{N}G_k = \frac{1}{N} \frac{1 - e^{-10}}{1 - e^{-(10+i2\pi k)/N}} \; . \tag{2.16}$$

In order to compare (2.15) and (2.16), we use the approximation

$$e^{-x} \approx 1 - x \tag{2.17}$$

which is valid for $|x| \ll 1$. For $|k|/N \ll 1$ we will have, using $(10 + i2\pi k)/N$ in place of x in (2.17),

$$e^{-(10+i2\pi k)/N} \approx 1 - (10 + i2\pi k)/N \; . \tag{2.18}$$

Using (2.18) in (2.16) we get

$$\frac{1}{N}G_k \approx \frac{1}{N} \frac{1 - e^{-10}}{(10 + i2\pi k)/N} = \frac{1 - e^{-10}}{10 + i2\pi k} = c_k \; . \tag{2.19}$$

Hence,

$$\frac{1}{N}G_k \approx c_k \tag{2.20}$$

provided $|k| \ll N$. Our rule of thumb that $|k| \leq N/8$ is not in contradiction with $|k| \ll N$.

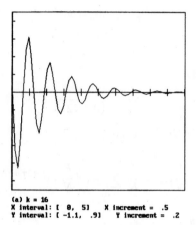

(a) k = 16
X interval: [0, 5]   X increment = .5
Y interval: [-1.1, .9]   Y increment = .2

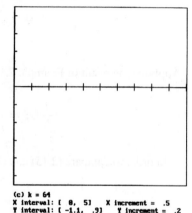

(b) k = 32
X interval: [0, 5]   X increment = .5
Y interval: [-1.1, .9]   Y increment = .2

(c) k = 64
X interval: [0, 5]   X increment = .5
Y interval: [-1.1, .9]   Y increment = .2

FIGURE 2.1
Graphs of $\mathrm{Im}\,(e^{-x}e^{-i2\pi kx/10})$ **for** $N = 128$ **points and** $k = 16, 32,$ **and** 64.

There is another consideration involved in demanding that $|k| \leq N/8$. In Figure 2.1 we show graphs of part of $e^{-x}e^{-i2\pi kx/10}$ obtained by connecting the sampled values [at $x_j = j(10/N)$] by line segments. These graphs are of the *imaginary part*

$$\mathrm{Im}\left[e^{-x}e^{-i2\pi kx/10}\right] = -e^{-x}\sin\frac{2\pi kx}{10} \tag{2.21}$$

but similar graphs could be drawn for the *real part*

$$\mathrm{Re}\left[e^{-x}e^{-i2\pi kx/10}\right] = e^{-x}\cos\frac{2\pi kx}{10}. \tag{2.22}$$

The graphs reveal (for the case of $N = 128$) that when $k = N/2$ the sampled version of the function in (2.21) is *indistinguishable* from the sampled version of the constant function 0. In this case, the 0-function is called an *alias* of the function in (2.21) for $k = N/2$.

The origin of our *rule of thumb* for taking $|k| \le N/8$ is based on the *qualitative* observation that, for such values of k, the sampled versions of our functions generally have no aliases with other samplings at lower frequencies. In general, it is difficult to give precise estimates on the size of k which will ensure a given degree of error in Approximation (2.3). In any case, to pursue this matter further would obscure our attempt at introducing the fundamentals of DFTs. (For further discussion, see [Br-H, Chapter 6].) ∎

2.4 Relation of the DFT to Sampled Fourier Series

An important application of the DFT is to calculations with sampled Fourier series. We will begin our treatment of this topic in this section and expand on it in Chapter 4 when we discuss filtering of Fourier series.

Using the approximation of Fourier coefficients by the DFT described in (2.3), we can write the M-harmonic partial sum S_M of the Fourier series for a function g as

$$S_M(x) \approx \sum_{k=-M}^{M} \frac{1}{N} G_k e^{i2\pi kx/P} \tag{2.23}$$

where $\{G_k\}$ is the N-point DFT of the sequence of samples $\{g(jP/N)\}_{j=0}^{N-1}$ of g. In (2.23) we assume that $M \le N/8$ and P is the period for the Fourier series expansion of g.

If we substitute for x the same sampling points $\{jP/N\}_{j=0}^{N-1}$, which we used to get G_k, then we obtain from (2.23)

$$S_M\left(j\frac{P}{N}\right) \approx \frac{1}{N} \sum_{k=-M}^{M} G_k e^{i2\pi jk/N}. \tag{2.24}$$

Formula (2.24) can be expressed as an N-point inverse DFT with weight $W = e^{i2\pi/N}$ if we make the following algebraic manipulations. First, since $W^N = 1$, we have $W^{N-k} = W^{-k}$. And, from (2.13), we have $G_{N-k} = G_{-k}$ (using $\{G_k\}$ in place of $\{H_k\}$). Thus, (2.24) can be rewritten as

$$S_M\left(j\frac{P}{N}\right) \approx \frac{1}{N}\sum_{k=0}^{M} G_k e^{i2\pi jk/N} + \frac{1}{N}\sum_{k=1}^{M} G_{-k} e^{-i2\pi jk/N}$$

$$= \frac{1}{N}\sum_{k=0}^{M} G_k e^{i2\pi jk/N} + \frac{1}{N}\sum_{k=1}^{M} G_{N-k} e^{i2\pi j(N-k)/N}$$

$$= \frac{1}{N}\sum_{k=0}^{M} G_k e^{i2\pi jk/N} + \frac{1}{N}\sum_{k=N-M}^{N-1} G_k e^{i2\pi jk/N}. \qquad (2.25)$$

To see that the last sums can be rewritten as a single DFT sum, we define $\{H_k\}_{k=0}^{N-1}$ by

$$H_k = \begin{cases} G_k & \text{for } k = 0, 1, \ldots, M \\ 0 & \text{for } k = M+1, \ldots, N-M-1 \\ G_k & \text{for } k = N-M, \ldots, N-1. \end{cases} \qquad (2.26)$$

We can interpret $\{H_k\}$ as a lopping off to 0 of $\{G_k\}$ in an interval about the index value $k = N/2$. Based on Formula (2.26) we can write

$$S_M\left(j\frac{P}{N}\right) \approx \frac{1}{N}\sum_{k=0}^{N-1} H_k e^{i2\pi jk/N} \qquad (2.27)$$

which expresses the sampled Fourier series partial sum values $\{S_M(jP/N)\}_{j=0}^{N-1}$ as the N-point inverse DFT (weight $W = e^{i2\pi/N}$) of $\{H_k\}$, multiplied by $1/N$. When graphing S_M the software uses the method of *linear interpolation*; in other words, it connects the sampled values of S_M with straight line segments.

To summarize our results, we compute an approximation of a Fourier series partial sum S_M for a function g as follows:

Step 1. Sample the given function g over the interval $[0, P]$, using N evenly spaced sample points, obtaining $\{g(jP/N)\}_{j=0}^{N-1}$.

Step 2. Compute the N-point DFT $\{G_k\}$ of the sequence $\{g(jP/N)\}_{j=0}^{N-1}$ using the weight $W = e^{-i2\pi/N}$.

Step 3. Lop off to 0 the DFT $\{G_k\}$ for $M+1 \le k \le N-M-1$ where $M \le N/8$, obtaining a new sequence $\{H_k\}$.

Step 4. Compute the N-point inverse DFT of $\{H_k\}$, using weight $W = e^{i2\pi/N}$, and multiply each number of this DFT by $1/N$. The result is the approximation to the sampled Fourier series partial sum $\{S_M(jP/N)\}_{j=0}^{N-1}$. *Note:* by periodicity, $S_M(P) = S_M(0)$.

Step 5. Connect the points of the sequence $\{S_M(jP/N)\}_{j=0}^{N}$ by straight line segments, yielding an approximation to $S_M(x)$ over the interval $0 \le x \le P$.

REMARK 2.3: When the interval used is $[-P/2, P/2]$, then *FAS* makes use of periodicity. It just converts to the interval $[0, P]$, applies the method described above, and then converts back to the interval $[-P/2, P/2]$. ∎

2.5 Discrete Sine and Cosine Transforms

Discrete sine and cosine transforms can be derived from the formulas for sine and cosine series described in Chapter 1, Section 1.6. The kth sine coefficient B_k of a real valued function g over the interval $[0, L]$ is defined by

$$B_k = \frac{2}{L} \int_0^L g(x) \sin \frac{k\pi x}{L} \, dx \tag{2.28}$$

while the kth cosine coefficient A_k is defined by

$$A_k = \frac{2}{L} \int_0^L g(x) \cos \frac{k\pi x}{L} \, dx. \tag{2.29}$$

To derive the discrete sine transform we approximate the integral in (2.28) using a uniform left-endpoint Riemann sum

$$B_k \approx \frac{2}{L} \sum_{j=0}^{N-1} g\left(j\frac{L}{N}\right) \sin \frac{k\pi j L/N}{L} \frac{L}{N}. \tag{2.30}$$

Therefore,

$$B_k \approx \frac{2}{N} \sum_{j=0}^{N-1} g\left(j\frac{L}{N}\right) \sin \frac{\pi j k}{N}.$$

Or, since the $j = 0$ term above is 0, we have

$$B_k \approx \frac{2}{N} \sum_{j=1}^{N-1} g\left(j\frac{L}{N}\right) \sin \frac{\pi j k}{N}. \tag{2.31}$$

In a similar way, we find that

$$A_k \approx \frac{2}{N} \sum_{j=0}^{N-1} g\left(j\frac{L}{N}\right) \cos \frac{\pi j k}{N}. \tag{2.32}$$

Based on Formulas (2.31) and (2.32), we make the following definition.

DEFINITION 2.3:   *For a real sequence* $\{h_j\}_{j=1}^{N-1}$ *the **Discrete Sine Transform** (DST), $\{H_k^S\}$, is defined by*

$$H_k^S = \sum_{j=1}^{N-1} h_j \sin \frac{\pi j k}{N} .$$

For a real sequence $\{h_j\}_{j=0}^{N-1}$ *the **Discrete Cosine Transform** (DCT), $\{H_k^C\}$, is defined by*

$$H_k^C = \sum_{j=0}^{N-1} h_j \cos \frac{\pi j k}{N} .$$

From Formula (2.31), we see that the kth Fourier sine coefficient B_k is approximated by $2/N$ times the DST of the sequence $\{g(jL/N)\}_{j=1}^{N-1}$. And from Formula (2.32), we see that the kth Fourier cosine coefficient A_k is approximated by $2/N$ times the DCT of the sequence $\{g(jL/N)\}_{j=0}^{N-1}$.

REMARK 2.4:   Just as we discussed in the previous sections for the DFT and Fourier series coefficients, the Approximations (2.31) and (2.32) are only valid for $k \ll N$. *We shall assume from now on that $k \le N/4$, for reasons similar to the ones given in Section 2.3.* ∎

A Fourier sine series and a Fourier cosine series can be approximated using a DST and a DCT, respectively. Let's begin with a Fourier sine series. Suppose that a function g is expanded in a Fourier sine series over $[0, L]$:

$$g \sim \sum_{k=1}^{\infty} B_k \sin \frac{k\pi x}{L} , \qquad \left(B_k = \frac{2}{L} \int_0^L g(x) \sin \frac{k\pi x}{L} \, dx \right). \qquad (2.33)$$

If $S_M(x)$ is used to denote the M-harmonic partial sum of the Fourier sine series in (2.33), then we have

$$S_M(x) = \sum_{k=1}^{M} B_k \sin \frac{k\pi x}{L} , \qquad \left(B_k = \frac{2}{L} \int_0^L g(x) \sin \frac{k\pi x}{L} \, dx \right). \qquad (2.34)$$

Assuming that $M \le N/4$ (see Remark 2.4), we approximate each B_k, using $2/N$

times the DST of $\{g(jL/N)\}$. Thus,

$$S_M(x) \approx \frac{2}{N} \sum_{k=1}^{M} G_k^S \sin \frac{k\pi x}{L} \tag{2.35}$$

where $\{G_k^S\}$ is the DST of $\{g(jL/N)\}$. Replacing x by the sample values jL/N, we have

$$S_M\left(j\frac{L}{N}\right) \approx \frac{2}{N} \sum_{k=1}^{M} G_k^S \sin \frac{\pi jk}{N}, \qquad (j = 1, \ldots, N-1). \tag{2.36}$$

Defining the sequence $\{H_k\}$ by

$$H_k = \begin{cases} G_k^S & \text{for } k = 1, \ldots, M \\ 0 & \text{for } k = M+1, \ldots, N-1 \end{cases} \tag{2.37}$$

we see that

$$S_M\left(j\frac{L}{N}\right) \approx \frac{2}{N} \sum_{k=1}^{N-1} H_k \sin \frac{\pi jk}{N}, \qquad (j = 1, \ldots, N-1). \tag{2.38}$$

Formula (2.38) shows that the sampled values of the M-harmonic Fourier sine series partial sum, $\{S_M(jL/N)\}$, are approximated by multiplying each element of the DST of the sequence $\{H_k\}$, defined in (2.37), by $2/N$. If these sample values are connected by line segments [including $S_M(0) = S_M(L) = 0$], then $S_M(x)$ can be approximated for $0 \le x \le L$. This, of course, is exactly what *FAS* does when it graphs a Fourier sine series.

Similarly, we can approximate a Fourier cosine series using a DCT. Suppose that the function g is expanded in a Fourier cosine series over the interval $[0, L]$:

$$g \sim \frac{1}{2}A_0 + \sum_{k=1}^{\infty} A_k \cos \frac{k\pi x}{L}, \qquad \left(A_k = \frac{2}{L} \int_0^L g(x) \cos \frac{k\pi x}{L} \, dx\right). \tag{2.39}$$

If $S_M(x)$ is used to denote the M-harmonic partial sum of the Fourier cosine series in (2.39), then

$$S_M(x) = \frac{1}{2}A_0 + \sum_{k=1}^{M} A_k \cos \frac{k\pi x}{L}. \tag{2.40}$$

Assuming that $M \leq N/4$ (see Remark 2.4), we approximate each A_k, using $2/N$ times the DCT of $\{g(jL/N)\}$. Thus,

$$S_M(x) \approx \frac{2}{N} \left[\frac{1}{2} G_0^C + \sum_{k=1}^{M} G_k^C \cos \frac{k\pi x}{L} \right] \tag{2.41}$$

where $\{G_k^C\}$ is the DCT of $\{g(jL/N)\}$. Replacing x by the sample values jL/N, we have

$$S_M\left(j\frac{L}{N}\right) \approx \frac{2}{N} \left[\frac{1}{2} G_0^C + \sum_{k=1}^{M} G_k^C \cos \frac{\pi jk}{N} \right]. \tag{2.42}$$

Defining the sequence $\{H_k\}$ by

$$H_k = \begin{cases} \frac{1}{2} G_0^C & \text{for } k = 0 \\ G_k^C & \text{for } k = 1, \ldots, M \\ 0 & \text{for } k = M+1, \ldots, N-1 \end{cases} \tag{2.43}$$

we see that

$$S_M\left(j\frac{L}{N}\right) \approx \frac{2}{N} \sum_{k=0}^{M} H_k \cos \frac{\pi jk}{N}, \qquad (j = 0, 1, \ldots, N-1). \tag{2.44}$$

Formula (2.44) shows that the sampled values of the M-harmonic Fourier cosine series partial sums $\{S_M(jL/N)\}$ are approximated by multiplying each element of the DCT of the sequence $\{H_k\}$, defined in (2.43), by $2/N$. If these sample values are connected by line segments [including $S_M(L) \approx (2/N) \sum_{k=0}^{N-1} H_k(-1)^k$, which must be calculated separately], then $S_M(x)$ can be approximated for $0 \leq x \leq L$. This, of course, is just what *FAS* does when it graphs a Fourier cosine series.

References

For more information on DFTs, see [Br,2], [Ra-R], [Bri], and [Op-S]. An illuminating treatment can be found in [Br-H].

Exercises

Section 1

2.1 Compute the N-point DFT of the *constant sequence* $\{1\}_{j=0}^{N-1}$.

2.2 Compute the 16-point DFT of the sequence

$$1\ 1\ 1\ 1\ 0\ 0\ 0\ 0\ 0\ 0\ 0\ 0\ 1\ 1\ 1\ 1\ .$$

2.3 Compute the 16-point DFT of the sequence

$$1\ 1\ 1\ 1\ 0\ 0\ 0\ 0\ 0\ 0\ 0\ 0\ 0\ 1\ 1\ 1\ .$$

2.4 Suppose that a *right-endpoint, uniform Riemann sum* (using points $x_j = jP/N$ for $j = 1, 2, \ldots, N$) is used to approximate the integral for c_k in (2.1). Show that c_k is approximated by an N-point DFT, using the sequence

$$g(P), \ g\left(\frac{P}{N}\right), \ g\left(2\frac{P}{N}\right), \ \ldots, \ g\left((N-1)\frac{P}{N}\right).$$

2.5 Show that by averaging approximations with left- and right-endpoint sums, the Fourier coefficient c_k in (2.1) can be approximated by an N-point DFT, using the sequence

$$\frac{g(0) + g(P)}{2}, \ g\left(\frac{P}{N}\right), \ g\left(2\frac{P}{N}\right), \ \ldots, \ g\left((N-1)\frac{P}{N}\right).$$

Section 2

2.6 Prove Parseval's Equality (2.14), and prove the linearity property (a) in Theorem 2.1.

2.7 Suppose that a sequence $\{h_j\}_{j=0}^{N-1}$ is modified as follows: N zeroes are appended to the sequence, to obtain a new sequence $\{g_j\}_{j=0}^{2N-1}$, where $g_j = h_j$ for $j = 0, 1, \ldots, N-1$, and $g_j = 0$ for $j = N, \ldots, 2N-1$. Show that for the $2N$-point DFT $\{G_k\}$ of $\{g_j\}$,

$$G_{2k} = H_k, \qquad k = 0, 1, \ldots, N-1$$

where $\{H_k\}$ is the N-point DFT of $\{h_j\}$.

Section 3

2.8 Using *FAS*, graph $\sin(k\pi x)$ over the interval $[-1, 1]$, using 1024 points, for $k = 32$, 64, 128, 256, 512, and 1024. When does aliasing begin (in a noticeable way) and how does this compare with the rule of thumb $|k| \leq 128$?

2.9 Repeat Exercise 2.8 for $\cos(k\pi x)$.

2.10 Repeat Exercises 2.8 and 2.9 for $N = 2048$ points and $k = 64, 128, 256, 512$, 1024, and 2048.

Section 4

2.11 For the function $f(x) = |x|$, graph the 35-harmonic partial sum of its Fourier series over $[-\pi, \pi]$, using 1024 points, in two ways. First, using the *Fourier series* procedure of *FAS*. Second, by graphing the partial sum directly, using the following formula:

$$f(x) = \pi/2 - (4/\pi)\text{sumk}(\cos((2k - 1)x)/(2k - 1) \wedge 2) \backslash k{=}1,18.$$

Compare these two graphs by finding the Sup-Norm difference between them (i.e., the maximum absolute value of their difference for each point in $[-\pi, \pi]$.)

2.12 Repeat Exercise 2.11 for $f(x) = (x^2 - \pi^2)^2$ over $[-\pi, \pi]$. [*Hint:* the Fourier series for $f(x)$ is $8\pi^4/15 - 48 \sum_{k=1}^{\infty}((-1)^k \cos(kx)/k^4)$.]

REMARK 2.5: The previous two exercises show that there is little difference between an exact calculation of a Fourier series partial sum and the *FAS* approximated partial sum (at least when the series is uniformly convergent). The DFT method, actually the FFT method, trades a slight amount of inaccuracy for a *massive* increase in speed of calculation. ∎

2.13 Repeat Exercise 2.11 for

$$f(x) = \begin{cases} 1 & \text{for } 0 < x \leq \pi \\ 0 & \text{for } -\pi \leq x < 0. \end{cases}$$

Why do you think there is significantly more discrepancy between the two graphs in this case than in Exercises 2.11 and 2.12? [The discrepancy is still small enough, however, for *FAS* to produce an acceptable *graphical sketch* of the Fourier series partial sum.]

2.14 Repeat Exercise 2.13, but now use 4096 points. The discrepancy between the two graphs should now be less.

Section 5

2.15 Compute a 41-harmonic Fourier sine series for the function $f(x) = x^2 - 2x$ over the interval $[0, 2]$ using 1024 points. Do this computation in two ways. First, using the *Fourier sine series* procedure of *FAS*. Second, by graphing the explicitly entered sum consisting of the 41 terms of the Fourier sine series. Using Sup-Norm differences, find the maximum difference between these two graphs. The discrepancy should be relatively small. (The fast sine transform method exchanges a slight inaccuracy for a *massive* increase in speed.)

2.16 Repeat Exercise 2.15, but use Fourier cosine series instead.

2.17 Repeat Exercise 2.15, but use $f(x) = x$ instead.

2.18 Repeat Exercise 2.15, but use $f(x) = x$ and Fourier cosine series. Why do you think the discrepancy is less than in Exercise 2.17?

2.19 Repeat Exercise 2.15 for

$$f(x) = \begin{cases} 1 & \text{for } 0 \leq x < 2 \\ 0 & \text{for } 2 < x \leq 3 \end{cases}$$

using 1024 points and interval $[0, 3]$.

2.20 Repeat Exercise 2.19, but use cosine series instead.

2.21 Repeat Exercises 2.17 and 2.19, but use 4096 points. The discrepancies should be less now.

3

The Fast Fourier Transform (FFT)

A direct calculation of an N-point DFT requires $(N - 1)^2$ multiplications and $N(N - 1)$ additions. For large N, say $N > 1000$, this would require too much computer time. The term *Fast Fourier Transform* (FFT) is used to describe a computer algorithm that reduces the amount of computer time enormously. The FFT that we describe in this chapter reduces the calculation time by a factor of 200 when $N = 1024$. The FFT is one of the greatest contributions to numerical analysis made in this century.

Besides the FFT we shall also describe a Fast Sine Transform (FST) and a Fast Cosine Transform (FCT).

See Appendix B for computer programs related to the material in this chapter.

3.1 Decimation in Time, Radix 2, FFT

In this section we shall describe one of the most widely used FFT algorithms, the decimation in time, radix 2, FFT. *Throughout this chapter we will assume that N is a power of* 2, *say* $N = 2^R$, *where R is a positive integer.* Also, we will continue with the notation introduced in the last chapter for the N-point DFT

$$H_k = \sum_{j=0}^{N-1} h_j W^{jk} \tag{3.1}$$

where W stands for either $e^{i2\pi/N}$ or $e^{-i2\pi/N}$. The base 2, for $N = 2^R$, is often called radix 2. In this book we shall work with radix 2 exclusively. Other types of radices, including mixed radices, are discussed in the references.

We begin our FFT algorithm by halving the N-point DFT in (3.1) into two sums, each of which is an $N/2$-point DFT:

$$H_k = \sum_{j=0}^{\frac{1}{2}N-1} h_{2j}(W^2)^{jk} + \sum_{j=0}^{\frac{1}{2}N-1} h_{2j+1}(W^2)^{jk} W^k. \qquad (3.2)$$

Based on (3.2) we write H_k as

$$H_k = H_k^0 + W^k H_k^1$$

$$H_k^0 = \sum_{j=0}^{\frac{1}{2}N-1} h_{2j}(W^2)^{jk}$$

$$H_k^1 = \sum_{j=0}^{\frac{1}{2}N-1} h_{2j+1}(W^2)^{jk}, \qquad (k = 0, 1, \ldots, N-1). \qquad (3.3)$$

Notice that the $N/2$-point DFTs $\{H_k^0\}$ and $\{H_k^1\}$ use weight W^2 (not W). The periods of $\{H_k^0\}$ and $\{H_k^1\}$ are $N/2$; they are the $N/2$-point DFTs of $\{h_0, h_2, \ldots, h_{N-2}\}$ and $\{h_1, h_3, \ldots, h_{N-1}\}$, respectively. It is important to realize that the binary (base 2) expansions of the indices of $h_0, h_2, \ldots, h_{N-2}$ all end with 0 (since the indices are even). This is the reason for the superscript 0 in $\{H_k^0\}$. Similarly, the binary expansions of the indices of $h_1, h_3, \ldots, h_{N-1}$ all end with 1 and that is the reason for the superscript 1 in $\{H_k^1\}$. Using the prefix DFT to stand for "DFT of" we can represent this as

$$\{H_k^0\} = DFT\{h_{\ldots 0}\}, \qquad \{H_k^1\} = DFT\{h_{\ldots 1}\}. \qquad (3.4)$$

Since $N = 2^R$ we can divide N evenly by 2, and we have

$$W^{N/2} = -1. \qquad (3.5)$$

Using (3.5), and the fact that $\{H_k^0\}$ and $\{H_k^1\}$ both have period $N/2$, we will write the first equation in (3.3) as

$$H_k = H_k^0 + W^k H_k^1, \qquad H_{k+N/2} = H_k^0 - W^k H_k^1 \qquad (3.6)$$

for $k = 0, 1, \ldots, N/2 - 1$. The calculations in (3.6) can be diagrammed as

$$H_k^0 \quad \longrightarrow \quad H_k^0 + W^k H_k^1$$

$$\times$$

$$H_k^1 \quad \longrightarrow \quad H_k^0 - W^k H_k^1, \qquad \left(k = 0, 1, \ldots, \frac{1}{2}N - 1\right). \qquad (3.7)$$

Diagram (3.7) is called a *butterfly*. There are $N/2$ butterflies for this stage of the FFT and each butterfly requires one multiplication, W^k times H_k^1. Figure 3.1 illustrates the butterflies needed for this stage when $N = 16$.

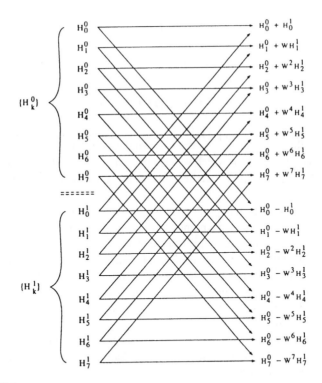

FIGURE 3.1
First reduction in FFT algorithm for $N = 16$.

The splitting of $\{H_k\}$ into two half-size DFTs, $\{H_k^0\}$ and $\{H_k^1\}$, can be repeated on $\{H_k^0\}$ and $\{H_k^1\}$ themselves. We get

$$H_k^0 = H_k^{00} + (W^2)^k H_k^{01}, \qquad H_{k+\frac{1}{4}N}^0 = H_k^{00} - (W^2)^k H_k^{01}$$

$$H_k^1 = H_k^{10} + (W^2)^k H_k^{11}, \qquad H_{k+\frac{1}{4}N}^1 = H_k^{10} - (W^2)^k H_k^{11} \qquad (3.8)$$

for $k = 0, 1, \ldots, N/4 - 1$. In (3.8), $\{H_k^{00}\}$ is the $N/4$-point DFT of $\{h_0, h_4, h_8, \ldots, h_{N-4}\}$, $\{H_k^{01}\}$ is the $N/4$-point DFT of $\{h_2, h_6, \ldots, h_{N-2}\}$, $\{H_k^{10}\}$ is the $N/4$-point DFT of $\{h_1, h_5, \ldots, h_{N-3}\}$, $\{H_k^{11}\}$ is the $N/4$-point DFT of $\{h_3, h_7, \ldots, h_{N-1}\}$. Or, in terms of the binary expansions of the indices of h_j, we have

$$\{H_k^{00}\} = DFT\{h_{\ldots 00}\},$$

$$\{H_k^{01}\} = DFT\{h_{\ldots 10}\},$$

$$\{H_k^{10}\} = DFT\{h_{\ldots 01}\},$$

$$\{H_k^{11}\} = DFT\{h_{\ldots 11}\}.$$

Notice that there is a *reversal* of the last two digits (*bits*) in the binary expansions of the indices j in $\{h_j\}$ for which we calculate the $N/4$-point DFTs $\{H_k^{00}\}$, $\{H_k^{01}\}$, $\{H_k^{10}\}$, $\{H_k^{11}\}$. In Figure 3.2 we have diagrammed the two stages of the FFT that we have described.

If we continue with this process of halving the order of the DFTs, then after $R = \log_2 N$ stages we reach a point where we are performing N one-point DFTs. A one-point DFT of a number h_j is just the identity $h_j \longrightarrow h_j$. Moreover, since the process of indexing the smaller size DFTs by superscripts written *by reversing the order of the bits* in the indices of $\{h_j\}$ will continue, by the time the one-point stage is reached *all* bits in the binary expansion of j will be arranged in reverse order. See Figure 3.3. Therefore, to begin the FFT calculation one must first *rearrange* $\{h_j\}$ so it is listed in *bit reverse order*.

In Figure 3.4 we have shown the calculations needed for performing an 8-point FFT. The result, of course, matches with the definition of an 8-point DFT. In the exercises the reader is asked to perform a similar calculation of a 16-point FFT. Ultimately, this work is essential to understanding the FFT algorithm.

The FFT results in an enormous savings in computing time. For each of the $\log_2 N$ stages there are $N/2$ multiplications, hence there are $(N/2) \log_2 N$ multiplications needed for the FFT. This takes much less time than the $(N - 1)^2$ multiplications needed for a direct DFT calculation. When $N = 1024$, the FFT requires 5120 multiplications, while the DFT requires $1,046,529$ multiplications. Therefore, for $N = 1024$, the FFT results in a savings, in multiplication time, by

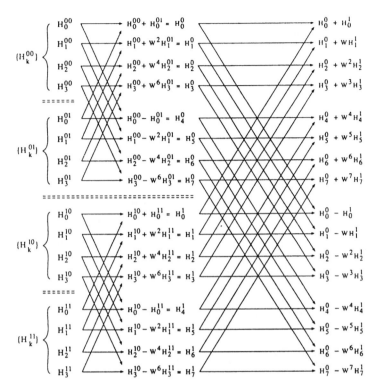

FIGURE 3.2
Two stages of FFT, $N = 16$.

a factor of almost 200. Actually, even more savings can be achieved, since multiplication by i (or $-i$) does not need to be programmed as multiplication, and we do not need to count as multiplications any butterflies that only involve additions and subtractions (see Exercise 3.1). Similar dramatic savings hold in regard to the number of additions needed for the FFT; we leave it as an exercise for the reader to calculate how many additions are needed for the FFT vs. the DFT. Throughout the remainder of the book we shall concentrate on counting multiplications, leaving the addition counts to the reader as exercises.

In the next few sections we shall analyze in detail the following components of the FFT:

(a) bit reversal reordering of the initial data

(b) rotations involved in butterfly computations

(c) computation of the sines and tangents needed to perform (b).

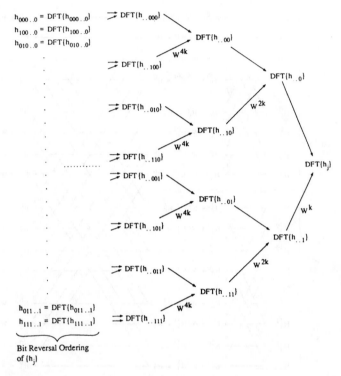

FIGURE 3.3
Reductions of DFTs and bit reversal. The indices are in binary form.

REMARK 3.1:   In this section we have described a decimation in time FFT. The other major type of FFT is known as a *decimation in frequency* FFT. This type of FFT is described in [Wa, Chapter 7.13].   ∎

3.2   Bit Reversal

In this section we shall examine in detail the first component of the FFT, bit reversal reordering of the initial data. We shall describe two methods of performing this bit reversal permutation. These methods are *Buneman's algorithm,* and another, faster, algorithm.

Buneman's algorithm is the simplest method for performing the bit reversal

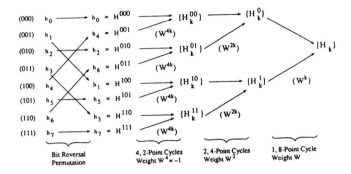

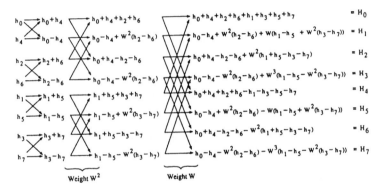

FIGURE 3.4
Decimation in Time FFT for $N = 8$. [*Note:* $W^4 = -1$ and $W^8 = 1$.]

permutation. It is based on a simple pattern. If we have permuted the N numbers

$$\{0, 1, \ldots, N - 1\}$$

by bit reversing their binary expansions, then the permutation of the $2N$ numbers

$$\{0, 1, \ldots, 2N - 1\}$$

is obtained by doubling the numbers in the permutation of $\{0, 1, \ldots, N - 1\}$ to get the first N numbers and then adding 1 to these doubled numbers to get the last N numbers. See Figure 3.5.

That Buneman's algorithm works is easily proved by looking at binary expansions. Suppose m has the following binary expansion

$$m = (a_1 a_2 \ldots a_R)_{\text{base 2}} \tag{3.9}$$

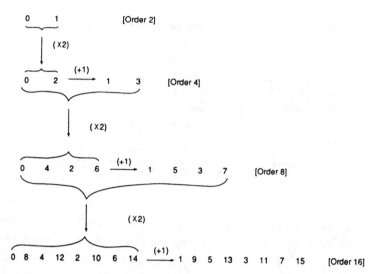

FIGURE 3.5
Buneman's algorithm for generating lists of bit reversed numbers.

where each a_j is either 0 or 1. The number m is mapped to $P_N(m)$, its bit reversed image, hence

$$P_N(m) = (a_R \ldots a_2 a_1)_{\text{base 2}} . \tag{3.10}$$

If we double $P_N(m)$, then

$$2P_N(m) = (a_R \ldots a_2 a_1 0)_{\text{base 2}} \tag{3.11}$$

and we see that $2P_N(m)$ is the bit reversed image of

$$m = (0 a_1 a_2 \ldots a_R)_{\text{base 2}} \tag{3.12}$$

where m is considered as an element of $\{0, 1, \ldots, 2N - 1\}$. Note that (3.12) describes the first N numbers in the set $\{0, 1, \ldots, 2N - 1\}$. Furthermore,

$$2P_N(m) + 1 = (a_R \ldots a_2 a_1 1)_{\text{base 2}} \tag{3.13}$$

is the bit reversal of

$$m = (1 a_1 a_2 \ldots a_R)_{\text{base 2}} \tag{3.14}$$

which accounts for the last N numbers in the list $\{0, 1, \ldots, 2N - 1\}$.

While Buneman's method is admirably simple, it has the defect that it performs *unnecessary* swaps. It can be improved upon by more efficient algorithms that perform only those swaps that are absolutely necessary. We will now describe one such algorithm, the one that *FAS* uses.

To see how this algorithm works, let's suppose that $N = 2^R$ where $R = 2Q$ is *even*. (The case of R odd will be discussed afterwards.) In this case, a number m from the list $\{0, 1, \ldots, N - 1\}$ has a binary expansion

$$m = \left(a_Q \ldots a_1 b_Q \ldots b_1 \right)_{\text{base 2}} \tag{3.15}$$

where every a_j and every b_j is either 0 or 1. The number $P_N(m)$ then has as its binary expansion

$$P_N(m) = \left(b_1 \ldots b_Q a_1 \ldots a_Q \right)_{\text{base 2}}. \tag{3.16}$$

If we let $M = 2^Q = N^{1/2}$, and $K = (a_Q \ldots a_1)_{\text{base 2}}$, and $L = (b_Q \ldots b_1)_{\text{base 2}}$, then Equations (3.15) and (3.16) can be written as

$$m = KM + L$$

$$P_N(m) = P_M(L)M + P_M(K) \tag{3.17}$$

for $K = 0, 1, \ldots, M - 1$ and $L = 0, 1, \ldots, M - 1$.

The equations in (3.17) show how the bit reversal permutation P_N can be split into two bit reversal permutations *of square root size*, P_M, where $M = N^{1/2}$.

Using the facts that $P_M \circ P_M$ is the identity permutation and that P_M permutes $\{0, 1, 2, \ldots, M - 1\}$ we can express the set of numbers described in (3.17) in the following new way:

$$n = P_M(K)M + L$$

$$P_N(n) = P_M(L)M + K \tag{3.18}$$

for $K = 0, 1, \ldots, M - 1$ and $L = 0, 1, \ldots, M - 1$. The equations in (3.18) describe all the bit reversals, $n \longrightarrow P_N(n)$. However, each swap of data,

$$h_n \iff h_{P_N(n)} \tag{3.19}$$

covers two of the bit reversals described in (3.18). Namely, $n \longrightarrow P_N(n)$ and $P_N(n) \longrightarrow n$. Furthermore, for those n for which $P_N(n) = n$, we do not need to perform the swaps in (3.19). If we were to make a table of K vs. $P_M(L)$, then the unnecessary swaps would be characterized by the *diagonal elements*, where

$L = K$. For, in this case, (3.18) becomes

$$n = P_M(K)M + K$$

$$P_N(n) = P_M(K)M + K \tag{3.20}$$

so $P_N(n) = n$. The *lower triangle* of this table, where $L < K$, corresponds to

$$n = P_M(K)M + L$$

$$P_N(n) = P_M(L)M + K \tag{3.21}$$

for $K = 1, 2, \ldots, M - 1$ and $L = 0, \ldots, K - 1$. The equations in (3.21) describe $(1/2)M(M - 1) = (1/2)N^{1/2}(N^{1/2} - 1)$ distinct swaps. This follows from the fact that there are no repeats in the list

$$\{n\} = \{P_M(K)M + L\}_{K=1, L=0}^{M-1, K-1} . \tag{3.22}$$

There are no repeats in the list in (3.22) because $n = P_M(K)M + L$ can be rewritten as an expression of the number n in base M, i.e., $n = (P_M(K), L)_{\text{base } M}$. Any other expression of n in this base M notation would look like $n = (P_M(K'), L')_{\text{base } M}$. But then, *the uniqueness of base M expressions* would imply that $L = L'$ and $P_M(K) = P_M(K')$. Consequently, $K = K'$ because P_M is a permutation. Hence, $K = K'$ and $L = L'$, so there are no repeats in the list in (3.22).

Therefore, we have shown the equations in (3.21) describe, *in a one-to-one fashion*, $(1/2)N^{1/2}(N^{1/2} - 1)$ swaps. But, that is precisely the number of swaps needed to perform the bit reversal permutation P_N. This last assertion follows from the fact that there are $N^{1/2}$ numbers of the form

$$(a_1 a_2 \ldots a_Q a_Q \ldots a_2 a_1)_{\text{base } 2} \tag{3.23}$$

which do *not* require swaps, since they are their own bit reversals. Subtracting these $N^{1/2}$ elements from the N total elements and multiplying by $1/2$ (since there is just one swap needed for each pair swapped) we get $(1/2)(N - N^{1/2}) = (1/2)N^{1/2}(N^{1/2} - 1)$. Thus, *the equations in (3.21) completely describe all bit reversals when the power R in $N = 2^R$ is even.*

When the power R is odd, a general bit reversal will look like

$$n = (a_Q \ldots a_1 c b_Q \ldots b_1)_{\text{base } 2} \longrightarrow (b_1 \ldots b_Q c a_1 \ldots a_Q)_{\text{base } 2} = P_N(n) \tag{3.24}$$

where c equals 0 or 1. For $M = (N/2)^{1/2}$, we can express (3.24) as

$$n = P_M(K)(2M) + L \longrightarrow P_N(n) = P_M(L)(2M) + K$$

$$n + M \longrightarrow P_N(n+M) = P_N(n) + M \qquad (3.25)$$

for $K = 1, 2, \ldots, M - 1$ and $L = 0, 1, \ldots, K - 1$.

To accomplish a bit reversal permutation one must program the equations in (3.21) for the power R being even, and program the transformations in (3.25) for the power R being odd, the swaps in each case being carried out according to (3.19). A computer procedure which accomplishes this, called BITREV, can be found in Appendix B. Part of the program uses Buneman's method to compute the *lower order* permutation P_M.

While Buneman's method is simple and relatively fast, the algorithm programmed in BITREV runs faster by a considerable percentage.

3.3 Rotations in FFTs

The second major component of the FFT is the succession of butterflies performed during each stage. Each butterfly calculation requires a multiplication of the form

$$W^m Z \qquad (3.26)$$

where Z is a complex number and so is W^m.

We now show that (3.26) can be interpreted as a rotation. The complex number Z can be written as

$$Z = A + iB \qquad (3.27)$$

where A and B are real numbers. The weight factor W^m can be written, if $W = e^{i2\pi/N}$, as

$$W^m = \cos\left(\frac{2\pi m}{N}\right) + i \sin\left(\frac{2\pi m}{N}\right), \qquad (m = 0, 1, \ldots, N/2 - 1). \quad (3.28)$$

For simplicity of notation we define $C(m)$ and $S(m)$ by

$$C(m) = \cos\left(\frac{2\pi m}{N}\right), \qquad S(m) = \sin\left(\frac{2\pi m}{N}\right) \qquad (3.29)$$

and, for later use, we define $T(m)$ by

$$T(m) = \tan\left(\frac{\pi m}{N}\right). \tag{3.30}$$

Using (3.29), Formula (3.28) becomes

$$W^m = C(m) + i S(m). \tag{3.31}$$

And, using (3.27) and (3.31), we express $W^m Z$ in (3.26) as

$$W^m Z = [C(m)A - S(m)B] + i [S(m)A + C(m)B]. \tag{3.32}$$

Or, in terms of real and imaginary parts,

$$\mathrm{Re}\,(W^m Z) = C(m)A - S(m)B, \qquad \mathrm{Im}\,(W^m Z) = S(m)A + C(m)B. \tag{3.33}$$

Because of (3.29), we can intepret (3.33) as describing a *rotation* of the vector (A, B) through the angle $2\pi m/N$.

Formula (3.33) requires 4 real multiplications. An algorithm due to Buneman reduces the number of multiplications needed to 3, while increasing the number of additions from 2 to 3. Since multiplications take up a large part of FFT computation time, usually taking much more time than the additions, this reduction of multiplications yields a considerable time savings. [*Note:* this last statement is not always true. In some PCs, for example, considerable effort has been made to speed up multiplications so that they are almost as fast as additions.]

Buneman's algorithm is based on the following trigonometric identity,

$$\tan\frac{1}{2}\theta = \frac{1 - \cos\theta}{\sin\theta} \tag{3.34}$$

which follows from using the double angle identities for cosine and sine, where $\theta = 2(\theta/2)$, on the right side of (3.34). From (3.34), with a little algebra, we get

$$\cos\theta = 1 - \sin\theta \tan\frac{1}{2}\theta. \tag{3.35}$$

And, multiplying both sides of (3.34) by $1 + \cos\theta$, yields (after applying a trigonometric identity)

$$\sin\theta = (1 + \cos\theta) \tan\frac{1}{2}\theta. \tag{3.36}$$

Replacing θ by $2\pi m/N$ in (3.35) and (3.36) yields

$$C(m) = 1 - S(m)T(m) \tag{3.37}$$

and

$$S(m) = [1 + C(m)]T(m). \tag{3.38}$$

Using (3.37) in the first equation in (3.33) we get

$$\text{Re}\,(W^m Z) = A - S(m)[B + T(m)A] \tag{3.39}$$

If we define the auxiliary quantity V by

$$V = B + T(m)A \tag{3.40}$$

then (3.39) becomes

$$\text{Re}\,(W^m Z) = A - S(m)V. \tag{3.41}$$

Using (3.38) and (3.37) in the second equation in (3.33) we get

$$\text{Im}\,(W^m Z) = [1 + C(m)]T(m)A + [1 - S(m)T(m)]B$$

$$= [B + T(m)A] + [C(m)A - S(m)B]T(m)$$

$$= V + \text{Re}\,(W^m Z)T(m). \tag{3.42}$$

Summarizing these results, we have from Formulas (3.40) to (3.42) a three-step process for computing $\text{Re}\,(W^m Z)$ and $\text{Im}\,(W^m Z)$ [when $W = e^{i2\pi/N}$]

$$V = B + T(m)A$$

$$\text{Re}\,(W^m Z) = A - S(m)V$$

$$\text{Im}\,(W^m Z) = V + \text{Re}\,(W^m Z)T(m). \tag{3.43}$$

The three-step process in (3.43) requires only 3 real multiplications instead of the 4 real multiplications required for (3.33).

We saw in Section 3.1 that the FFT requires a total of $(N/2)\log_2 N$ complex multiplications. There are $2N\log_2 N$ *real* multiplications needed if Formula (3.33) is used. Using Formula (3.43) requires only $(3N/2)\log_2 N$ real multiplications. This represents a savings of about 25% when a computer is used that takes *significantly longer to perform multiplications than additions* [Formula (3.43) requires 3 additions while (3.33) requires 2]. Such a savings represents a significant reduction in computation time.

3.4   Computation of Sines and Tangents

In the previous section we saw that, because of Formula (3.43), the butterflies in the FFT require the computation of $S(m)$ and $T(m)$ for $m = 0, \ldots, N/2 - 1$, where $S(m)$ and $T(m)$ are defined by

$$S(m) = \sin\left(\frac{2\pi m}{N}\right),$$

$$T(m) = \tan\left(\frac{\pi m}{N}\right), \qquad \left(m = 0, \ldots, \frac{1}{2}N - 1\right). \qquad (3.44)$$

Buneman's method requires that we compute the quantities $S(m)$ and $T(m)$ in advance of performing the butterflies in the FFT. This saves time because the computer does not have to calculate a particular weight factor over and over again from one stage of the FFT to the next. There is also a time savings when several FFTs of the same size (N points) are computed in a row; the same values of $S(m)$ and $T(m)$ can be used again and again. In fact, this is what happens in *FAS* when several partial sums for Fourier series, using the same number of points, are computed in a row.

In this section we will show that only the values of $S(m)$ and $T(m)$ for $m = 0$, $\ldots, N/4 - 1$ need to be precalculated, and we will discuss Buneman's method for efficient computation of these values. They are *not* computed by repeatedly invoking sine and tangent functions; there is a much more efficient way of computing these very special values.

First, we show why only the values of $S(m)$ and $T(m)$ for $m = 0, \ldots, N/4 - 1$ need to be computed. This reduction saves memory storage. If $N = 1024$, for example, then we only need two 256-point auxiliary arrays to store these values of $S(m)$ and $T(m)$. This is a tolerable increase in extra memory.

As it stands from the discussion in Sections 3.1 and 3.3, the multiplication $W^m Z$, performed in any butterfly, is computed using [see Formula (3.43)]:

$$V = B + T(m)A, \qquad\qquad (Z = A + iB)$$

$$\mathrm{Re}\,(W^m Z) = A - S(m)V$$

$$\mathrm{Im}\,(W^m Z) = V + \mathrm{Re}\,(W^m Z)T(m) \qquad (3.45)$$

for $m = 0, 1, 2, \ldots, N/2 - 1$. However, rather than apply (3.45) for all of these values of m, we notice that if $N/4 \le m \le N/2 - 1$ then

$$W^m = W^{m-\frac{1}{4}N} W^{\frac{1}{4}N}$$

$$= W^{m-\frac{1}{4}N} i$$

Therefore, if $Z = A + iB$, then

$$W^m Z = W^{m-\frac{1}{4}N}(-B + iA).$$

Consequently, by replacing A by $-B$ and B by A, the equations in (3.45) become

$$V = A - T\left(m - \frac{1}{4}N\right)B$$

$$\mathrm{Re}\,(W^m Z) = -B - S\left(m - \frac{1}{4}N\right)V$$

$$\mathrm{Im}\,(W^m Z) = V + \mathrm{Re}\,(W^m Z)T\left(m - \frac{1}{4}N\right) \qquad (3.46)$$

for $m = N/4, \ldots, N/2 - 1$. Thus, if we use the formulas in (3.45) for $m = 0$, $1, \ldots, N/4 - 1$, and the formulas in (3.46) for $m = N/4, \ldots, N/2 - 1$, it is evident that *only values from the sets* $\{S(m)\}_{m=0}^{N/4-1}$ *and* $\{T(m)\}_{m=0}^{N/4-1}$ *are needed to compute the multiplications in the butterflies of the FFT.*

We will now discuss Buneman's algorithm for efficiently calculating values of $S(m)$ and $T(m)$ for $m = 0, 1, \ldots, N/4$. As we can see from (3.44), we do not need to compute arbitrary values of the sine function. Rather, we must compute $\sin(2\pi m/N)$ for $m = 0, 1, \ldots, N/4$. Buneman devised a clever method for inductively constructing these values.

The method is based on the following trigonometric identity:

$$\sin\theta = \frac{\sin(\theta + \phi) + \sin(\theta - \phi)}{2\cos\phi}. \qquad (3.47)$$

Formula (3.47) follows from the sine addition and subtraction identities applied to the numerator on the right side of (3.47). Using (3.47) it is possible to compute $S(m)$ for $m = 0, 1, \ldots, N/4$ by an inductive process. The first step is to put $S(0) = 0$ and $S(N/4) = 1$. Then, we put $\phi_1 = \pi/4$ and define $z_1 = \sqrt{2}$.

Using (3.47) we have, since $z_1 = 2\cos\phi_1$,

$$S\left(\frac{1}{8}N\right) = \frac{S(\frac{1}{4}N) + S(0)}{z_1}$$

The second step is to multiply ϕ_1 by $1/2$ and put $\phi_2 = \phi_1/2$. Then, by the cosine half-angle formula

$$2\cos\phi_2 = 2\cos\left(\frac{1}{2}\phi_1\right) = (2 + 2\cos\phi_1)^{\frac{1}{2}}$$

$$= (2 + z_1)^{\frac{1}{2}}. \tag{3.48}$$

Defining z_2 to be $2\cos\phi_2$, we rewrite (3.48) as

$$z_2 = (2 + z_1)^{\frac{1}{2}}. \tag{3.49}$$

Combining (3.48) and (3.49), we get from (3.47) that

$$S\left(\frac{1}{16}N\right) = \frac{S(\frac{1}{8}N) + S(0)}{z_2}, \qquad S\left(\frac{3}{16}N\right) = \frac{S(\frac{1}{4}N) + S(\frac{1}{8}N)}{z_2}.$$

The third step is to define z_3 by

$$z_3 = (2 + z_2)^{\frac{1}{2}}. \tag{3.50}$$

Using z_3 in place of $2\cos\phi$ in (3.47), with $\phi_3 = \phi_2/2$ in place of ϕ, we obtain

$$S\left(\frac{1}{32}N\right) = \frac{S(\frac{1}{16}N) + S(0)}{z_3}, \qquad S\left(\frac{3}{32}N\right) = \frac{S(\frac{1}{8}N) + S(\frac{1}{16}N)}{z_3},$$

$$S\left(\frac{5}{32}N\right) = \frac{S(\frac{3}{16}N) + S(\frac{1}{8}N)}{z_3}, \qquad S\left(\frac{7}{32}N\right) = \frac{S(\frac{1}{4}N) + S(\frac{3}{16}N)}{z_3}.$$

This completes the first three stages of Buneman's method. We have shown how to compute $S(jN/32)$ for $j = 0, 1, \ldots, 8$. If $N = 32$, then we are done. If $N = 64$ we would need to go through one more stage. Buneman's method is to continue the process we have begun. For $N = 2^R$, Buneman's method takes $R - 2$

stages to compute all the sines $S(m)$ for $m = 0, 1, \ldots, N/4$. For each new stage we compute

$$\phi_{k+1} = \frac{1}{2}\phi_k, \qquad z_{k+1} = (2 + z_k)^{\frac{1}{2}} \tag{3.51}$$

and use (3.47) in the form

$$\sin \theta = \frac{\sin(\theta + \phi_{k+1}) + \sin(\theta - \phi_{k+1})}{z_{k+1}}$$

to compute the values of $S(m)$ where the angles $2\pi m/N$ lie midway between the angles previously computed. Once we have the values of $\{S(m)\}_{m=0}^{N/4}$, then the values of $\{T(m)\}_{m=0}^{N/4}$ are easily computed using Formula (3.34) [see Exercise 3.10].

A computer procedure, called SINESTANS, which executes the method just described, can be found in Appendix B.

Bracewell has shown that Buneman's methods of handling trigonometric operations in the FFT, discussed in this section and the previous section, result in significant time savings over more conventional methods (see [Br]).

We have now finished our discussion of the basic elements of the FFT algorithm. See Appendix B for computer programs that implement this algorithm.

3.5 Computing Two Real FFTs Simultaneously

In this section and the next, we shall discuss FFTs of *real* data. Most FFTs are performed on real valued data sequences. For real data there are some useful devices which reduce the number of calculations involved and, consequently, the computation time.

In this section, we show how to compute two real FFTs simultaneously by performing one complex FFT. This time-saving device will be used in the next section to reduce the amount of calculation needed to perform an FFT of real data by about one half. It will also be used in Chapter 4 to facilitate the computation of the convolution of two real data sequences.

Suppose $\{f_j\}_{j=0}^{N-1}$ and $\{g_j\}_{j=0}^{N-1}$ are two sequences of N *real* numbers. Define the sequence of N complex numbers $\{h_j\}$ by

$$h_j = f_j + ig_j, \qquad (j = 0, 1, \ldots, N-1). \tag{3.52}$$

Taking the N-point FFT of $\{h_j\}$, we get $\{H_k\}$, where

$$H_k = \sum_{j=0}^{N-1}(f_j + ig_j)W^{jk}, \qquad (k = 0, 1, \ldots, N-1) \qquad (3.53)$$

and $W = e^{i2\pi/N}$ (or $W = e^{-i2\pi/N}$). Using algebra we rewrite (3.53) as

$$H_k = F_k + iG_k, \qquad (k = 0, 1, \ldots, N-1) \qquad (3.54)$$

where

$$F_k = \sum_{j=0}^{N-1} f_j W^{jk}, \qquad G_k = \sum_{j=0}^{N-1} g_j W^{jk} \qquad (3.55)$$

are N-point FFTs of $\{f_j\}$ and $\{g_j\}$, respectively.

There is a crucial symmetry property for the DFT of a real valued sequence that we need to use. It is an analog of the symmetry relation for Fourier series coefficients of a real valued function. We take the complex conjugate F_{N-k}^* of F_{N-k} and we get

$$F_{N-k}^* = \left[\sum_{j=0}^{N-1} f_j W^{j(N-k)}\right]^* = \sum_{j=0}^{N-1} f_j^*(W^{jN})^*(W^{-jk})^*. \qquad (3.56)$$

Because f_j is real we have $f_j^* = f_j$ and, because $W^N = 1$, we have $(W^{jN})^* = 1$. Finally, because $W^{-jk} = (W^{jk})^*$ we have $(W^{-jk})^* = W^{jk}$. From these facts, we observe that (3.56) simplifies to

$$F_{N-k}^* = F_k \qquad (\text{or } F_{N-k} = F_k^*). \qquad (3.57)$$

Similarly, we have

$$G_{N-k}^* = G_k \qquad (\text{or } G_{N-k} = G_k^*) \qquad (3.58)$$

because $\{g_j\}$ are real numbers, too. Using (3.57) and (3.58), we obtain from (3.54)

$$H_{N-k}^* = F_k - iG_k. \qquad (3.59)$$

By combining (3.54) and (3.59) we get

$$F_k = \frac{1}{2}\left[H_{N-k}^* + H_k\right], \qquad G_k = \frac{i}{2}\left[H_{N-k}^* - H_k\right]. \qquad (3.60)$$

The formulas in (3.60) show us how to obtain $\{F_k\}_{k=0}^{N-1}$ and $\{G_k\}_{k=0}^{N-1}$.

We have shown how to obtain the FFTs of two real valued N-point sequences $\{f_j\}$ and $\{g_j\}$. The method consists of the following three steps:

Step 1. Form the sequence $h_j = f_j + ig_j$ for $j = 0, 1, \ldots, N-1$.

Step 2. Perform the N-point FFT of $\{h_j\}$ to get $\{H_k\}$.

Step 3. Use Formula (3.60) to get F_k and G_k for $k = 0, 1, \ldots, N-1$.

3.6 Computing a Real FFT

In this section we shall apply the method of the previous section, along with Formula (3.57), to compute the FFT of a real data sequence. This method will reduce the number of computations involved, from just computing an FFT in the form described in Section 3.1, by about one half.

We begin by considering again the first reduction involved in computing the FFT of the sequence of N *real* points $\{f_j\}$,

$$F_k = \sum_{j=0}^{\frac{1}{2}N-1} f_{2j}(W^2)^{jk} + \sum_{j=0}^{\frac{1}{2}N-1} f_{2j+1}(W^2)^{jk}W^k$$

$$= F_k^0 + F_k^1 W^k \tag{3.61}$$

where $\{F_k^0\}$ and $\{F_k^1\}$ are the $N/2$-point FFTs of the real sequences $\{f_{2j}\}$ and $\{f_{2j+1}\}$, respectively. To calculate *simultaneously* the $N/2$-point FFTs of $\{f_{2j}\}$ and $\{f_{2j+1}\}$ we use the method summarized at the end of Section 3.5. First, we form the sequence $\{h_j\}$ where $h_j = f_{2j} + if_{2j+1}$. Second, we calculate $\{F_k^0\}$ and $\{F_k^1\}$ by

$$F_k^0 = \frac{1}{2}\left[H_{\frac{1}{2}N-k}^* + H_k\right]$$

$$F_k^1 = \frac{i}{2}\left[H_{\frac{1}{2}N-k}^* - H_k\right], \qquad \left(k = 0, 1, \ldots, \frac{1}{2}N - 1\right). \tag{3.62}$$

Now, for $k = 0, 1, \ldots, N/2 - 1$, we obtain F_k from (3.61):

$$F_k = F_k^0 + F_k^1 W^k, \qquad \left(k = 0, 1, \ldots, \frac{1}{2}N - 1\right). \tag{3.63}$$

To get F_k for $k = N/2 + 1, \ldots, N - 1$ we make use of (3.57). Namely,

$$F_{N-k} = F_k^*, \qquad \left(k = 1, \ldots, \frac{1}{2}N - 1\right). \tag{3.64}$$

To obtain $F_{N/2}$ we note that $\{F_k^0\}$ and $\{F_k^1\}$ have period $N/2$. Therefore, $F_{N/2}^0 = F_0^0$ and $F_{N/2}^1 = F_0^1$. Furthermore, $W^{N/2} = -1$, hence we obtain from (3.61)

$$F_{\frac{1}{2}N} = F_0^0 - F_0^1. \tag{3.65}$$

Formulas (3.63) to (3.65) give a complete description of $\{F_k\}_{k=0}^{N-1}$.

The method we have just described for computing an FFT of a real sequence is considerably faster than just applying the FFT described in Section 3.1. For example, the method of this section requires $(1/2)(N/2)\log_2(N/2) + N/2$ *complex* multiplications. This results in $(3N/4)\log_2(N/2) + 3N/2$ *real* multiplications if we use Buneman's method for performing butterflies (which was described in Section 3.3). This represents a reduction by approximately one half of the number of real multiplications, $(3N/2)\log_2 N$, needed for the complex FFT described in Section 3.1. This reduction by one half is what we should expect based on the symmetry inherent in Formula (3.57).

A computer procedure, called REALFFT, which performs the algorithm above, can be found in Appendix B.

We can also apply the symmetry in Formula (3.57) to reduce the number of calculations needed for an inverse FFT of the FFT of a real sequence. We will describe our approach to this problem after the next section, since it involves fast sine and cosine transforms.

3.7   Fast Sine and Cosine Transforms

To compute discrete sine transforms (DSTs) and discrete cosine transforms (DCTs) efficiently, we make use of the work of the previous section and we use some important symmetry properties of finite sequences. An efficient method for computing a DST is called a *Fast Sine Transform* (FST). After discussing an FST we will then discuss an efficient method for computing a DCT, which is called a *Fast Cosine Transform* (FCT).

We will need to make use of some fundamental symmetry properties of finite sequences. These symmetry properties are closely related to the properties of evenness and oddness of functions. A sequence $\{C_j\}_{j=1}^{N-1}$ is called *even about* $N/2$

if

$$C_{N-j} = C_j, \qquad (j = 1, \ldots, N-1). \qquad (3.66)$$

The sequence is called *odd about* $N/2$ if

$$C_{N-j} = -C_j, \qquad (j = 1, \ldots, N-1). \qquad (3.67)$$

If a sequence is even about $N/2$, then

$$\sum_{j=1}^{N-1} C_j = 2 \sum_{j=1}^{\frac{1}{2}N-1} C_j + C_{\frac{1}{2}N}. \qquad (3.68)$$

If a sequence is odd about $N/2$, then

$$\sum_{j=1}^{N-1} C_j = 0. \qquad (3.69)$$

The proofs of (3.68) and (3.69) are left to the reader as exercises.

Here are a few examples. The sequence $\{\sin(j\pi/N)\}_{j=1}^{N-1}$ is even about $N/2$. For *fixed* k, the sequence $\{\sin(2\pi jk/N)\}_{j=1}^{N-1}$ is odd about $N/2$. And, for fixed k, the sequence $\{\cos(2\pi jk/N)\}_{j=1}^{N-1}$ is even about $N/2$.

The algebra of even and odd sequences about $N/2$ is the same as the algebra of even and odd functions from calculus. For example, if $\{A_j\}_{j=1}^{N-1}$ is odd about $N/2$ and $\{S_j\}_{j=1}^{N-1}$ is even about $N/2$, then $\{A_j S_j\}_{j=1}^{N-1}$ is odd about $N/2$. The general pattern is (even)(even) = even, (odd)(odd) = even, and (odd)(even) = odd.

Now, to compute an FST of a real sequence $\{h_j\}_{j=1}^{N-1}$, we define a new sequence $\{f_j\}_{j=0}^{N-1}$ by

$$f_0 = 0$$

$$f_j = s_j + a_j, \qquad (j = 1, \ldots, N-1) \qquad (3.70)$$

where

$$s_j = \left[h_j + h_{N-j}\right] \sin \frac{j\pi}{N},$$

$$a_j = \frac{1}{2}\left[h_j - h_{N-j}\right]. \qquad (3.70a)$$

The sequence $\{s_j\}_{j=1}^{N-1}$ defined in (3.70a) is even about $N/2$, while the sequence $\{a_j\}_{j=1}^{N-1}$ is odd about $N/2$. We now compute an FFT (using weight $W = e^{i2\pi/N}$) of $\{f_j\}$. This FFT $\{F_k\}$ has real parts $\{R_k\}$ defined by

$$R_k = \sum_{j=0}^{N-1} f_j \cos \frac{2\pi jk}{N} = \sum_{j=1}^{N-1} f_j \cos \frac{2\pi jk}{N}. \qquad (3.71)$$

The imaginary parts $\{I_k\}$ of $\{F_k\}$ are defined by

$$I_k = \sum_{j=0}^{N-1} f_j \sin \frac{2\pi jk}{N} = \sum_{j=1}^{N-1} f_j \sin \frac{2\pi jk}{N}. \qquad (3.72)$$

From the symmetry relation (3.69) we get [using (3.71), (3.70), and (3.70a)]

$$R_k = \sum_{j=1}^{N-1} s_j \cos \frac{2\pi jk}{N} + \sum_{j=1}^{N-1} a_j \cos \frac{2\pi jk}{N}$$

$$= \sum_{j=1}^{N-1} s_j \cos \frac{2\pi jk}{N}$$

$$= \sum_{j=1}^{N-1} h_j \sin \frac{j\pi}{N} \cos \frac{2\pi jk}{N} + \sum_{j=1}^{N-1} h_{N-j} \sin \frac{j\pi}{N} \cos \frac{2\pi jk}{N}. \qquad (3.73)$$

By substituting $N-j$ in place of j we find that, because $\{\sin(j\pi/N)\cos(2\pi jk/N)\}$ is even about $N/2$,

$$\sum_{j=1}^{N-1} h_{N-j} \sin \frac{j\pi}{N} \cos \frac{2\pi jk}{N} = \sum_{j=1}^{N-1} h_j \sin \frac{j\pi}{N} \cos \frac{2\pi jk}{N}. \qquad (3.74)$$

Substituting from (3.74) back into (3.73) we get

$$R_k = \sum_{j=1}^{N-1} 2h_j \sin \frac{j\pi}{N} \cos \frac{2\pi jk}{N}. \qquad (3.75)$$

Applying the trigonometric identity $2 \sin \phi \cos \theta = \sin(\theta + \phi) - \sin(\theta - \phi)$ we obtain from (3.75)

$$R_k = \sum_{j=1}^{N-1} h_j \left[\sin\left(\frac{2k+1}{N} j\pi \right) - \sin\left(\frac{2k-1}{N} j\pi \right) \right]$$

$$= H_{2k+1}^S - H_{2k-1}^S. \tag{3.76}$$

Similarly, for I_k we have

$$I_k = \sum_{j=1}^{N-1} s_j \sin \frac{2\pi jk}{N} + \sum_{j=1}^{N-1} a_j \sin \frac{2\pi jk}{N}$$

$$= \sum_{j=1}^{N-1} a_j \sin \frac{2\pi jk}{N}$$

$$= \sum_{j=1}^{N-1} \frac{1}{2} h_j \sin \frac{2\pi jk}{N} - \sum_{j=1}^{N-1} \frac{1}{2} h_{N-j} \sin \frac{2\pi jk}{N}$$

$$= \sum_{j=1}^{N-1} h_j \sin \frac{2\pi jk}{N} .$$

Thus,

$$I_k = H_{2k}^S. \tag{3.77}$$

Formula (3.77) says that the *even-indexed* numbers H_{2k}^S are given by I_k, the imaginary part of F_k, for $k = 0, 1, \ldots, N/2 - 1$. The *odd-indexed* numbers H_{2k+1}^S are gotten from the following recursion formula

$$H_{2k+1}^S = H_{2k-1}^S + R_k, \qquad \left(k = 1, \ldots, \frac{1}{2}N - 1 \right), \tag{3.78}$$

obtained from (3.76), where R_k is the real part of F_k. To get the recursion in (3.78) started, the value of H_1^S is needed. But,

$$H_1^S = \sum_{j=1}^{N-1} h_j \sin \frac{j\pi}{N} \tag{3.79}$$

can be calculated first (for instance, when creating $\{f_j\}$), and then used to begin the recursion in (3.78).

The calculation of a FCT follows a pattern similar to that of the FST. In this case, we define the sequence $\{f_j\}$ by

$$f_0 = h_0$$

$$f_j = s_j + a_j, \qquad (j = 1, \ldots, N-1) \tag{3.80}$$

where

$$s_j = \frac{1}{2} \left[h_j + h_{N-j} \right],$$

$$a_j = \left[h_{N-j} - h_j \right] \sin \frac{j\pi}{N}. \tag{3.80a}$$

We leave it as an exercise for the reader to show that

$$H_{2k}^C = R_k, \qquad \left(k = 0, 1, \ldots, \frac{1}{2}N - 1 \right) \tag{3.81}$$

and

$$H_{2k+1}^C = H_{2k-1}^C + I_k, \qquad \left(k = 1, \ldots, \frac{1}{2}N - 1 \right)$$

$$H_1^C = \sum_{j=0}^{N-1} h_j \cos \frac{j\pi}{N} \tag{3.82}$$

where $\{R_k\}$ and $\{I_k\}$ are the real and imaginary parts of the FFT of $\{f_j\}$ using weight $W = e^{i2\pi/N}$.

Computer procedures, called COSTRAN and SINETRAN, which perform the algorithms described in this section can be found in Appendix B.

3.8   Inversion of Discrete Sine and Cosine Transforms

In this section we will describe how to invert DSTs and DCTs, thereby recovering the original sequences before they were transformed. First, we show that a DST has the nice property that *it is its own inverse* (up to multiplication by a constant).

More precisely, we will show that the inversion of a DST $\{H_k^S\}$ of the sequence $\{h_j\}$ is given by

$$\frac{2}{N} \sum_{k=1}^{N-1} H_k^S \sin \frac{\pi jk}{N} = h_j, \qquad (j = 1, \ldots, N-1). \qquad (3.83)$$

Formula (3.83) follows from

$$\sum_{k=1}^{N-1} \sin \frac{\pi mk}{N} \sin \frac{\pi jk}{N} = \begin{cases} \frac{1}{2}N & \text{if } m = j \\ 0 & \text{if } m \neq j. \end{cases} \qquad (3.84)$$

We will prove (3.84) in a moment. First, we show that (3.83) follows from it.
The left side of (3.83) can be rewritten as

$$\frac{2}{N} \sum_{k=1}^{N-1} \left[\sum_{m=1}^{N-1} h_m \sin \frac{\pi mk}{N} \right] \sin \frac{\pi jk}{N} = \sum_{m=1}^{N-1} h_m \left[\frac{2}{N} \sum_{k=1}^{N-1} \sin \frac{\pi mk}{N} \sin \frac{\pi jk}{N} \right]$$

which, by applying (3.84), yields (3.83).
Now, to prove (3.84), we note that for *fixed* j and m, the sequence

$$\left\{ \sin \frac{\pi mk}{N} \sin \frac{\pi jk}{N} \right\}_{k=1}^{2N-1}$$

is even about N. Hence, from Formula (3.68), using N in place of $N/2$ and k in place of j, and noting that the middle element $\sin(\pi mN/N) \sin(\pi jN/N)$ equals 0, we have

$$\sum_{k=1}^{N-1} \sin \frac{\pi mk}{N} \sin \frac{\pi jk}{N} = \frac{1}{2} \sum_{k=0}^{2N-1} \sin \frac{\pi mk}{N} \sin \frac{\pi jk}{N}$$

$$= \frac{1}{2} \sum_{k=0}^{2N-1} \sin \frac{2\pi mk}{2N} \sin \frac{2\pi jk}{2N}$$

$$= \frac{1}{4} \sum_{k=0}^{2N-1} \cos \frac{2\pi(m-j)k}{2N}$$

$$- \frac{1}{4} \sum_{k=0}^{2N-1} \cos \frac{2\pi(m+j)k}{2N}. \qquad (3.85)$$

Moreover,

$$\sum_{k=0}^{2N-1} \cos \frac{2\pi (m-j)k}{2N} = \text{Re}\left[\sum_{k=0}^{2N-1} e^{i2\pi(m-j)k/2N}\right]$$

$$= \begin{cases} 0 & \text{for } m \neq j \\ 2N & \text{for } m = j \end{cases} \qquad (3.86)$$

where the last equality follows from Formula (2.7), Chapter 2, with $2N$ in place of N. [The argument being similar to the one used to prove Theorem 2.1(c) in Chapter 2.]

Similarly, we have

$$\sum_{k=0}^{2N-1} \cos \frac{2\pi (m+j)k}{2N} = 0. \qquad (3.87)$$

Substituting from (3.86) and (3.87) back into (3.85) we obtain Formula (3.84).

There is also an inversion procedure for DCTs. It is, however, more complicated than the inversion procedure for DSTs. First, we need to prove the following formula:

$$\sum_{k=0}^{N-1} \cos \frac{\pi jk}{N} \cos \frac{\pi mk}{N} = \begin{cases} N & \text{for } m = j = 0 \\ \frac{1}{2}N & \text{for } m = j \neq 0 \\ \frac{1}{2} - \frac{1}{2}(-1)^{j+m} & \text{for } m \neq j. \end{cases} \qquad (3.88)$$

Since, for fixed j and m, the sequence $\{\cos(\pi jk/N)\cos(\pi mk/N)\}_{k=1}^{2N-1}$ is even about N, we have

$$\sum_{k=0}^{N-1} \cos \frac{\pi jk}{N} \cos \frac{\pi mk}{N} = \frac{1}{2} \sum_{k=0}^{2N-1} \cos \frac{\pi jk}{N} \cos \frac{\pi mk}{N} + \frac{1}{2} - \frac{1}{2}(-1)^{j+m}. \quad (3.89)$$

That (3.89) implies (3.88) is shown by a similar argument to the one used above to show (3.84). The details are left to the reader as an exercise.

From (3.88) we find that

$$\frac{2}{N} \sum_{k=0}^{N-1} H_k^C \cos \frac{\pi jk}{N} = \frac{2}{N} \sum_{m=0}^{N-1} h_m \left[\sum_{k=0}^{N-1} \cos \frac{\pi jk}{N} \cos \frac{\pi mk}{N}\right]$$

$$= \begin{cases} 2h_0 + \frac{1}{N}\sum_{m=1}^{N-1} h_m - \frac{1}{N}\sum_{m=1}^{N-1} h_m(-1)^m & \text{for } j = 0 \\ \\ h_j + \frac{1}{N}\sum_{m=0}^{N-1} h_m - \frac{1}{N}\sum_{m=0}^{N-1} h_m(-1)^{m+j} & \text{for } j \neq 0. \end{cases} \quad (3.90)$$

Consequently, since

$$\frac{1}{N}\sum_{m=0}^{N-1} h_m - \frac{1}{N}\sum_{m=0}^{N-1} h_m(-1)^{m+j} = \frac{1}{N}\sum_{m=0}^{N-1} h_m \left[1 - (-1)^{m+j}\right]$$

we obtain from (3.90)

$$\frac{2}{N}\sum_{k=0}^{N-1} H_k^C \cos\frac{\pi jk}{N} = \begin{cases} h_j + \frac{2}{N}\sum_{m \text{ odd}}^{N-1} h_m & \text{for } j \text{ even} \neq 0 \\ \\ h_j + \frac{2}{N}\sum_{m \text{ even}}^{N-2} h_m & \text{for } j \text{ odd} \\ \\ 2h_0 + \frac{2}{N}\sum_{m \text{ odd}}^{N-1} h_m & \text{for } j = 0. \end{cases} \quad (3.91)$$

Using (3.91) it is possible to recover $\{h_j\}$ from $\{H_k^C\}$ *provided* the sums

$$s_1 = \frac{2}{N}\sum_{m \text{ odd}}^{N-1} h_m \quad \text{and} \quad s_2 = \frac{2}{N}\sum_{m \text{ even}}^{N-2} h_m$$

are known. To find these unknown sums, we introduce the sequence $\tilde{h}_j$ defined by

$$\tilde{h}_j = \frac{2}{N}\sum_{k=0}^{N-1} H_k^C \cos\frac{\pi jk}{N}, \quad (j = 0, 1, \ldots, N-1).$$

Then, from (3.91) we obtain

$$\sum_{j \text{ even}}^{N-2} \tilde{h}_j = h_0 + \sum_{j=0}^{N-1} h_j \quad (3.92)$$

and

$$\sum_{j \text{ odd}}^{N-1} \tilde{h}_j = \sum_{j=0}^{N-1} h_j. \quad (3.93)$$

From (3.92) and (3.93) it follows that

$$h_0 = \sum_{j \text{ even}}^{N-2} \tilde{h}_j - \sum_{j \text{ odd}}^{N-1} \tilde{h}_j. \quad (3.94)$$

Hence, since (3.91) implies

$$\tilde{h}_0 = 2h_0 + \frac{2}{N} \sum_{m \text{ odd}}^{N-1} h_m$$

we get

$$s_1 = \tilde{h}_0 - 2 \left[\sum_{j \text{ even}}^{N-2} \tilde{h}_j - \sum_{j \text{ odd}}^{N-1} \tilde{h}_j \right]. \tag{3.95}$$

Furthermore, it also follows from (3.93), and the original definitions of s_1 and s_2, that

$$s_2 = \frac{2}{N} \sum_{j \text{ odd}}^{N-1} \tilde{h}_j - s_1. \tag{3.96}$$

Thus, we can invert $\{H_k^C\}$ to recover $\{h_j\}$ by the following steps:

Step 1. Form the sequence $\{\tilde{h}_j\}$ defined by

$$\tilde{h}_j = \frac{2}{N} \sum_{k=0}^{N-1} H_k^C \cos \frac{\pi j k}{N}, \qquad (j = 0, 1, \ldots, N-1).$$

Step 2. Calculate the quantities s_1 and s_2 by

$$s_1 = \tilde{h}_0 - 2 \left[\sum_{j \text{ even}}^{N-2} \tilde{h}_j - \sum_{j \text{ odd}}^{N-1} \tilde{h}_j \right]$$

$$s_2 = \frac{2}{N} \sum_{j \text{ odd}}^{N-1} \tilde{h}_j - s_1.$$

Step 3. The sequence $\{h_j\}_{j=0}^{N-1}$ is then obtained by

$$h_j = \begin{cases} \frac{1}{2}[\tilde{h}_0 - s_1] & \text{for } j = 0 \\ \tilde{h}_j - s_1 & \text{for } j \text{ even and } \neq 0 \\ \tilde{h}_j - s_2 & \text{for } j \text{ odd.} \end{cases}$$

3.9 Inversion of the FFT of a Real Sequence

In this section we show how the symmetry property (3.57) of the FFT of a real sequence can be used to reduce by about one half the computations involved in inverting the FFT. Although there are other methods of doing this than the method described below, this method is of interest because it uses no extra sine nor tangent values other than the sets $\{S(m)\}_{m=0}^{N/4}$ and $\{T(m)\}_{m=0}^{N/4}$ which were used for performing the FFT itself.

We will assume in this section that $W = e^{i2\pi/N}$. Let R_k and I_k stand for the real and imaginary parts of the FFT $\{F_k\}$. That is,

$$F_k = R_k + i I_k, \qquad (k = 0, 1, \ldots, N-1) \tag{3.97}$$

where F_k is defined by

$$F_k = \sum_{j=0}^{N-1} f_j W^{jk}.$$

Then, by the formula for DFT inversion

$$f_j = \frac{1}{N} g_j, \qquad (j = 0, 1, \ldots, N-1) \tag{3.98}$$

where

$$g_j = \sum_{k=0}^{N-1} F_k (W^{jk})^*. \tag{3.99}$$

Using $W = e^{i2\pi/N}$ and (3.97) we have

$$g_j = \sum_{j=0}^{N-1} (R_k + i I_k) e^{-i2\pi jk/N}$$

$$= \sum_{k=0}^{N-1} (R_k + i I_k)\left(\cos\frac{2\pi jk}{N} - i \sin\frac{2\pi jk}{N}\right). \tag{3.100}$$

Since $\{f_j\}$ consists of real numbers, so does $\{g_j\}$ because of (3.98). Consequently, only the real part of the sum in (3.100) is non-zero. Therefore, we must have

$$g_j = \sum_{k=0}^{N-1} R_k \cos\frac{2\pi jk}{N} + \sum_{k=0}^{N-1} I_k \sin\frac{2\pi jk}{N}. \tag{3.101}$$

Because of (3.57) we have

$$R_{N-k} = R_k, \qquad I_{N-k} = -I_k, \qquad (k = 1, \ldots, N - 1). \qquad (3.102)$$

It follows from (3.102) that, for each fixed j, the sequences

$$\left\{ R_k \cos \frac{2\pi jk}{N} \right\}_{k=1}^{N-1} \qquad \text{and} \qquad \left\{ I_k \sin \frac{2\pi jk}{N} \right\}_{k=1}^{N-1}$$

are even about $N/2$. From Formula (3.68) we obtain

$$\sum_{k=1}^{N-1} R_k \cos \frac{2\pi jk}{N} = R_{\frac{1}{2}N}(-1)^j + 2 \sum_{k=1}^{\frac{1}{2}N-1} R_k \cos \frac{2\pi jk}{N} \qquad (3.103)$$

and

$$\sum_{k=1}^{N-1} I_k \sin \frac{2\pi jk}{N} = 2 \sum_{k=1}^{\frac{1}{2}N-1} I_k \sin \frac{2\pi jk}{N}. \qquad (3.104)$$

Using (3.103) and (3.104) in (3.101) yields

$$g_j = R_0 + R_{\frac{1}{2}N}(-1)^j + 2 \sum_{k=1}^{\frac{1}{2}N-1} R_k \cos \frac{\pi jk}{\frac{1}{2}N} + 2 \sum_{k=1}^{\frac{1}{2}N-1} I_k \sin \frac{\pi jk}{\frac{1}{2}N}. \qquad (3.105)$$

Formula (3.105) shows that $\{g_j\}_{j=0}^{N-1}$, and consequently $\{f_j\}_{j=0}^{N-1}$, can be generated from an $N/2$-point FCT of

$$\{0, 2R_1, 2R_2, \ldots, 2R_{\frac{1}{2}N-1}\}$$

and an $N/2$-point FST of

$$\{2I_1, 2I_2, \ldots, 2I_{\frac{1}{2}N-1}\}.$$

These FCTs and FSTs are even and odd about $N/2$, respectively. Taking this symmetry into account, Formula (3.105) applies to $j = 0, 1, \ldots, N - 1$.

As we noted above, an interesting feature of (3.105) is that to compute the FCTs and FSTs, *of order* $N/2$, by the methods of Section 3.7, one needs only the *same* sines $\{S(m)\}_{m=0}^{N/4}$ and tangents $\{T(m)\}_{m=0}^{N/4}$ generated to calculate the FFT $\{F_k\}_{k=0}^{N-1}$. Therefore, inverting the FFT using (3.105) requires only these same sines and tangents. This is a useful memory savings.

A computer procedure, called InvRFFT, which carries out the algorithm described in this section can be found in Appendix B.

References

For further discussion of DFTs and FFTs, see [Bri], [El-R], [Nu], [Op-S], [PFTV], [Ra], and [Ra-G]. An important set of foundational papers is collected in [Ra-R]. FFT programs in various computer languages can be found in [PFTV].

Exercises

Section 1

3.1 Make a diagram, like the one in Figure 3.4, for the 16-point FFT. How many multiplications are required?

3.2 Construct tables of bit reversed numbers for $N = 4, 8, 16$, and 32.

3.3 Suppose that $N = 4^R$. Derive an FFT based on successive divisions of N by 4. [*Remark:* This is called a radix 4 FFT.]

Section 2

3.4 Construct tables of base 4 digit reversals for $N = 4$ and $N = 16$. [This is needed because of Exercise 3.3.]

3.5 Describe the analog of Buneman's algorithm for digit reversal, base 4. [If $n = 1\,3\,2\,0\,1$ (base 4), then the digit reversal of n is $1\,0\,2\,3\,1$.]

3.6 Repeat Exercise 3.5 for base r, where r is an integer greater than 1.

3.7 Generalize the second algorithm described in this section, using base 4.

3.8 Generalize the second algorithm described in this section, but use base r.

Section 3

3.9 Show that $S(N/2 - m) = S(m)$ for $m = 0, 1, \ldots, N/4$.

3.10 Show that, for $m = 0, 1, \ldots, N/4$

$$C(m) = S\left(\frac{1}{4}N - m\right), \qquad C\left(\frac{1}{2}N - m\right) = -S\left(\frac{1}{4}N - m\right).$$

Using these identities and Formula (3.34), show that the values of $\{T(m)\}_{m=0}^{N/4}$ can be computed using the values $\{S(m)\}_{m=0}^{N/4}$.

3.11 What sines and tangents are needed for performing rotations in a radix 4 FFT $(N = 4^R)$?

Section 4

3.12 Use the first three stages of Buneman's method, along with a hand calculator, to find $\sin(2\pi j/32)$ for $j = 0, 1, \ldots, 8$. Check your results by computing the same values directly on a calculator.

3.13 Develop the fourth stage of Buneman's method to find $\sin(2\pi j/64)$ for $j = 0, 1, \ldots, 16$.

3.14 Suppose that $\{F_k\}$ is the N-point DFT of $\{f_j\}$, defined by

$$F_k = \sum_{j=0}^{N-1} f_j W^{jk}.$$

Prove that, for each $j = 0, 1, \ldots, N-1$,

$$f_j = \frac{1}{N} \left[\sum_{k=0}^{N-1} F_k^* W^{jk} \right]^*. \tag{3.106}$$

3.15 Explain why Formula (3.106) in Exercise 3.14 shows that only the values $\{S(j)\}_{j=0}^{N/4}$ and $\{T(j)\}_{j=0}^{N/4}$ are needed for inverting an FFT, when that FFT used weight $W = e^{i2\pi/N}$. (See Sample Program 2 in Appendix B.)

3.16 Prove (3.68) and (3.69).

3.17 Show that (3.81) and (3.82) are true.

Section 8

3.18 Show that (3.88) follows from (3.89).

3.19 Prove that, for $m, k = 0, 1, \ldots, N-1$,

$$\sum_{n=0}^{N-1} \cos \frac{(n+\frac{1}{2})(m+\frac{1}{2})\pi}{N} \cos \frac{(n+\frac{1}{2})(k+\frac{1}{2})\pi}{N} = \begin{cases} \frac{1}{2}N & \text{if } m = k \\ 0 & \text{if } m \neq k. \end{cases} \tag{3.107}$$

Prove also that a formula like (3.107) holds if sines are used instead of cosines.

3.20 Using the results of Exercise 3.19, formulate a definition of an N-point DCT *that is (up to a constant multiple) its own inverse.* Do the same for sines instead of cosines. How would you program a fast version of this discrete cosine (sine) transform?

4

Some Applications of Fourier Series

In this chapter we will describe computer modeling of solutions to a few fundamental equations from mathematical physics. We shall discuss one-dimensional
versions of the heat equation, the wave equation, and the Schrödinger wave equation for an enclosed, freely moving particle. We shall also examine some of the
modifications of Fourier series that are used in signal processing.

4.1   Heat Equation

In this section we shall describe computer modeling of a simple problem in heat
conduction. Let the function $u(x, t)$, where $0 \leq x \leq L$ and $t \geq 0$, be interpreted
as the temperature of a thin insulated rod, with a temperature of 0 at the ends, and
let the function $g(x)$ be interpreted as the *initial temperature* $u(x, 0) = g(x)$. In
[Wa, Chapter 3.1] it is shown that

$$\frac{\partial u}{\partial t} = a^2 \frac{\partial^2 u}{\partial x^2} \qquad \text{(heat equation)}$$

$$u(0, t) = 0, \quad u(L, t) = 0 \qquad \text{(boundary conditions)}$$

$$u(x, 0) = g(x) \qquad \text{(initial condition)}. \qquad (4.1)$$

The constant a^2 is called the *diffusion constant* for the material of the rod. Its units
are commonly cm^2/sec. Copper, for example, has a diffusion constant of 1.14,
while granite has a diffusion constant of 0.011.

Problem (4.1) is a classic problem in mathematical physics. It was first solved
by Fourier (see [Fo]). Its derivation and solution by Fourier series is described in
[Wa], [Ch-B], and [We].

A second interpretation of Problem (4.1) is that the function u describes the relative concentration of a solvent within a surrounding solute. The function $g(x)$ describes the initial relative concentration. This interpretation is described in many physical chemistry textbooks.

Usually, the method of separation of variables is used to solve (4.1). Since this method is so well known, we will describe another method. We begin by expanding $u(x, t)$ in a Fourier series in the spatial variable x. Since all of the functions in the system $\{\sin(n\pi x/L)\}_{n=1}^{\infty}$ satisfy the *zero boundary conditions* at $x = 0$ and $x = L$, we expand $u(x, t)$ in a *Fourier sine series*

$$u(x, t) = \sum_{n=1}^{\infty} B_n(t) \sin \frac{n\pi x}{L},$$

$$\left(B_n(t) = \frac{2}{L} \int_0^L u(x, t) \sin \frac{n\pi x}{L} \, dx \right). \tag{4.2}$$

To find the functions $B_n(t)$ we substitute into the heat equation in (4.1) and, differentiating term by term, get

$$\sum_{n=1}^{\infty} B_n'(t) \sin \frac{n\pi x}{L} = \sum_{n=1}^{\infty} -\left(\frac{n\pi a}{L}\right)^2 B_n(t) \sin \frac{n\pi x}{L}. \tag{4.3}$$

Expressing (4.3) in terms of one sine series yields

$$\sum_{n=1}^{\infty} \left[B_n'(t) + \left(\frac{n\pi a}{L}\right)^2 B_n(t) \right] \sin \frac{n\pi x}{L} = 0. \tag{4.4}$$

Since the sine coefficients of the 0-function are all 0 we infer from (4.4) that

$$B_n'(t) + \left(\frac{n\pi a}{L}\right)^2 B_n(t) = 0, \qquad (n = 1, 2, 3, \ldots). \tag{4.5}$$

Substituting $t = 0$ into the formula for $B_n(t)$ in (4.2) we also have

$$B_n(0) = \frac{2}{L} \int_0^L u(x, 0) \sin \frac{n\pi x}{L} \, dx, \qquad (n = 1, 2, 3, \ldots). \tag{4.6}$$

The right side of (4.6) is a constant for each n; it represents the sine coefficients of the initial temperature $u(x, 0)$. If we define b_n by

$$b_n = \frac{2}{L} \int_0^L u(x, 0) \sin \frac{n\pi x}{L} \, dx, \qquad (n = 1, 2, 3, \ldots) \tag{4.7}$$

then (4.6) becomes

$$B_n(0) = b_n, \qquad (n = 1, 2, 3, \ldots). \tag{4.8}$$

For each n, the solution to the differential equation (4.5) with initial condition (4.8) is

$$B_n(t) = b_n e^{-(n\pi a/L)^2 t}. \tag{4.9}$$

Combining (4.9) with the first equation in (4.2) yields the following Fourier sine series solution to Problem (4.1),

$$u(x, t) = \sum_{n=1}^{\infty} \left[e^{-(n\pi a/L)^2 t} \right] b_n \sin \frac{n\pi x}{L} \tag{4.10a}$$

where

$$u(x, 0) = \sum_{n=1}^{\infty} b_n \sin \frac{n\pi x}{L}. \tag{4.10b}$$

In other words, *the temperature $u(x, t)$ for $t > 0$ is obtained by putting damping factors*

$$e^{-(n\pi a/L)^2 t} \qquad \text{(Gaussian damping factor)}$$

on the coefficients of the Fourier sine series for the initial temperature $u(x, 0)$.

Example 4.1:

Suppose that $a^2 = 1.14$ cm^2/sec, the diffusion constant of copper, and $L = 10$ cm. Graph $u(x, t)$ for $t = 0.1, 0.5, 1.0,$ and 2.0 sec; given an initial temperature of

$$u(x, 0) = \begin{cases} 0 & \text{for } 0 \leq x \leq 4 \text{ and } 6 \leq x \leq 10 \\ 20x - 80 & \text{for } 4 < x < 5 \\ 120 - 20x & \text{for } 5 \leq x < 6. \end{cases}$$

SOLUTION   In Figure 4.1(b) we show the graph of the 25 harmonic partial sum for the Fourier sine series of $u(x, 0)$ using 512 points. We will use this partial sum as an approximation of $u(x, 0)$. Although the graph of this partial sum appears noticeably different than the graph of the initial function $u(x, 0)$ in Figure 4.1(a), we shall see below that it will still provide good approximations for the positive values of t that we are using.

Now, for the times given, we have damping factors $e^{-(n\pi a/L)^2 t}$. *Using **FAS**, these factors can be put as coefficients on the terms of the* 25 *harmonic partial sum by choosing to do a filter and then choosing **Gauss** as the type of filter. For*

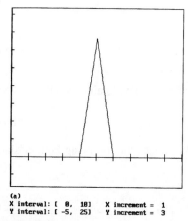
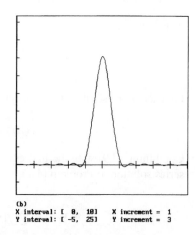

(a)
X interval: [0, 10]   X increment = 1
Y interval: [-5, 25]   Y increment = 3

(b)
X interval: [0, 10]   X increment = 1
Y interval: [-5, 25]   Y increment = 3

FIGURE 4.1
(a) Graph of the initial temperature $u(x,0)$ in Example 4.1. (b) Graph of the 25 harmonic Fourier sine series partial sum for this initial temperature.

$t = 0.1$ the damping constant T asked for by *FAS* is $T = a^2 0.1 = 0.114$. And, for $t = 0.5$, 1.0, and 2.0 sec, we use $T = 0.57$, 1.14, and 2.28, respectively. The graphs of $u(x, t)$ for these values of t are shown in Figure 4.2.

We will now show that using a 25 harmonic partial sum gives good approximations to $u(x, t)$ for the times $t = 0.1$, 0.5, 1.0, and 2.0 that we used above. First, we observe that the sine coefficients b_n satisfy

$$|b_n| = |\frac{1}{5} \int_0^{10} u(x,0) \sin \frac{n\pi x}{10} \, dx| \leq \frac{1}{5} \int_0^{10} |u(x,0)| \, dx.$$

Thinking of the integral of $|u(x, 0)|$ as an area, we get

$$|b_n| \leq 4, \qquad (n = 1, 2, 3, \ldots). \tag{4.11}$$

From (4.11) we obtain (putting $a^2 t = T$)

$$\left| u(x,t) - \sum_{n=1}^{25} b_n e^{-n^2(\pi/10)^2 T} \sin \frac{n\pi x}{10} \right| = \left| \sum_{n=26}^{\infty} b_n e^{-n^2(\pi/10)^2 T} \sin \frac{n\pi x}{10} \right|$$

$$\leq \sum_{n=26}^{\infty} |b_n| e^{-n^2(\pi/10)^2 T}$$

$$\leq \sum_{n=26}^{\infty} 4 e^{-n^2(\pi/10)^2 T}.$$

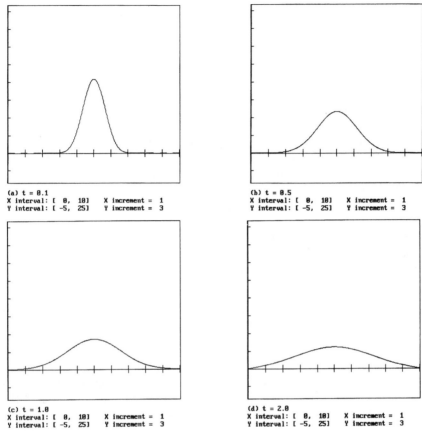

(a) t = 0.1
X interval: [0, 10]   X increment = 1
Y interval: [-5, 25]   Y increment = 3

(b) t = 0.5
X interval: [0, 10]   X increment = 1
Y interval: [-5, 25]   Y increment = 3

(c) t = 1.0
X interval: [0, 10]   X increment = 1
Y interval: [-5, 25]   Y increment = 3

(d) t = 2.0
X interval: [0, 10]   X increment = 1
Y interval: [-5, 25]   Y increment = 3

FIGURE 4.2
**Graphs of the temperature $u(x, t)$ evolving from the initial temperature shown
in Figure 4.1(a).**

Now, if we write out the terms of the last series, we have

$$\sum_{n=26}^{\infty} 4e^{-n^2(\pi/10)^2 T} = 4e^{-26^2(\pi/10)^2 T} + 4e^{-27^2(\pi/10)^2 T} + 4e^{-28^2(\pi/10)^2 T} + \cdots$$

$$\leq 4e^{-26^2(\pi/10)^2 T}\left[1 + e^{-53(\pi/10)^2 T} + e^{-(2)53(\pi/10)^2 T} + \cdots\right]$$

$$= 4e^{-26^2(\pi/10)^2 T}\sum_{k=0}^{\infty} e^{-k53(\pi/10)^2 T} = \frac{4e^{-26^2(\pi/10)^2 T}}{1 - e^{-53(\pi/10)^2 T}}.$$

Combining this last result with our previous estimates yields

$$\left| u(x,t) - \sum_{n=1}^{25} b_n e^{-n^2(\pi/10)^2 T} \sin \frac{n\pi x}{10} \right| \leq \frac{4e^{-26^2(\pi/10)^2 T}}{1 - e^{-53(\pi/10)^2 T}}. \tag{4.12}$$

Substituting the values of $T = 0.114$, 0.57, 1.14, and 2.28 into the right side of (4.12), we see that the maximum differences between $u(x,t)$ and the 25 harmonic partial sums used for each $T = a^2 t$ are no more than about 0.00443, 1.284×10^{-16}, 3.725×10^{-33}, and 3.451×10^{-66}, respectively. This shows what *excellent* approximations the 25 harmonic partial sums are to the actual temperature $u(x,t)$ for the last three values of T. For $T = 0.114$ more harmonics must be used to obtain such precision. ∎

We will end this section with a brief description of how *FAS* graphs the filtered sine series partial sums just described. First, the Fourier sine coefficients b_n are approximated by a uniform, left endpoint, sum:

$$b_n \approx \frac{2}{N} \sum_{j=1}^{N-1} u(jL/N, 0) \sin \frac{\pi j n}{N}. \tag{4.13}$$

Thus, at least for $1 \leq n \leq N/4$, we will have b_n approximated by $2/N$ times the discrete sine transform (DST) of $\{u(jL/N, 0)\}_{j=1}^{N-1}$. Denoting this DST by $\{U_n\}$, we substitute into an M-harmonic partial sum of the series for $u(x,t)$ in (4.10a) and we obtain (putting $a^2 t = T$)

$$u(x,t) \approx \frac{2}{N} \sum_{n=1}^{M} \left[U_n e^{-(n\pi/L)^2 T} \right] \sin \frac{n\pi x}{L}, \qquad (T = a^2 t). \tag{4.14}$$

Assuming that $M \leq N/4$ (as in Chapter 2, Section 2.5) we substitute $x = mL/N$, for $m = 0, 1, \ldots, N - 1$, into (4.14), obtaining

$$u(mL/N, t) \approx \frac{2}{N} \sum_{n=1}^{N-1} V_n \sin \frac{\pi nm}{N} \tag{4.15a}$$

where

$$V_n = \begin{cases} U_n e^{-(n\pi/L)^2 T} & \text{for } n = 1, 2, \ldots, M \\ 0 & \text{for } n = M + 1, \ldots, N - 1. \end{cases} \tag{4.15b}$$

Based on (4.15a) and (4.15b) we have $u(x,t)$ approximated by the following steps:

Step 1. A DST $\{U_n\}_{n=1}^{N-1}$ is computed from the set of sample values $\{u(jL/N, 0)\}_{j=1}^{N-1}$ of the initial function $u(x, 0)$.

Step 2. The DST $\{U_n\}$ is lopped off by setting to 0 the values for the indices $n = M + 1, \ldots, N - 1$ where $M \leq N/4$. The remaining values of U_n are multiplied by the damping factors $e^{-(n\pi/L)^2 T}$ where T is a non-negative constant (called the damping constant) entered by the user. For solving (4.1), T should be given the value $a^2 t$ for each specific time t. The new sequence obtained by this procedure is called $\{V_n\}$.

Step 3. An inverse DST is applied to V_n to obtain approximations to the function values $\{u(mL/N, t)\}_{m=1}^{N-1}$. Since $u(x, t)$ equals the sine series in (4.10a) we also have $u(0, t) = 0$ and $u(L, t) = 0$.

Step 4. The values of $\{u(mL/N, t)\}_{m=0}^{N}$ are connected by straight line segments to give an approximation of $u(x, t)$ for $0 \leq x \leq L$.

4.2 The Wave Equation

In this section we describe the computer modeling of solutions to the problem of describing the motion of a vibrating string.

Consider a string of length L with both its ends fixed; for example, a guitar string or a violin string. It can be shown (see [Wa, Chapter 3.1]) that in the absence of any externally applied force, the position $y(x, t)$ of the string for $0 \leq x \leq L$ and $t \geq 0$ satisfies

$$c^2 \frac{\partial^2 y}{\partial x^2} = \frac{\partial^2 y}{\partial t^2} \qquad \text{(wave equation)}$$

$$y(0, t) = 0, \quad y(L, t) = 0 \qquad \text{(boundary conditions)}$$

$$y(x, 0) = f(x), \quad \frac{\partial y}{\partial t}(x, 0) = g(x) \qquad \text{(initial conditions)}. \qquad (4.16)$$

The function $f(x)$ specifies the *initial position* of the string, while the function $g(x)$ specifies the *initial velocity* of the string.

The system of units involved is the CGS system. The constant c^2 is equal to T/ρ where T is the constant tension along the string and ρ is the constant linear density of the string. In the CGS system c has the units of *velocity*, cm/sec.

Expanding $y(x, t)$ in a Fourier sine series,

$$y(x, t) = \sum_{n=1}^{\infty} B_n(t) \sin \frac{n\pi x}{L},$$

$$\left(B_n(t) = \frac{2}{L} \int_0^L y(x, t) \sin \frac{n\pi x}{L} \, dx \right) \tag{4.17}$$

and substituting this series for $y(x, t)$ into the wave equation in (4.16) and into the initial conditions in (4.16), we obtain

$$\sum_{n=1}^{\infty} \left[B_n''(t) + (nc\pi/L)^2 B_n(t) \right] \sin \frac{n\pi x}{L} = 0$$

$$B_n(0) = \frac{2}{L} \int_0^L f(x) \sin \frac{n\pi x}{L} dx \qquad (n = 1, 2, 3, \ldots)$$

$$B_n'(0) = \frac{2}{L} \int_0^L g(x) \sin \frac{n\pi x}{L} dx \qquad (n = 1, 2, 3, \ldots). \tag{4.18}$$

Letting a_n and b_n stand for the nth sine coefficients of f and g, respectively, we obtain from (4.18)

$$B_n''(t) + (nc\pi/L)^2 B_n(t) = 0, \qquad B_n(0) = a_n, \qquad B_n'(0) = b_n \tag{4.19}$$

for $n = 1, 2, 3, \ldots$. For each n the solution to (4.19) is

$$B_n(t) = a_n \cos (nc\pi/L)t + b_n \frac{\sin(nc\pi/L)t}{nc\pi/L}. \tag{4.20}$$

Using the notation

$$\text{sinc } v = \frac{\sin \pi v}{\pi v}, \qquad (\text{sinc } 0 = 1) \tag{4.21}$$

we can express (4.20) as

$$B_n(t) = a_n \cos (nc\pi/L)t + b_n t \text{ sinc } (nc/L)t. \tag{4.22}$$

Combining (4.22) with (4.17) we have found the following series solution to Problem (4.16):

$$y(x, t) = \sum_{n=1}^{\infty} [a_n \cos (nc\pi/L)t + b_n t \text{ sinc } (nc/L)t] \sin \frac{n\pi x}{L}$$

$$a_n = \frac{2}{L} \int_0^L y(x, 0) \sin \frac{n\pi x}{L} dx,$$

$$b_n = \frac{2}{L} \int_0^L \frac{\partial y}{\partial t}(x, 0) \sin \frac{n\pi x}{L} dx. \tag{4.23}$$

Example 4.2:
Suppose that $L = 50\,\text{cm}$, $c = 100\,\text{cm/sec}$, the initial position is $f(x) = 0.1x(50 - x)$, and the initial velocity g is zero. Graph $y(x, t)$ for the times $t = 0.1, 0.2$, and 0.3 sec.

SOLUTION   Since $g(x) = 0$ for $0 \le x \le L$, it follows that $b_n = 0$ for $n = 1,$ $2, 3, \ldots$. Using a 200 harmonic partial sum for $y(x, t)$ yields

$$y(x, t) \approx \sum_{n=1}^{200} a_n \cos(n\pi ct/L) \sin \frac{n\pi x}{L} \tag{4.24}$$

where $L = 50$ and $c = 100$. The sine series partial sum in (4.24) is obtained by applying *filter coefficients*

$$\cos(n\pi ct/L), \quad (n = 1, 2, 3, \ldots) \tag{4.25}$$

to the terms of the 200 harmonic partial sum of the sine series for $y(x, 0) = f(x)$. This can be done by first calculating the sine series partial sum, using 1024 points. Then, after pressing y in response to the question about using a filter, one presses the *Tab* key to display the second list of filter choices. The filter needed in this case is the *Cos* filter. You will be asked for the value of c, called the *wave constant,* which is 100 for this example. Then you will be asked for the value of t, called the *time constant.* After performing these steps and entering the three values of t for the time constant, you will obtain graphs like the ones shown in Figure 4.3(a). ∎

The approximation in (4.24) is a good one. Calculating the sine coefficients $\{a_n\}$

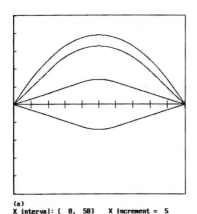

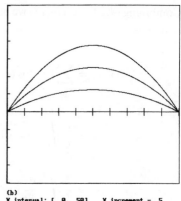

(a)
X interval: [0, 50]   X increment = 5
Y interval: [-80, 80]   Y increment = 16

(b)
X interval: [0, 50]   X increment = 5
Y interval: [-40, 60]   Y increment = 10

FIGURE 4.3
(a) Graphs of the string positions for Example 4.2. (b) Graphs of the string positions for Example 4.3.

for $f(x) = 0.1x(50 - x)$, we get

$$a_n = \frac{1000}{n^3\pi^3}\left[1 - (-1)^n\right], \qquad (n = 1, 2, 3, \ldots). \qquad (4.26)$$

Therefore,

$$\left| y(x, t) - \sum_{n=1}^{200} a_n \cos(n\pi ct/L) \sin\frac{n\pi x}{L} \right| = \left| \sum_{n=201}^{\infty} a_n \cos(n\pi ct/L) \sin\frac{n\pi x}{L} \right|$$

$$\leq \sum_{n=201}^{\infty} |a_n| = \sum_{k=101}^{\infty} \frac{2000}{(2k-1)^3\pi^3}.$$

The last sum above can be estimated using the integral test from calculus

$$\sum_{k=101}^{\infty} \frac{2000}{(2k-1)^3\pi^3} \leq \int_{100}^{\infty} \frac{2000}{(2x-1)^3\pi^3}\, dx = \frac{500}{(199)^2\pi^3}$$

$$< 0.41 \times 10^{-3}.$$

Thus, the approximation in (4.24) involves an error of less than 0.41×10^{-3} for $0 \leq x \leq 50$ and all time $t \geq 0$.

REMARK 4.1: Since Formula (4.26) gives us an exact expression for the Fourier sine series coefficients, it is possible to graph exact partial sums (instead of the Fast Sine Transform [FST] approximations used in the example above). The gain in accuracy, however, is insignificant (see Exercise 4.12) and certainly does not justify the huge increase in time needed to calculate sums of 200 terms for every point plotted. Of course, for most functions we cannot obtain exact expressions for the Fourier sine series coefficients so we *must* use FST approximations. ∎

Example 4.3:

Suppose that L and c have the same values as in the previous example, but that the initial velocity is described by $20x(50 - x)$ while the initial position is zero. Graph $y(x, t)$ for times $t = 0.001, 0.002$, and 0.003 sec.

SOLUTION In this case, the coefficients $\{a_n\}$ are all 0. Again using 200 harmonics in a partial sum for $y(x, t)$ we have

$$y(x, t) \approx \sum_{n=1}^{200} [t \text{ sinc } (nct/L)] b_n \sin \frac{n\pi x}{L} . \qquad (4.27)$$

In this example, one computes the filter factors

$$t \text{ sinc } (nct/L), \qquad (n = 1, 2, 3, \ldots). \qquad (4.28)$$

One chooses the filter *Sinc* on the second filter menu (press the *Tab* key when the first filter menu appears). You are asked to enter the wave constant, which is $c = 100$ as in the previous example, and the time constants (0.001, 0.002, and 0.003). This is how one obtains the graphs shown in Figure 4.3(b). ∎

The approximation in (4.27) is more accurate than the previous example. Since the values of b_n are 200 times the values of a_n in the previous example (since $g = 200f$ where f is as in the previous example), we obtain

$$\left| y(x, t) - \sum_{n=1}^{200} t \text{ sinc } \frac{nct}{L} b_n \sin \frac{n\pi x}{L} \right|$$

$$= \left| \sum_{n=201}^{\infty} \frac{200 \sin(nc\pi t/L)}{nc\pi/L} \frac{1000[1 - (-1)^n]}{n^3\pi^3} \sin \frac{n\pi x}{L} \right|$$

$$\leq \sum_{k=101}^{\infty} \frac{200\,000}{(2k-1)^4 \pi^4} < 4.34 \times 10^{-5}.$$

Therefore, the approximation in (4.27) involves an error of no more than 4.34×10^{-5} for $0 \leq x \leq 50$ and all time $t \geq 0$.

REMARK 4.2:   The filtered Fourier sine series partial sums described in this section can be used to produce *simulations* of vibrating strings in the form of *computer animations*. For an example of such a computer animation, view the graphbook *VIBR_STR* which can be found on the computer disk accompanying this book. ∎

4.3   Schrödinger's Equation for a Free Particle

In this section we shall discuss a problem involving the one-dimensional Schrödinger's equation:

$$\frac{-\hbar}{i} \frac{\partial \psi}{\partial t} = \frac{-\hbar^2}{2m} \frac{\partial^2 \psi}{\partial x^2} + V\psi.$$

We shall assume that the potential term, V, equals 0. Hence, the equation simplifies to

$$\frac{\partial \psi}{\partial t} = \frac{i\hbar}{2m} \frac{\partial^2 \psi}{\partial x^2}.$$

We shall now describe computer modeling of the filtered Fourier series solution to the following problem:

$$\frac{\partial \psi}{\partial t} = \frac{i\hbar}{2m} \frac{\partial^2 \psi}{\partial x^2} \qquad \text{(Schrödinger's equation)}$$

$$\psi(x, t) = 0, \qquad \text{for } x \leq 0 \text{ or } x \geq L \qquad \text{(boundary conditions)}$$

$$\psi(x, 0) = f(x), \qquad \text{for } 0 \leq x \leq L \qquad \text{(initial condition)}. \qquad (4.29)$$

The function ψ is interpreted in quantum mechanics as a generator of a probability density function, $|\psi|^2 \ (= \psi \psi^*)$, governing the position x of a particle of mass m

at time $t \geq 0$. Later, we will show that

$$\|\psi\|_2^2 = \int_0^L |\psi(x, t)|^2 dx = 1 \tag{4.30}$$

provided that the given initial function f satisfies

$$\|f\|_2^2 = \int_0^L |f(x)|^2 dx = 1. \tag{4.31}$$

Formulas (4.30) and (4.31) are the *normalization conditions* that are imposed on the function ψ whose evolution from f is governed by Schrödinger's equation. The constant $\hbar$ is Planck's constant divided by 2π. The value of $\hbar$ is 1.054×10^{-27} erg-sec. By comparison, an electron has mass 0.911×10^{-27} g. Because of the absence of a potential term in Schrödinger's equation in (4.29), the particle is said to move *freely*. The boundary conditions in (4.29) say that the particle is constrained to move along the x-axis between 0 and L. Problem (4.29) is a one-dimensional version of the *particle in a box* problem.

To find the solution ψ to (4.29) we proceed as in the previous two sections. We expand $\psi(x, t)$ and $f(x)$ in a Fourier sine series in x:

$$\psi(x, t) = \sum_{n=1}^{\infty} C_n(t) \sin \frac{n\pi x}{L}, \qquad \left[C_n(t) = \frac{2}{L} \int_0^L \psi(x, t) \sin \frac{n\pi x}{L} dx \right]$$

$$f(x) = \sum_{n=1}^{\infty} c_n \sin \frac{n\pi x}{L}, \qquad \left[c_n = \frac{2}{L} \int_0^L f(x) \sin \frac{n\pi x}{L} dx \right]. \tag{4.32}$$

Substituting the series for $\psi(x, t)$ into Schrödinger's equation in (4.29), differentiating term by term, and combining the two series into one, yields

$$\sum_{n=-\infty}^{\infty} \left[C_n'(t) + \frac{i\hbar}{2m} \left(\frac{n\pi}{L} \right)^2 C_n(t) \right] \sin \frac{n\pi x}{L} = 0. \tag{4.33}$$

Setting $t = 0$ in the formula for $C_n(t)$ in (4.32) yields [by comparison to the formula for c_n]

$$C_n(0) = c_n, \qquad (n = 0, \pm 1, \pm 2, \ldots). \tag{4.34}$$

Setting each coefficient in (4.33) equal to 0, we get

$$C_n'(t) = \frac{-i\hbar}{2m} \left(\frac{n\pi}{L} \right)^2 C_n(t), \qquad (n = 0, \pm 1, \pm 2, \ldots). \tag{4.35}$$

Solving (4.35), subject to the initial condition (4.34), for each n yields

$$C_n(t) = c_n e^{\frac{-i\hbar}{2m}\left(\frac{n\pi}{L}\right)^2 t}, \qquad (n = 0, \pm 1, \pm 2, \ldots). \qquad (4.36)$$

Combining (4.36) with (4.32), we have our solution to (4.29):

$$\psi(x, t) = \sum_{n=1}^{\infty} \left[e^{\frac{-i\hbar}{2m}\left(\frac{n\pi}{L}\right)^2 t} \right] c_n \sin \frac{n\pi x}{L} \qquad (4.37a)$$

where

$$\psi(x, 0) = \sum_{n=1}^{\infty} c_n \sin \frac{n\pi x}{L}. \qquad (4.37b)$$

Our solution to (4.29) expresses $\psi(x, t)$ as a *filtered Fourier sine series* obtained from the Fourier series for the initial function $\psi(x, 0)$. The filter factors are the complex, time-dependent exponentials given in Equation (4.36).

Example 4.4:
Suppose that the initial function $\psi(x, 0)$ is given by

$$\psi(x, 0) = \begin{cases} 0.5 & \text{for } 30 < x < 34 \\ 0 & \text{for } 0 < x < 30 \text{ or } 34 < x < 64 \end{cases}$$

(see Figure 4.4) and the particle has a mass of 0.911×10^{-27} g (mass of an electron). Graph $\psi(x, t)$ for $t = 0.05$ and 0.15 sec.

SOLUTION   The initial function $\psi(x, 0)$ is *real valued*. Consequently, the Fourier series for $\psi(x, t)$ can be split into real and imaginary parts

$$\psi(x, t) = \psi^R(x, t) - i\psi^I(x, t) \qquad (4.38)$$

where

$$\psi^R(x, t) = \sum_{n=1}^{\infty} \cos\left[\frac{\hbar}{2m} \left(\frac{n\pi}{L}\right)^2 t \right] c_n \sin \frac{n\pi x}{L}$$

$$\psi^I(x, t) = \sum_{n=1}^{\infty} \sin\left[\frac{\hbar}{2m} \left(\frac{n\pi}{L}\right)^2 t \right] c_n \sin \frac{n\pi x}{L}. \qquad (4.39)$$

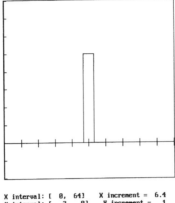

X interval: [0, 64]   X increment = 6.4
Y interval: [-.2, .8]   Y increment = .1

FIGURE 4.4
Graph of the initial function in Example 4.4.

FAS can be used to graph partial sum approximations to ψ^R and ψ^I, and use these approximations to approximate the probability density function

$$|\psi|^2 = (\psi^R)^2 + (\psi^I)^2. \tag{4.40}$$

To graph a 2000 harmonic partial sum approximation to $\psi^R(x, t)$, call it $S_{2000}^R(x, t)$, one first computes a 2000 harmonic partial sum for $\psi(x, 0)$, using *FAS* (8192 points were used for these examples). Then, one chooses the *FresCos* filter from the second filter menu. The filter procedure asks you to input the value of $\hbar/2m$, which it calls the *wave constant*. For an electron, the wave constant has a value of 0.578. The *FresCos* procedure then asks you for the *time constant,* for which you enter one of the values of t. *FAS* then computes filter factors

$$\cos\left[\frac{\hbar}{2m}\left(\frac{n\pi}{L}\right)^2 t\right], \qquad (n = 0, \pm1, \pm2, \ldots)$$

and graphs S_{2000}^R, the 2000 harmonic partial sum approximation to ψ^R.

To graph the 2000 harmonic partial sum approximation to $\psi^I(x, t)$, call it $S_{2000}^I(x, t)$, you choose *FresSin* from the filter menu, and then enter the same wave constant and the same time constants as for ψ^R. If you graph S_{2000}^R and S_{2000}^I simultaneously for $t = 0.05$ and 0.15 sec, then you obtain graphs like the ones shown in Figure 4.5. In Figure 4.6 we show graphs of

$$|S_{2000}(x, t)|^2 = [S_{2000}^R(x, t)]^2 + [S_{2000}^I(x, t)]^2$$

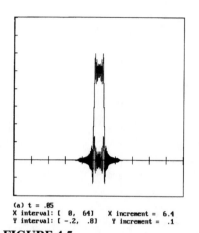
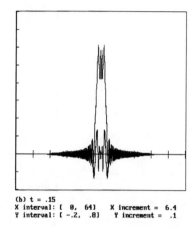

(a) t = .05
X interval: [0, 64]   X increment = 6.4
Y interval: [-.2, .8]   Y increment = .1

(b) t = .15
X interval: [0, 64]   X increment = 6.4
Y interval: [-.2, .8]   Y increment = .1

FIGURE 4.5
Approximations of the real and imaginary parts of $\psi(x, t)$ in Example 4.4 for $t = 0.05$ and 0.15 sec.

which is the 2000 harmonic approximation to $|\psi(x, t)|^2$. These graphs were obtained by choosing *Graphs* on the display menu and then entering the formula

$$f(x) = g1(x) \wedge 2 + g2(x) \wedge 2$$

This sequence of steps is performed after both S_{2000}^{R} and S_{2000}^{I} are displayed simultaneously.

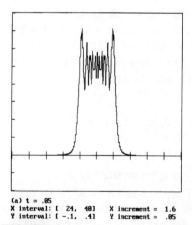
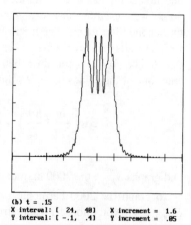

(a) t = .05
X interval: [24, 40]   X increment = 1.6
Y interval: [-.1, .4]   Y increment = .05

(b) t = .15
X interval: [24, 40]   X increment = 1.6
Y interval: [-.1, .4]   Y increment = .05

FIGURE 4.6
Approximations of $|\psi(x, t)|^2$ in Example 4.4 for $t = 0.05$ and 0.15 sec.

The similarity between these approximations of $|\psi(x, t)|^2$ and one-dimensional Fresnel diffraction patterns from optics will be obvious to anyone familiar with such

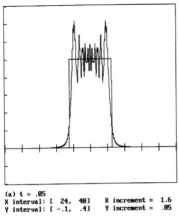 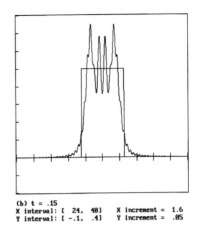

(a) t = .05
X interval: [24, 40]   X increment = 1.6
Y interval: [-.1, .4]   Y increment = .05

(b) t = .15
X interval: [24, 40]   X increment = 1.6
Y interval: [-.1, .4]   Y increment = .05

FIGURE 4.7
Comparison between $|\psi(x,0)|^2$ and $|\psi(x,t)|^2$ for $t = 0.05$ and 0.15 sec.

diffraction patterns. (See Chapter 6, Section 6.2.) It is interesting to compare these approximations of $|\psi(x,t)|^2$ with the graph of the initial distribution $|\psi(x,0)|^2$; see Figure 4.7.   ∎

It is interesting to see what happens if the initial probability density is narrowed, as in our next example.

Example 4.5:
Suppose that the initial function $\psi(x,0)$ is

$$\psi(x,0) = \begin{cases} \sqrt{2} & \text{for } 31.75 < x < 32.25 \\ 0 & \text{for } 0 < x < 31.75 \text{ and } 32.25 < x < 64. \end{cases}$$

Graph approximations to $|\psi(x,t)|^2$ for the same times as in the previous example and for the same mass, too.

SOLUTION   The graphs of $|S_{2000}|^2$ ($\approx |\psi(x,t)|^2$) are shown in Figure 4.8. The similarity between the graphs for $t = 0.05$ and 0.15 sec and *Fraunhofer diffraction patterns* from optics will be obvious to anyone familiar with such diffraction patterns. (See Chapter 6, Section 6.3.)   ∎

We close this section by showing that, given (4.31), Equation (4.30) must hold. And, we also show that the 2000 harmonic partial sum approximations used in the

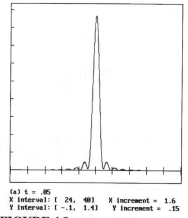
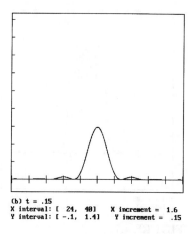

(a) t = .05
X interval: [24, 40]   X increment = 1.6
Y interval: [-.1, 1.4]   Y increment = .15

(b) t = .15
X interval: [24, 40]   X increment = 1.6
Y interval: [-.1, 1.4]   Y increment = .15

FIGURE 4.8
Approximations of $|\psi(x, t)|^2$ in Example 4.5 for $t = 0.05$ and 0.15 sec.

last two examples are fairly good if the 2-Norm

$$\|g\|_2 = \left[\int_0^L |g(x)|^2 \, dx \right]^{\frac{1}{2}} \tag{4.41}$$

is used to measure the magnitudes of the differences between functions.

We need the following theorem, whose proof is beyond the scope of this text.

THEOREM 4.1:
The series

$$\sum_{n=1}^{\infty} a_n \sin \frac{n\pi x}{L}$$

*is a Fourier sine series for a function g, where $\|g\|_2 < \infty$, if and only if $\sum_{n=1}^{\infty} |a_n|^2$ converges. Moreover, when $\|g\|_2 < \infty$, then **Parseval's equality** holds*

$$\frac{L}{2} \sum_{n=1}^{\infty} |a_n|^2 = \int_0^L |g(x)|^2 \, dx = \|g\|_2^2$$

*and we have the **completeness relation***

$$\lim_{M \to \infty} \|g - S_M\|_2 = \lim_{M \to \infty} \left[\frac{L}{2} \sum_{n > M}^{\infty} |a_n|^2 \right]^{\frac{1}{2}} = 0$$

where S_M is the M-harmonic partial sum of the Fourier sine series for g.

Theorem 4.1 is a corollary of the following theorem (proved in [Ru, Chapter 4]). In this theorem, $\|g\|_2$ stands for $[\int_{-L}^{L} |g(x)|^2 \, dx]^{1/2}$.

THEOREM 4.2: COMPLETENESS OF FOURIER SERIES

The series

$$\sum_{n=-\infty}^{\infty} a_n e^{i\pi nx/L}$$

is a Fourier series for a function g, where $\|g\|_2 < \infty$, if and only if $\sum_{n=-\infty}^{\infty} |a_n|^2$ converges. Moreover, when $\|g\|_2 < \infty$, then **Parseval's equality** *holds*

$$2L \sum_{n=-\infty}^{\infty} |a_n|^2 = \int_{-L}^{L} |g(x)|^2 \, dx = \|g\|_2^2$$

and we have the **completeness relation**

$$\lim_{M \to \infty} \|g - S_M\|_2 = \lim_{M \to \infty} \left[2L \sum_{|n|>M}^{\infty} |a_n|^2 \right]^{\frac{1}{2}} = 0$$

where S_M is the M-harmonic partial sum of the Fourier series for g.

Using Theorem 4.1 we have, based on (4.31) and the second equation in (4.32),

$$1 = \|f\|_2^2 = \frac{L}{2} \sum_{n=1}^{\infty} |c_n|^2. \tag{4.42}$$

However, we also have

$$\left| e^{i \frac{\hbar}{2m} \left(\frac{n\pi}{L}\right)^2 t} c_n \right| = |c_n|$$

so (4.42) implies

$$1 = \frac{L}{2} \sum_{n=1}^{\infty} \left| e^{i \frac{\hbar}{2m} \left(\frac{n\pi}{L}\right)^2 t} c_n \right|^2. \tag{4.43}$$

Using Theorem 4.1 again, we conclude from (4.43) that the series in the first equation in (4.32) is the Fourier series for $\psi(x, t)$ and (4.30) holds.

Theorem 4.1 can also be used to show that the 2000 harmonic partial sums in the examples above are good approximations. Letting $S_M(x, t)$ stand for the M-harmonic partial sum of the series in (4.37a), it follows that $\psi(x, t) - S_M(x, t)$ has the Fourier sine series expansion

$$\psi(x, t) - S_M(x, t) = \sum_{n>M}^{\infty} \left[e^{i \frac{\hbar}{2m} \left(\frac{n\pi}{L}\right)^2 t} c_n \right] \sin \frac{n\pi x}{L} \tag{4.44}$$

and, for $t = 0$ in particular,

$$\psi(x, 0) - S_M(x, 0) = \sum_{n > M}^{\infty} c_n \sin \frac{n\pi x}{L}. \qquad (4.45)$$

By Theorem 4.1 applied to (4.44) and (4.45) we obtain, by the same method as above,

$$\|\psi(x, 0) - S_M(x, 0)\|_2 = \left[\frac{L}{2} \sum_{|n| > M}^{\infty} |c_n|^2 \right]^{\frac{1}{2}}$$

$$= \|\psi(x, t) - S_M(x, t)\|_2. \qquad (4.46)$$

Equation (4.46) says that the 2-Norm measure of the magnitude of the difference between $\psi(x, t)$ and $S_M(x, t)$ is a constant for all $t \geq 0$. Using the choices *Graphs* (on the display menu) and *Norm diff* (entering 2 for the *Power*), one can estimate the 2-Norm magnitude of the difference between the initial function $\psi(x, 0)$ and its M-harmonic Fourier series $S_M(x, 0)$ *when both graphs are displayed.* When $M = 2000$, the magnitude of the 2-Norm difference obtained for Example 4.4 is approximately 0.0039 while for Example 4.5 it is approximately 0.11. In both cases, $S_{2000}(x, t)$ is a reasonable approximation, in terms of 2-Norm, to $\psi(x, t)$. This is especially true if the 2-Norm difference is thought of as measuring the distance between two vectors of length 1 (corresponding to an *angular,* or more precisely, *chordal,* distance).

4.4   Filters Used in Signal Processing

In each of the preceding sections an initial Fourier sine series was *filtered* by multiplying its coefficients by *filter factors*. These factors were time dependent. In this section we will give an introduction to some of the common *time-independent* filters, most of which are used in signal processing.

Cesàro Filter

Cesàro filtering is also known as the method of *arithmetic means*. To motivate this procedure, consider Figure 4.9. In parts (b) to (d) of Figure 4.9, we can see how the successive partial sums of the Fourier series seem to *interlace* around the graph of the step function. By interlace we mean not only that the partial sums oscillate about the function, but that *at most points* they also change from being

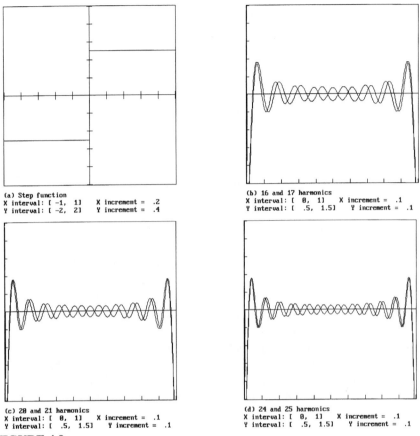

(a) Step function
X interval: [−1, 1]   X increment = .2
Y interval: [−2, 2]   Y increment = .4

(b) 16 and 17 harmonics
X interval: [0, 1]   X increment = .1
Y interval: [.5, 1.5]   Y increment = .1

(c) 20 and 21 harmonics
X interval: [0, 1]   X increment = .1
Y interval: [.5, 1.5]   Y increment = .1

(d) 24 and 25 harmonics
X interval: [0, 1]   X increment = .1
Y interval: [.5, 1.5]   Y increment = .1

FIGURE 4.9
Behavior of Fourier series partial sums for a step function.

above the step function to below the step function (or from below to above) as one
passes from one partial sum to the next. It makes sense then to form an average,
an arithmetic mean, of partial sums in order to better approximate the function.

DEFINITION 4.1:   *Given a function with Fourier series partial sums* $\{S_n\}_{n=0}^{\infty}$,
the Mth ***arithmetic mean,*** *or* ***Cesàro filtered Fourier series using*** M ***harmonics,***
is denoted by σ_M *where*

$$\sigma_M = \frac{1}{M}\left[S_0 + S_1 + \ldots + S_{M-1}\right].$$

By replacing S_k by $\sum_{j=-k}^{k} c_j e^{i\pi jx/L}$ we get

$$\sigma_M(x) = \frac{1}{M} \sum_{k=0}^{M-1} \left[\sum_{j=-k}^{k} c_j e^{i\pi jx/L} \right]. \tag{4.47}$$

By fixing a value of n, for some $n = 0, \pm 1, \pm 2, \ldots, \pm M$, and counting how often c_n appears in the sums in (4.47), we obtain

$$\sigma_M(x) = \sum_{n=-M}^{M} \left(1 - \left| \frac{n}{M} \right| \right) c_n e^{i\pi nx/L}. \tag{4.48}$$

If we compare σ_M in (4.48) to S_M, where

$$S_M(x) = \sum_{n=-M}^{M} c_n e^{i\pi nx/L} \tag{4.49}$$

we see that σ_M is obtained from S_M by multiplying the coefficients in S_M by the *filter factors* $\{1 - |n/M|\}_{n=-M}^{n=M}$. These factors are often called *convergence factors* since they will usually help improve the pointwise convergence of the Cesàro filtered Fourier series to the original function. One form of this improved convergence is *suppression of Gibbs' phenomenon* (see Figure 4.10). Another aspect of Cesàro filtering is that the filter coefficients are near 0 for $|n|$ near M,

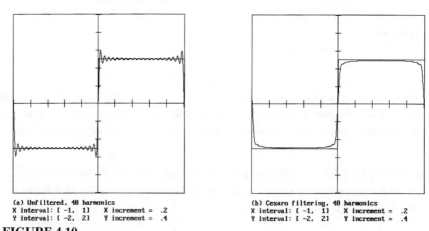

(a) Unfiltered, 40 harmonics
X interval: [-1, 1]   X increment = .2
Y interval: [-2, 2]   Y increment = .4

(b) Cesaro filtering, 40 harmonics
X interval: [-1, 1]   X increment = .2
Y interval: [-2, 2]   Y increment = .4

FIGURE 4.10
Suppression of Gibbs' phenomenon by Cesàro filtering.

consequently the *higher frequency harmonics in S_M are damped down in σ_M.* This is a common feature of many of the filtering procedures used in signal processing.

de le Vallée Poussin Filter

Another filter that is closely related to the Cesàro filter is the de la Vallée Poussin filter (dlVP filter, for short). To motivate the use of the dlVP filter, we note that for small values of n the Fourier series partial sum S_n is often not a good approximation to the original function. For example, graphs of the 2 and 3 harmonic Fourier series partial sums for the step function shown in Figure 4.11 are not close to that

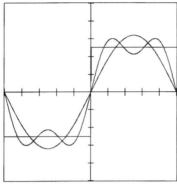

```
X interval: [ -1,  1]    X increment =  .2
Y interval: [ -2,  2]    Y increment =  .4
```

FIGURE 4.11
Graphs of 2 and 3 harmonic Fourier series partial sums for a step function.

step function at all. Putting this another way, notice that in Figure 4.9, we chose partial sums that were intertwined about the step function *and that this intertwining becomes denser with larger number of harmonics.* Therefore, it might be more advantageous to average only the upper half of the partial sums. If we have an even number of harmonics, say $2M$, then we define V_{2M} by

$$V_{2M} = \frac{1}{M} \sum_{k=M}^{2M-1} S_k. \tag{4.50}$$

This function V_{2M} is called the *dlVP filtered partial sum using $2M$ harmonics.*
We now show that

$$V_{2M}(x) = \sum_{n=-2M}^{2M} v_n c_n e^{i\pi nx/L} \tag{4.51}$$

where

$$v_n = \begin{cases} 1 & \text{if } |n| \le M \\ 2(1 - |\dfrac{n}{2M}|) & \text{if } M < |n| \le 2M. \end{cases} \tag{4.52}$$

First, we rewrite (4.50) using the definition of S_k:

$$V_{2M}(x) = \frac{1}{M} \sum_{k=M}^{2M-1} \left[\sum_{j=-k}^{k} c_j e^{i\pi jx/L} \right]. \tag{4.53}$$

Second, we find the coefficient of $e^{i\pi nx/L}$ for each $n = 0, \pm 1, \ldots, \pm 2M$. There are two cases to consider.

Case 1. ($|n| \leq M$). For this case c_n appears in each term in brackets in (4.53) for $k = M$ to $2M - 1$. Therefore, the exponential $e^{i\pi nx/L}$ has coefficient

$$\frac{1}{M} \sum_{k=M}^{2M-1} c_n = c_n \left[\frac{1}{M} \sum_{k=M}^{2M-1} 1 \right] = c_n.$$

Hence $v_n = 1$ as in (4.52).

Case 2. ($M < |n| \leq 2M$). For this case, c_n appears in only those terms in the brackets in (4.53) where $k = |n|, \ldots, 2M - 1$. Therefore, the coefficient of $e^{i\pi nx/L}$ is

$$\frac{1}{M} \sum_{k=|n|}^{2M-1} c_n = c_n \left[\frac{1}{M} \sum_{k=|n|}^{2M-1} 1 \right] = c_n \frac{2M - |n|}{M}$$

$$= \left(2 - |\frac{n}{M}| \right) c_n = 2 \left(1 - |\frac{n}{2M}| \right) c_n.$$

Hence, $v_n = 2(1 - |n/2M|)$ as in (4.52).

In *FAS*, the dlVP filter is found on the second filter menu. *FAS* will calculate the dlVP filtered Fourier series

$$V_M(x) = \sum_{n=-M}^{M} v_n c_n e^{i\pi nx/L} \tag{4.54}$$

where

$$v_n = \begin{cases} 1 & \text{if } 2|n| \leq M \\ 2\left(1 - \frac{|n|}{M}\right) & \text{if } M \leq 2|n| \leq 2M \end{cases} \tag{4.55}$$

which *for 2M in place of M*, matches the description of V_{2M} in (4.51) and (4.52). Formula (4.54) has the advantage that it allows the calculation of a dlVP filter for an odd number of harmonics as well as an even number of harmonics.

In Figure 4.12 we have graphed a dlVP filtered Fourier series partial sum for

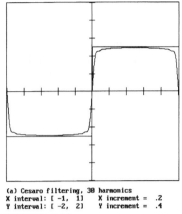
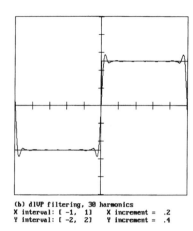

(a) Cesaro filtering, 30 harmonics
X interval: [-1, 1]   X increment = .2
Y interval: [-2, 2]   Y increment = .4

(b) dlVP filtering, 30 harmonics
X interval: [-1, 1]   X increment = .2
Y interval: [-2, 2]   Y increment = .4

FIGURE 4.12
Comparison of Cesàro and dlVP filtering.

a step function. For this example we can see that dlVP filtering gives a closer approximation to the original step function than Cesàro filtering does.

Hamming and Hanning Filters

We close this section by mentioning two other filters frequently used in signal processing. When the function $H_M(x)$ is computed by the formula

$$H_M(x) = \sum_{n=-M}^{M} A_n c_n e^{i\pi nx/L} \tag{4.56}$$

where

$$A_n = 0.5 + 0.5 \cos \frac{n\pi}{M} \tag{4.57}$$

then the Fourier series

$$\sum_{n=-\infty}^{\infty} c_n e^{i\pi nx/L}$$

is said to be *hanning filtered*. The function $H_M(x)$ is called the *hanning filtered Fourier series using M harmonics.*

If, instead of Formula (4.57), one uses

$$A_n = 0.54 + 0.46 \cos \frac{n\pi}{M} \tag{4.58}$$

then the Fourier series is said to be *Hamming filtered.* And the function $H_M(x)$ is called the *Hamming filtered Fourier series using M harmonics.*

REMARK 4.3: (a) The similarity between the standard names for these filters, hanning and Hamming, is a cause for confusion. The hanning filter (note the small h) is named in honor of the Austrian mathematician von Hann. The Hamming filter (note the capital H) is named after Richard W. Hamming, who is from the U.S. (b) The underlying motivation for hanning and Hamming filtering is not as easily explained as our two previous examples. Their description requires an examination of *point spread functions*, which we discuss in Section 4.6. ∎

4.5 Designing Filters

All of the filters described in Section 4.4 fit into one common framework. Understanding this framework will allow you to design your own filters, using the function creation procedure of *FAS*.

Each filtering procedure used a modified Fourier series partial sum

$$\sum_{n=-M}^{M} F\left(\frac{n}{M}\right) c_n e^{i\pi nx/L} \tag{4.59}$$

where the *unfiltered* Fourier series partial sum was

$$\sum_{n=-M}^{M} c_n e^{i\pi nx/L}. \tag{4.60}$$

In (4.59) we shall assume that the function F is a continuous, even function over the interval $[-1, 1]$. For three of the filters described in Section 4.4, we have the following formulas for $F(n/M)$ and $F(x)$:

Filter	$F(n/M)$	$F(x)$
Cesàro	$1 - |n/M|$	$1 - |x|$
hanning	$0.5 + 0.5\cos(\frac{\pi n}{M})$	$0.5 + 0.5\cos(\pi x)$
Hamming	$0.54 + 0.46\cos(\frac{\pi n}{M})$	$0.54 + 0.46\cos(\pi x)$

Since $F(x)$ is even, it really only needs to be defined over the interval $[0, 1]$. In which case, the functions above can be expressed as

Filter	$F(x)$, x in $[0, 1]$
Cesàro	$1 - x$
hanning	$0.5 + 0.5\cos(\pi x)$
Hamming	$0.54 + 0.46\cos(\pi x)$

The *User* choice on the filter menu of *FAS* allows you to design you own filter function $F(x)$ over the interval $[0, 1]$. *FAS* then automatically takes care of computing $\{F(n/M)\}$, using the even extension of F for negative n, for $n = 0, \pm1, \pm2, \ldots, \pm M$ and using these filter coefficients in (4.59). Here is an example.

Example 4.6:
If you choose *User* on the filter menu, you can create the following function:

$$F(x) = \begin{cases} 1 & \text{if } 0 \le x \le 0.2 \\ \cos[\pi(x - 0.2)/1.6] & \text{if } 0.2 < x \le 1 \end{cases}$$

by entering the formula

$$f(x) = (0 \le x \le 0.2) + (0.2 < x \le 1)\cos[\pi(x - 0.2)/1.6]. \qquad (4.61)$$

If you create this filter function for $M = 30$ harmonics (using, say 512 points), then the graph shown in Figure 4.13(a) results when the initial function is $f(x) = x$

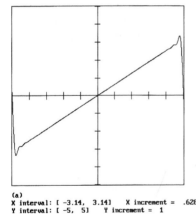
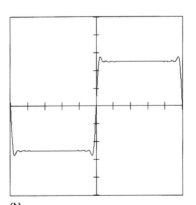

(a)
X interval: [-3.14, 3.14]   X increment = .628
Y interval: [-5, 5]   Y increment = 1

(b)
X interval: [-3.14, 3.14]   X increment = .628
Y interval: [-2, 2]   Y increment = .4

FIGURE 4.13
User-filtered Fourier series.

over $[-\pi, \pi]$. If the initial function is

$$f(x) = \begin{cases} -1 & \text{for } -\pi < x < 0 \\ 1 & \text{for } 0 < x < \pi \end{cases} \qquad (4.62)$$

then the graph shown in Figure 4.13(b) results for $M = 30$ harmonics.

Earlier in this chapter we described filtering of Fourier sine series. A filtered Fourier sine series partial sum has the form

$$\sum_{n=1}^{M} F\left(\frac{n}{M}\right) b_n \sin \frac{n\pi x}{L} \tag{4.63}$$

where $F(x)$ is a continuous function over the interval $[0, 1]$, and

$$\sum_{n=1}^{M} b_n \sin \frac{n\pi x}{L}$$

is the M-harmonic partial sum of the Fourier sine series. By choosing *User* on the filter menu when doing sine series, you can create your own filter function $F(x)$, just as for Fourier series. Here is an example of an application that requires the creation of such a filter.

Example 4.7: INHOMOGENEOUS WAVE EQUATION

It can be shown (see Chapter 3.1 of [Wa]) that when a force $F(x, t)$ *per unit length* is applied to an elastic string with fixed ends and constant tension, then the string height $y(x, t)$ satisfies (ρ is the linear density of the string, L is the length of the string):

$$c^2 \frac{\partial^2 y}{\partial x^2} + \frac{F(x, t)}{\rho} = \frac{\partial^2 y}{\partial t^2} \qquad \text{(inhomogenous wave equation)}$$

$$y(0, t) = 0, \quad y(L, t) = 0 \qquad \text{(boundary conditions)}$$

$$y(x, 0) = f(x), \quad \frac{\partial y}{\partial t}(x, 0) = g(x) \qquad \text{(initial conditions)}. \tag{4.64}$$

Problem (4.64) is solved by a method similar to the one used to solve (4.16). Expanding $y(x, t)$ and $F(x, t)/\rho$ in Fourier sine series for $0 \le x \le L$ we have

$$y(x, t) = \sum_{n=1}^{\infty} B_n(t) \sin \frac{n\pi x}{L},$$

$$\left(B_n(t) = \frac{2}{L} \int_0^L y(x, t) \sin \frac{n\pi x}{L} dx \right) \tag{4.65}$$

and

$$\frac{F(x,t)}{\rho} = \sum_{n=1}^{\infty} F_n(t) \sin \frac{n\pi x}{L},$$

$$\left(F_n(t) = \frac{2}{L} \int_0^L \frac{F(x,t)}{\rho} \sin \frac{n\pi x}{L} \, dx \right). \qquad (4.66)$$

Since $F(x,t)$ will be given to us, we assume that the functions $F_n(t)$ are also known. Substituting the series from (4.65) and (4.66) into the inhomogeneous wave equation in (4.64), differentiating term by term, and rearranging into a single series, yields

$$\sum_{n=1}^{\infty} \left[F_n(t) - \left(\frac{nc\pi}{L}\right)^2 B_n(t) - B_n''(t) \right] \sin \frac{n\pi x}{L} = 0. \qquad (4.67)$$

Since the 0-function has sine coefficients all equal to 0, we set the expression in brackets in (4.67) equal to 0, obtaining

$$B_n''(t) + \left(\frac{nc\pi}{L}\right)^2 B_n(t) = F_n(t), \qquad (n = 1, 2, 3, \ldots). \qquad (4.68)$$

For each n, we also have the initial conditions

$$B_n(0) = a_n, \qquad \left(a_n = \frac{2}{L} \int_0^L y(x,0) \sin \frac{n\pi x}{L} \, dx \right)$$

$$B_n'(0) = b_n, \qquad \left(b_n = \frac{2}{L} \int_0^L \frac{\partial y}{\partial t}(x,0) \sin \frac{n\pi x}{L} \, dx \right). \qquad (4.69)$$

For notational purposes, we define ω_n to be $nc\pi/L$. Our problem is then to solve, for each n, each of the problems

$$B_n''(t) + \omega_n^2 B_n(t) = F_n(t), \qquad \left(\omega_n = \frac{nc\pi}{L} \right)$$

$$B_n(0) = a_n, \qquad B_n'(0) = b_n \qquad (4.70)$$

for the unknown function $B_n(t)$.

For instance, suppose that we have a string of length $L = 10$ and that a force of $0.2\rho \sin(\omega t)$ is applied to the string between $x = 2$ and $x = 3$. Suppose also that

$c = 100$ and that the string is initially at rest in the horizontal position. *We will show how a filtered Fourier sine series can be used to describe the motion of the string.*

For this example, we have

$$y(x, 0) = 0, \qquad \frac{\partial y}{\partial t}(x, 0) = 0 \qquad\qquad (4.71)$$

since the string is initially horizontal and at rest. Consequently, using (4.71) in (4.69), we have

$$a_n = 0, \qquad b_n = 0, \qquad (n = 1, 2, 3, \ldots). \qquad (4.72)$$

Now, the function $F(x, t)/\rho$ has the form

$$\frac{F(x, t)}{\rho} = G(x) \sin \omega t$$

where

$$G(x) = \begin{cases} 0.2 & \text{for } 2 < x < 3 \\ 0 & \text{for } 0 < x < 2 \text{ or } 3 < x < 10. \end{cases} \qquad (4.73)$$

Consequently,

$$F_n(t) = (\sin \omega t) \frac{2}{L} \int_0^L G(x) \sin \frac{n\pi x}{L} \, dx$$

$$= K_n \sin \omega t, \qquad \left(K_n = \frac{2}{L} \int_0^L G(x) \sin \frac{n\pi x}{L} \, dx \right). \qquad (4.74)$$

We will see that for our computer work, the exact expression for K_n is *not* needed. The important thing is that $\{K_n\}$ is the set of Fourier sine coefficients for the function G in (4.73). That is,

$$G(x) = \sum_{n=1}^{\infty} K_n \sin \frac{n\pi x}{L},$$

$$\left(K_n = \frac{2}{L} \int_0^L G(x) \sin \frac{n\pi x}{L} \, dx \right). \qquad (4.75)$$

Now, after substituting the form for $F_n(t)$ from (4.74) and the values for a_n and b_n from (4.72) back into Problem (4.70), we have to solve

$$B_n''(t) + \omega_n^2 B_n(t) = K_n \sin \omega t$$

$$B_n(0) = 0, \quad B_n'(0) = 0. \tag{4.76}$$

If we assume that $\omega \neq \omega_n$ for any n, then the solution is found to be

$$B_n(t) = \frac{K_n}{\omega_n^2 - \omega^2} \sin \omega t - \frac{K_n \omega/\omega_n}{\omega_n^2 - \omega^2} \sin \omega_n t.$$

After simplifying this expression for $B_n(t)$ and substituting into the series for $y(x, t)$ in (4.65), we obtain

$$y(x, t) = \sum_{n=1}^{\infty} \left[\frac{\omega_n \sin \omega t - \omega \sin \omega_n t}{\omega_n (\omega_n^2 - \omega^2)} \right] K_n \sin \frac{n\pi x}{L} \tag{4.77}$$

where $L = 10$ and $\omega_n = nc\pi/L = 10n\pi$.

Formula (4.77) expresses $y(x, t)$ as a *time-dependent* filtered Fourier sine series, where the initial (unfiltered) series is the Fourier sine series for the function G in (4.75). Approximating the series in (4.77) using M harmonics yields

$$y(x, t) \approx \sum_{n=1}^{M} \left[\frac{\omega_n \sin \omega t - \omega \sin \omega_n t}{\omega_n (\omega_n^2 - \omega^2)} \right] K_n \sin \frac{n\pi x}{L}. \tag{4.78}$$

To create a filter function we need to express the filter coefficients

$$\frac{\omega_n \sin \omega t - \omega \sin \omega_n t}{\omega_n (\omega_n^2 - \omega^2)} \tag{4.79}$$

in the form $F(n/M)$ for a function $F(x)$ over the interval $[0, 1]$. The key thing to observe is that only ω_n depends on n. In fact, $\omega_n = 10n\pi = 10\pi M(n/M)$. Consequently, substituting $10\pi M x$ in place of ω_n in (4.79) yields the filter function

$$F(x) = \frac{(10\pi M x) \sin \omega t - \omega \sin[(10\pi M x)t]}{(10\pi M x)[(10\pi M x)^2 - \omega^2]}.$$

Or, more simply, using the sinc function defined in (4.21)

$$F(x) = \frac{\sin \omega t - \omega t \, \text{sinc}\,(10Mxt)}{(10\pi M x)^2 - \omega^2}. \tag{4.80}$$

Suppose we use 1024 points, $M = 200$ harmonics, $\omega = 29.9\pi$, and times
$t = 1.0, t = 1.01$, and $t = 1.03$ sec. By computing a 200 harmonic sine series for
$G(x) = 0.2(2 < x < 3)$, over the interval $[0, 10]$, and creating a filter function
using Formula (4.80) *with the values of M, ω, and each t that we have chosen,*
we obtain the graphs shown in Figure 4.14(a). Or, if we use $\omega = 19.9\pi$, then we
obtain the graphs shown in Figure 4.14(b) (using the same number of harmonics
and the same times).

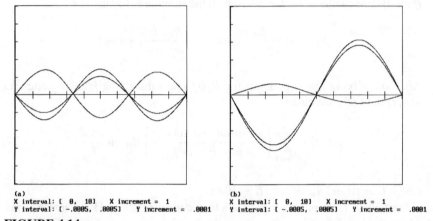

(a)
X interval: [0, 10]   X increment = 1
Y interval: [-.0005, .0005]   Y increment = .0001

(b)
X interval: [0, 10]   X increment = 1
Y interval: [-.0005, .0005]   Y increment = .0001

FIGURE 4.14
**String motion under driving force. (a) Frequency $\omega = 29.9\pi$, (b) frequency
$\omega = 19.9\pi$.**

4.6   Convolution and Point Spread Functions

In this section we continue our examination of filters. We will examine the
concepts of convolution and point spread functions, which are essential for a solid
understanding of filtered Fourier series. In the preceding sections we applied filters
to Fourier series in the following way. From an initial Fourier series for a function
f

$$\sum_{n=-\infty}^{\infty} c_n e^{i\pi nx/L}, \quad \left(c_n = \frac{1}{2L} \int_{-L}^{L} f(x) e^{-i\pi nx/L} \, dx \right) \tag{4.81}$$

we obtain a filtered Fourier series

$$\sum_{n=-\infty}^{\infty} c_n F_n e^{i\pi nx/L}. \tag{4.82}$$

To understand this filtering process, we need the following theorem.

THEOREM 4.3:   FOURIER SERIES CONVOLUTION THEOREM
If f and $\mathcal{P}$ have period 2L, and Fourier series expansions

$$f(x) = \sum_{n=-\infty}^{\infty} c_n e^{i\pi nx/L}, \qquad \left(c_n = \frac{1}{2L} \int_{-L}^{L} f(x)e^{-i\pi nx/L}\,dx\right) \quad (4.83a)$$

$$\mathcal{P}(x) = \sum_{n=-\infty}^{\infty} F_n e^{i\pi nx/L}, \qquad \left(F_n = \frac{1}{2L} \int_{-L}^{L} \mathcal{P}(x)e^{-i\pi nx/L}\,dx\right) \quad (4.83b)$$

then there is a function, denoted by $f * \mathcal{P}$, with Fourier series expansion

$$f * \mathcal{P}(x) = \sum_{n=-\infty}^{\infty} c_n F_n e^{i\pi nx/L}. \qquad (4.83c)$$

This function $f * \mathcal{P}$ is called the **convolution over** $[-L, L]$ **of f and** $\mathcal{P}$. *It is defined by*

$$f * \mathcal{P}(x) = \frac{1}{2L} \int_{-L}^{L} f(s)\mathcal{P}(x - s)\,ds. \qquad (4.83d)$$

PROOF   Given the existence of this function $f * \mathcal{P}$, we compute its nth Fourier coefficient (call it p_n)

$$p_n = \frac{1}{2L} \int_{-L}^{L} f * \mathcal{P}(x)e^{-i\pi nx/L}\,dx$$

$$= \frac{1}{2L} \int_{-L}^{L} \left[\frac{1}{2L} \int_{-L}^{L} f(s)\mathcal{P}(x - s)\,ds\right] e^{-i\pi nx/L}\,dx. \qquad (4.84)$$

Since $e^{-i\pi nx/L}$ is independent of s we bring it inside the integral in brackets and reverse the order of integration, obtaining

$$p_n = \frac{1}{2L} \int_{-L}^{L} f(s) \left[\frac{1}{2L} \int_{-L}^{L} \mathcal{P}(x - s)e^{-i\pi nx/L}\,dx\right] ds. \qquad (4.85)$$

Replacing x by $(x - s) + s$ in $e^{-i\pi nx/L}$ and doing some algebra, we get (after factoring out $e^{-in\pi s/L}$ from the inner integral)

$$p_n = \frac{1}{2L} \int_{-L}^{L} f(s)e^{-in\pi s/L} \left[\frac{1}{2L} \int_{-L}^{L} \mathcal{P}(x - s)e^{-in\pi(x-s)/L} \, dx \right] ds. \quad (4.86)$$

Changing variables, the inner integral in (4.86) becomes

$$\frac{1}{2L} \int_{-L}^{L} \mathcal{P}(x - s)e^{-in\pi(x-s)/L} \, dx = \frac{1}{2L} \int_{-L-s}^{L-s} \mathcal{P}(v)e^{-in\pi v/L} \, dv$$

$$= \frac{1}{2L} \int_{-L}^{L} \mathcal{P}(v)e^{-in\pi v/L} \, dv. \quad (4.87)$$

The second equality holds because $\mathcal{P}(v)e^{-in\pi v/L}$ has period $2L$.

The last integral in (4.87) is the nth Fourier coefficient for $\mathcal{P}$, which we called F_n in Equation (4.83b). Therefore, returning to (4.86), we have

$$p_n = \frac{1}{2L} \int_{-L}^{L} f(s)e^{-in\pi s/L} F_n \, ds = c_n F_n.$$

Thus, the nth Fourier coefficient for $f * \mathcal{P}$ is $c_n F_n$ so (4.83c) holds. ∎

Before we examine the application of this convolution theorem to filtering of Fourier series, we give an example of a convolution.

Example 4.8:
Compute $f * g$ over the interval $[-4, 4]$ where

$$f(x) = g(x) = \begin{cases} 1 & \text{for } |x| < 0.5 \\ 0 & \text{for } 0.5 < |x| < 4. \end{cases}$$

SOLUTION   Applying (4.83d) to compute $f * g$, with g in place of $\mathcal{P}$, we have

$$f * g(x) = \frac{1}{8} \int_{-4}^{4} f(s)g(x - s) \, ds = \frac{1}{8} \int_{-0.5}^{0.5} g(x - s) \, ds.$$

We leave it as an exercise for the reader to finish the calculation, showing that

$$f * g(x) = \begin{cases} (1 - |x|)/8 & \text{for } |x| < 1 \\ 0 & \text{for } 1 \leq |x| < 4. \end{cases} \tag{4.88}$$

With *FAS* we can easily check (4.88). First, choose *Convolution,* then choose the interval $[-4, 4]$ and 1024 points. *FAS* then asks you to define a first function and a second function for convolution. In both cases, you could enter the function rect(x). After doing its computations, *FAS* asks you if you want to divide by the length of the interval, and you should press y for this example (in Chapter 5, we will discuss another type of convolution for which we do not divide by the length of the interval). The result is shown in Figure 4.15. ∎

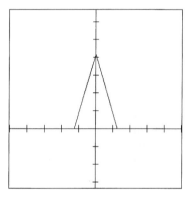

X interval: [−4, 4]   X increment = .8
Y interval: [−.1, .2]   Y increment = .03

FIGURE 4.15
The convolution described in Example 4.8.

REMARK 4.4:   It is an important fact that convolution is *commutative.* That is, if f and g have period $2L$, then

$$f * g = g * f.$$

The proof of this is not hard. We leave it to the reader as an exercise. ∎

The convolution theorem gives us a tool for studying filtering of Fourier series. If we look at a partial sum of the filtered Fourier series in (4.82), containing M harmonics, then we have

$$\sum_{n=-M}^{M} c_n F_n e^{i\pi nx/L}. \tag{4.89}$$

By Theorem 4.3, we then have

$$\sum_{n=-M}^{M} c_n F_n e^{i\pi nx/L} = f * \mathcal{P}_M(x) \qquad (4.90)$$

where

$$\mathcal{P}_M(x) = \sum_{n=-M}^{M} F_n e^{i\pi nx/L}. \qquad (4.91)$$

Indeed, (4.90) follows immediately from (4.83c) since the Fourier coefficients of $\mathcal{P}_M$ are all 0 for $|n| > M$.

We now give a name to this important function $\mathcal{P}_M$.

DEFINITION 4.2:   *Given a sequence $\{F_n\}$ of filter coefficients, the function $\mathcal{P}_M$, defined in (4.91), is called the **point spread function** (PSF) or **kernel** for the filter process.*

We will now describe how *FAS* plots filtered Fourier series and how it plots *PSFs*. This method is similar to the method of approximating sampled Fourier series that we described in Chapter 2, section 2.4.

First, since $f * \mathcal{P}_M$ has period $2L$, it can be sampled over the interval $[0, 2L]$. Replacing x in Formula (4.90) by $x_j = 2Lj/N$ for $j = 0, 1, \ldots, N - 1$, we get

$$f * \mathcal{P}_M(x_j) = \sum_{n=-M}^{M} c_n F_n e^{i2\pi nj/N}, \qquad (j = 0, 1, \ldots, N - 1). \qquad (4.92)$$

With some index shifting we can view (4.92) as an N-point DFT [provided $M < (1/2)N$]. Splitting the sum in (4.92) into two sums, one over positive indices and the other over negative indices, we get

$$f * \mathcal{P}_M(x_j) = \sum_{n=0}^{M} c_n F_n e^{i2\pi nj/N} + \sum_{n=1}^{M} c_{-n} F_{-n} e^{-i2\pi nj/N}. \qquad (4.93)$$

Hence, because $e^{-i2\pi nj/N} = e^{i2\pi(N-n)j/N}$,

$$f * \mathcal{P}_M(x_j) = \sum_{n=0}^{M} c_n F_n e^{i2\pi nj/N} + \sum_{n=1}^{M} c_{-n} F_{-n} e^{i2\pi(N-n)j/N}. \qquad (4.94)$$

Now, recall that the Fourier series coefficients of f are approximated by a DFT (see Remark 2.1[a] in Chapter 2). Denoting this DFT by $\{G_n\}_{n=0}^{N-1}$, we have (for

$M \leq N/8$)

$$c_n \approx G_n/N \qquad \text{and} \qquad c_{-n} \approx G_{N-n}/N. \qquad (4.95)$$

Using (4.95), we rewrite (4.94) as the following approximation:

$$f * \mathcal{P}_M(x_j) \approx \frac{1}{N} \sum_{n=0}^{M} G_n F_n e^{i2\pi nj/N}$$

$$+ \frac{1}{N} \sum_{n=1}^{M} G_{N-n} F_{-n} e^{i2\pi(N-n)j/N}. \qquad (4.96)$$

If we define the sequence $\{H_n\}$ by

$$H_n = \begin{cases} G_n F_n/N & \text{for } n = 0, 1, \ldots, M \\ 0 & \text{for } n = M+1, \ldots, N-M-1 \\ G_n F_{n-N}/N & \text{for } n = N-M, \ldots, N-1 \end{cases} \qquad (4.97)$$

then we can express (4.96) as

$$f * \mathcal{P}_M(x_j) \approx \sum_{n=0}^{N-1} H_n e^{i2\pi nj/N}. \qquad (4.98)$$

Formula (4.98) shows how we can approximate $f * \mathcal{P}_M$ at the points

$$x_j = j \frac{2L}{N}, \qquad (j = 0, 1, \ldots, N-1)$$

by doing an FFT, with weight $e^{i2\pi/N}$, on the sequence $\{H_n\}$ defined in (4.97). The value $f * \mathcal{P}_M(x_N) = f * \mathcal{P}_M(2L)$ can be obtained by periodicity:

$$f * \mathcal{P}_M(2L) = f * \mathcal{P}_M(0).$$

By connecting all of these values by line segments, $f * \mathcal{P}_M(x)$ is approximated for every x in the interval $[0, 2L]$. (When the interval used is $[-L, L]$, then *FAS* makes use of periodicity. It just converts to $[0, 2L]$, does the calculations described above, and converts back to $[-L, L]$.)

Using the method just described, it is also possible to graph the *PSF* $\mathcal{P}_M$. All we need for this is a sequence $\{\Delta_N(j)\}$ whose FFT $\{G_n\}$ is the constant sequence $\{N\}$, since then (4.97) will reduce to the Fourier coefficients of $\mathcal{P}_M$. If we define Δ_N by

$$\Delta_N(j) = \begin{cases} N & \text{if } j = 0 \\ 0 & \text{if } j = 1, 2, \ldots, N-1 \end{cases} \qquad (4.99)$$

then the N-point DFT of Δ_N is the constant sequence $\{N\}_{j=0}^{N-1}$. The function Δ_N is called a *discrete delta function*. The function Δ_N can be graphed by *FAS* as follows. Suppose, for example, that $N = 1024$. You then enter the function

$$f(x) = 1024\delta(x)$$

and Δ_{1024} will be graphed (see Figure 4.16[a]). If you then produce an M harmonic filtered Fourier series, you will obtain graphs of $\mathcal{P}_M$. See Figure 4.16(b) to (d) for graphs of $\mathcal{P}_{40}$ for Cesàro, hanning, and Hamming filtering.

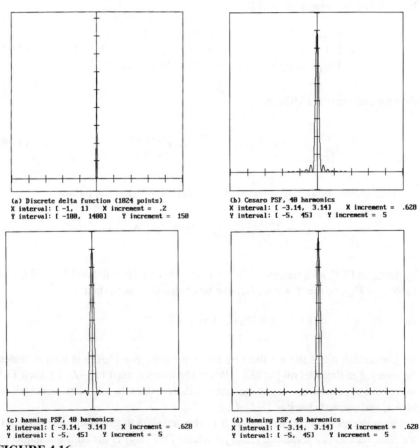

(a) Discrete delta function (1024 points)
X interval: [-1, 1]   X increment = .2
Y interval: [-100, 1400]   Y increment = 150

(b) Cesaro PSF, 40 harmonics
X interval: [-3.14, 3.14]   X increment = .628
Y interval: [-5, 45]   Y increment = 5

(c) hanning PSF, 40 harmonics
X interval: [-3.14, 3.14]   X increment = .628
Y interval: [-5, 45]   Y increment = 5

(d) Hamming PSF, 40 harmonics
X interval: [-3.14, 3.14]   X increment = .628
Y interval: [-5, 45]   Y increment = 5

FIGURE 4.16
Graphs of some point spread functions.

4.7 Discrete Convolutions Using FFTs

In the previous section we defined the concept of convolution of two functions. In this section we describe the method by which the computer can be used to approximate the convolution

$$f * g(x) = \frac{1}{2L} \int_{-L}^{L} f(s)g(x - s)\, ds \qquad (4.100)$$

where f and g both have period $2L$.

The method used to approximate (4.100) is very similar to the method described at the end of Section 4.6. It basically consists of multiplying Fourier coefficients for f and g and then doing an inverse FFT.

Dividing the interval $[-L, L]$ into N subintervals of *equal length* $2L/N$ with left endpoints

$$\{s_m\}_{m=0}^{N-1} = \left\{-L + m\frac{2L}{N}\right\}_{m=0}^{N-1}$$

and using a left-endpoint sum for the integral in (4.100) yields

$$f * g(x) \approx \frac{1}{2L} \sum_{m=0}^{N-1} f(s_m)g(x - s_m)\, \Delta s_m. \qquad (4.101)$$

Since $\Delta s_m = 2L/N$ for each m, we have

$$f * g(x) \approx \frac{1}{N} \sum_{m=0}^{N-1} f(s_m)g(x - s_m). \qquad (4.102)$$

Now, *since g has period $2L$*, if we extend the sequence $\{s_m\}$ beyond the interval $[-L, L]$ using the formula

$$s_m = -L + m\frac{2L}{N}, \qquad (m = 0, \pm 1, \pm 2, \ldots)$$

then we will have for all integers j and m

$$g(x_j - s_m) = g(s_{j-m}) \qquad (4.103)$$

where x_j is defined by

$$x_j = j\frac{2L}{N}, \qquad (j = 0, \pm 1, \pm 2, \ldots).$$

Using (4.103), we can rewrite (4.102) with x_j in place of x:

$$f * g(x_j) \approx \frac{1}{N} \sum_{m=0}^{N-1} f(s_m) g(s_{j-m}). \qquad (4.104)$$

The right side of (4.104) is a discrete convolution (multiplied by $1/N$).

DEFINITION 4.3:   *Let $\{u_j\}$ and $\{v_j\}$ be two sequences of numbers, each having period N. The **cyclic convolution** of $\{u_j\}$ and $\{v_j\}$ is the sequence $\{u * v_j\}$ defined by*

$$u * v_j = \sum_{m=0}^{N-1} u_m v_{j-m}.$$

For example, Formula (4.104) shows that the sequence $\{f * g(x_j)\}$ is approximated by $1/N$ times the cyclic convolution of the sequences $\{f(s_j)\}$ and $\{g(s_j)\}$.

A direct calculation of $u * v_j$ for $j = 0, 1, \ldots, N - 1$ would seem to require $4N^2$ real multiplications (if complex-valued sequences are used). The following theorem allows this number to be reduced to $(9/2)N \log_2 N + 6N$ real multiplications by taking advantage of the FFT. The FFT provides a very efficient way of performing cyclic convolutions.

THEOREM 4.4: DISCRETE CONVOLUTION
If $\{u_j\}$ and $\{v_j\}$ are sequences of period N with DFTs $\{U_k\}$ and $\{V_k\}$, respectively, then the DFT of $\{u * v_j\}$ is $\{U_k V_k\}$.

PROOF   If we put $W = e^{-i2\pi/N}$, then the DFT of $\{u * v_j\}$ is

$$\sum_{j=0}^{N-1} u * v_j \, W^{jk}.$$

Replacing $u * v_j$ by the sum that defines it, we have upon rearranging sums

$$\sum_{j=0}^{N-1} u * v_j W^{jk} = \sum_{j=0}^{N-1} \left[\sum_{m=0}^{N-1} u_m v_{j-m} \right] W^{jk}$$

$$= \sum_{m=0}^{N-1} u_m \left[\sum_{j=0}^{N-1} v_{j-m} W^{jk} \right].$$

Replacing W^{jk} by $W^{(j-m)k} W^{mk}$ in the last sum, yields

$$\sum_{j=0}^{N-1} u * v_j W^{jk} = \sum_{m=0}^{N-1} u_m W^{mk} \left[\sum_{j=0}^{N-1} v_{j-m} W^{(j-m)k} \right]$$

$$= \sum_{m=0}^{N-1} u_m W^{mk} \left[\sum_{p=-m}^{N-m-1} v_p W^{pk} \right]. \qquad (4.105)$$

Now, the last sum in brackets can be rewritten as follows:

$$\sum_{p=-m}^{N-m-1} v_p W^{pk} = \sum_{p=0}^{N-m-1} v_p W^{pk} + \sum_{p=-m}^{-1} v_p W^{pk}$$

$$= \sum_{p=0}^{N-m-1} v_p W^{pk} + \sum_{p=N-m}^{N-1} v_{p-N} W^{(p-N)k}.$$

Because $\{v_p\}$ has period N, we have $v_{p-N} = v_p$ for all p. Also, $W^{-N} = 1$. Therefore,

$$\sum_{p=-m}^{N-m-1} v_p W^{pk} = \sum_{p=0}^{N-1} v_p W^{pk} = V_k.$$

Hence (4.105) becomes

$$\sum_{j=0}^{N-1} u * v_j W^{jk} = \sum_{m=0}^{N-1} u_m W^{mk} V_k = U_k V_k$$

which proves that the DFT of $\{u * v_j\}$ is $\{U_k V_k\}$. ∎

Using Theorem 4.4 we can efficiently compute the cyclic convolution $\{u * v_j\}_{j=0}^{N-1}$. First, compute the FFTs of $\{u_j\}_{j=0}^{N-1}$ and $\{v_j\}_{j=0}^{N-1}$, which are $\{U_k\}_{k=0}^{N-1}$ and $\{V_k\}_{k=0}^{N-1}$. Then, multiply each element of the FFTs, obtaining $\{U_k V_k\}_{k=0}^{N-1}$. Finally, take the inverse FFT of $\{U_k V_k\}_{k=0}^{N-1}$ and divide by N. Since the two FFTs and the inverse FFT each take $(3/2)N \log_2 N$ real multiplications, and multiplying each element of the two FFTs takes $4N$ real multiplications, we find that the whole process takes $(9/2)N \log_2 N + 6N$ real multiplications. This indicates how efficient the FFT method is, as compared to a direct computation, which would take $4N^2$ real multiplications. For instance, when $N = 1024$, the FFT method takes 80 times less multiplications than a direct computation.

4.8   Kernels for Some Common Filters

In this section we will examine kernels (*PSFs*) for the Cesàro, dlVP, hanning, and Hamming filters.

To begin our discussion we must first examine the kernel for an unfiltered Fourier series,

$$\sum_{n=-\infty}^{\infty} c_n e^{inx}, \qquad \left(c_n = \frac{1}{2\pi} \int_{-\pi}^{\pi} f(x)e^{-inx}\, dx \right)$$

for a function f having period 2π. For simplicity, *we shall assume that all functions discussed in this section have period 2π;* this results in no loss of generality. The partial sum S_M for the Fourier series above is

$$S_M(x) = \sum_{n=-M}^{M} c_n e^{inx}. \tag{4.106}$$

This partial sum can be thought of as a filtering of a Fourier series, in which case its kernel D_M is defined by

$$D_M(x) = \sum_{n=-M}^{M} 1 \cdot e^{inx}. \tag{4.107}$$

Using this kernel, we have

$$S_M(x) = f * D_M(x) = \frac{1}{2\pi} \int_{-\pi}^{\pi} f(s)D_M(x-s)\, ds. \tag{4.108}$$

A closed form expression for D_M can be found. First, we split the sum for D_M in (4.107) into two sums

$$D_M(x) = \sum_{n=0}^{M} e^{inx} + \sum_{n=1}^{M} e^{-inx}. \tag{4.109}$$

Using Formula (2.7) from Chapter 2, with M in place of $N - 1$, e^{ix} in place of r for the first sum in (4.109), and e^{-ix} in place of r for the second sum in (4.109), we obtain

$$D_M(x) = \frac{1 - e^{i(M+1)x}}{1 - e^{ix}} + \frac{e^{-ix} - e^{-i(M+1)x}}{1 - e^{-ix}}$$

$$= \frac{-e^{i(M+1)x} + e^{iMx} - e^{-i(M+1)x} + e^{-iMx}}{(1 - e^{ix})(1 - e^{-ix})}$$

$$= \frac{2\cos Mx - 2\cos(M+1)x}{2 - 2\cos x}. \tag{4.110}$$

Thus,

$$D_M(x) = \frac{\cos Mx - \cos(M+1)x}{1 - \cos x}. \tag{4.111}$$

Formula (4.111) can be further simplified using trigonometric identities. First, since

$$\cos Mx = \cos\left(M + \frac{1}{2} - \frac{1}{2}\right)x$$

it follows that

$$\cos Mx = \cos\left(M + \frac{1}{2}\right)x\cos\frac{1}{2}x + \sin\left(M + \frac{1}{2}\right)x\sin\frac{1}{2}x$$

and, similarly,

$$\cos(M+1)x = \cos\left(M + \frac{1}{2}\right)x\cos\frac{1}{2}x - \sin\left(M + \frac{1}{2}\right)x\sin\frac{1}{2}x.$$

Hence, by subtraction

$$\cos Mx - \cos(M+1)x = 2\sin\left(M + \frac{1}{2}\right)x\sin\frac{1}{2}x. \tag{4.112}$$

For $M = 0$, Formula (4.112) becomes

$$1 - \cos x = 2\sin^2\frac{1}{2}x. \tag{4.113}$$

Combining Formulas (4.111) to (4.113) yields the following closed form for D_M:

$$D_M(x) = \frac{\sin (M + \frac{1}{2})x}{\sin \frac{1}{2}x}.$$ (4.114)

Formula (4.114) is the classic expression for *Dirichlet's kernel* D_M. Another expression that is more amenable to graphing with *FAS* is

$$D_M(x) = (2M + 1) \operatorname{sinc} [(M + 0.5)x/\pi]/ \operatorname{sinc} (0.5x/\pi)$$ (4.114a)

where sinc is the function defined in (4.21). Graphs of Dirichlet's kernel using Formula (4.114a) are shown in Figure 4.17.

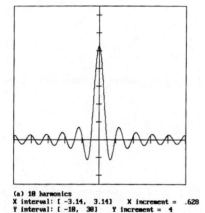

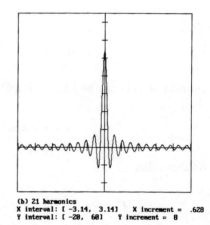

(a) 10 harmonics
X interval: [-3.14, 3.14]   X increment = .628
Y interval: [-10, 30]   Y increment = 4

(b) 21 harmonics
X interval: [-3.14, 3.14]   X increment = .628
Y interval: [-20, 60]   Y increment = 8

FIGURE 4.17
Dirichlet's kernel.

To derive the kernel for Cesàro filtering, we use Definition 4.1 and Formula (4.108) to obtain

$$\sigma_M(x) = \frac{1}{M} \sum_{k=0}^{M-1} f * D_k(x) = f * \left(\frac{1}{M} \sum_{k=0}^{M-1} D_k(x) \right).$$ (4.115)

Denoting the kernel for Cesàro filtering by C_M, we have from (4.115)

$$C_M(x) = \frac{1}{M} \sum_{k=0}^{M-1} D_k(x)$$ (4.116)

and

$$\sigma_M(x) = f * C_M(x).$$ (4.117)

To obtain a closed form for C_M we use (4.114) in (4.116), obtaining

$$C_M(x) = \frac{1}{M} \sum_{k=0}^{M-1} \frac{\sin(k + \frac{1}{2})x}{\sin \frac{1}{2}x}. \tag{4.118}$$

We now show that

$$C_M(x) = \frac{1}{M} \frac{\sin^2(\frac{1}{2}Mx)}{\sin^2(\frac{1}{2}x)}. \tag{4.119}$$

First, multiply (4.118) by $2 \sin^2(x/2)$, and get

$$2\sin^2\left(\frac{1}{2}x\right) C_M(x) = \frac{1}{M} \sum_{k=0}^{M-1} 2\sin\left(k + \frac{1}{2}\right) x \sin \frac{1}{2}x. \tag{4.120}$$

Using the trigonometric identity

$$2\sin\theta \sin\phi = \cos(\theta - \phi) - \cos(\theta + \phi) \tag{4.121}$$

we obtain from (4.120)

$$2\sin^2\left(\frac{1}{2}x\right) C_M(x) = \frac{1}{M} \sum_{k=0}^{M-1} [\cos kx - \cos(k+1)x]$$

$$= \frac{1}{M} [1 - \cos Mx]. \tag{4.122}$$

The last equality holding because of the telescoping of the finite series. Putting $\theta = \phi = (1/2)Mx$ in (4.121), we obtain

$$1 - \cos Mx = 2\sin^2\left(\frac{1}{2}Mx\right)$$

which used in (4.122) yields

$$2\sin^2\left(\frac{1}{2}x\right) C_M(x) = \frac{2}{M} \sin^2\left(\frac{1}{2}Mx\right). \tag{4.123}$$

Dividing (4.123) by $2\sin^2(x/2)$ yields (4.119), our desired formula.

Another form of C_M kernel, more amenable to graphing by *FAS*, is

$$C_M(x) = M[\text{sinc}(Mx/a)/\text{sinc}(x/a)] \wedge 2 \backslash a{=}2\pi \tag{4.124}$$

REMARK 4.5:   The kernel C_M is usually called *Fejér's kernel*. Sometimes we will refer to it as *Cesàro's kernel* in order to remind the reader that it is the kernel for Cesàro filtering.   ∎

We now obtain a closed form for the dlVP kernel using $2M$ harmonics. By Formula (4.51) the kernel V_{2M} for the dlVP filter must be (for $L = \pi$)

$$V_{2M}(x) = \sum_{n=-2M}^{2M} v_n e^{inx} \tag{4.125}$$

where

$$v_n = \begin{cases} 1 & \text{if } |n| \leq M \\ 2(1 - |\dfrac{n}{2M}|) & \text{if } M < |n| \leq 2M. \end{cases} \tag{4.126}$$

The second part of Formula (4.126) reminds us of the coefficients for the Cesàro kernel C_{2M}, multiplied by 2. Once we realize this, then with some algebra we obtain

$$V_{2M}(x) = 2C_{2M}(x) - C_M(x). \tag{4.127}$$

Using either Formula (4.119) or (4.124), Formula (4.127) gives a closed form expression for the dlVP kernel V_{2M}. For graphing with *FAS*, however, Formula (4.124) is the better choice (if one desires to use a formula, the method described at the end of Section 4.6 is much easier in this case).

We close this section by describing closed forms for the hanning and Hamming kernels. The *hanning kernel* h_M is defined by [see (4.57)]

$$h_M(x) = \sum_{n=-M}^{M} \left(0.5 + 0.5 \cos \frac{n\pi}{M}\right) e^{inx} \tag{4.128}$$

We leave it as an exercise for the reader to verify that

$$h_M(x) = \frac{1}{2} D_M(x) + \frac{1}{4} D_M\left(x + \frac{\pi}{M}\right) + \frac{1}{4} D_M\left(x - \frac{\pi}{M}\right) \tag{4.129}$$

where D_M is Dirichlet's kernel. In Figure 4.18(a) we show graphs of

$$\frac{1}{2} D_M(x) \qquad \text{and} \qquad \frac{1}{4} D_M\left(x + \frac{\pi}{M}\right) + \frac{1}{4} D_M\left(x - \frac{\pi}{M}\right)$$

for $M = 32$. The effect of adding the second function to the first is a large amount of cancellation of the oscillations of D_M. This is evident in Figure 4.18(b).

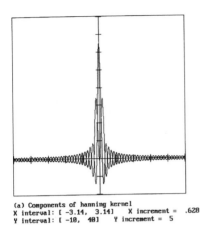

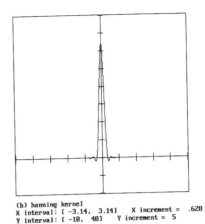

(a) Components of hanning kernel
X interval: [-3.14, 3.14]   X increment = .628
Y interval: [-10, 40]   Y increment = 5

(b) hanning kernel
X interval: [-3.14, 3.14]   X increment = .628
Y interval: [-10, 40]   Y increment = 5

FIGURE 4.18
Construction of hanning kernel.

The *Hamming kernel* H_M is defined by [see (4.58)]

$$H_M(x) = \sum_{n=-M}^{M} \left(0.54 + 0.46 \cos \frac{n\pi}{M}\right) e^{inx}. \qquad (4.130)$$

By comparing Formulas (4.107), (4.128), and (4.130) it follows that

$$H_M(x) = 0.08 D_M(x) + 0.92 h_M(x). \qquad (4.131)$$

In Figure 4.19(a) we have graphed $0.08 D_{32}(x)$ and $0.92 h_{32}(x)$. It can be seen that the similarity between the graphs lies in the nearly perfect match (with *opposite* signs) between the two side lobes just to the right and left of the two center lobes. Consequently, as can be seen in Figure 4.19(b), which is a graph of H_{32}, the effect of summing $0.08 D_{32}$ and $0.92 h_{32}$ is to *cancel out these sidelobes*. The main advantage of this is to *suppress Gibbs' phenomenon* in Fourier series expansions.

4.9 Convergence of Filtered Fourier Series

In this section we will prove a theorem that has wide applicability to the convergence of filtered Fourier series. It covers many (but not all) of the filters commonly used in mathematics, signal processing, and other areas.

We begin with the following definition of a summation kernel.

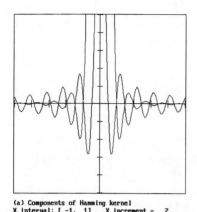

(a) Components of Hamming kernel
X interval: [-1, 1]   X increment = .2
Y interval: [-2, 2]   Y increment = .4

(b) Hamming kernel
X interval: [-3.14, 3.14]   X increment = .628
Y interval: [-2, 2]   Y increment = .4

FIGURE 4.19
Construction of Hamming kernel.

DEFINITION 4.4:   *A PSF, $\mathcal{P}_M$, defined on $[-L, L]$ is called a **summation***
***kernel** if it satisfies the following conditions:*

(a) *For each M,*

$$\frac{1}{2L} \int_{-L}^{L} \mathcal{P}_M(x)\, dx = 1.$$

(b) *There is a positive constant C for which*

$$\frac{1}{2L} \int_{-L}^{L} |\mathcal{P}_M(x)|\, dx \leq C \qquad \text{for all M.}$$

(c) *Suppose $\epsilon > 0$ is given, no matter how small. Then, given $\delta > 0$, satisfying*
     $0 < \delta < L$, we will have for $\delta \leq |x| \leq L$

$$|\mathcal{P}_M(x)| < \epsilon$$

***provided** M is chosen sufficiently large.*

REMARK 4.6:   Condition (c) in Definition 4.4 can also be stated as

$$\lim_{M \to \infty} \left[\sup_{\delta \leq |x| \leq L} |\mathcal{P}_M(x)| \right] = 0$$

for each $\delta > 0$.   ∎

The Cesàro, dlVP, and hanning kernels are all summation kernels over $[-\pi, \pi]$. We will show this for the Cesàro kernel, leaving the verification for the dlVP kernel to the reader as an exercise. Showing that the hanning kernel is a summation kernel is easier for us to do by an indirect argument utilizing the Fourier transform, so we will not do it until Section 5.10 of Chapter 5.

Example 4.9:
Show that the Cesàro kernel is a summation kernel.

SOLUTION We must check that the properties (a) to (c) in Definition 4.4 are satisfied.

(a) From Formula (4.116) we obtain

$$\frac{1}{2\pi} \int_{-\pi}^{\pi} C_M(x)\, dx = \frac{1}{M} \sum_{k=0}^{M-1} \frac{1}{2\pi} \int_{-\pi}^{\pi} D_k(x)\, dx. \qquad (4.132)$$

Using Formula (4.107) we have, because of the orthogonality of $\{e^{inx}\}$,

$$\frac{1}{2\pi} \int_{-\pi}^{\pi} D_k(x)\, dx = \sum_{n=-k}^{k} \frac{1}{2\pi} \int_{-\pi}^{\pi} e^{inx}\, dx$$

$$= \frac{1}{2\pi} \int_{-\pi}^{\pi} 1\, dx = 1.$$

Hence, Formula (4.132) becomes

$$\frac{1}{2\pi} \int_{-\pi}^{\pi} C_M(x)\, dx = \frac{1}{M} \sum_{k=0}^{M-1} 1 = 1 \qquad (4.133)$$

so (a) is true.

(b) From Formula (4.119) we see that $C_M(x) \geq 0$ for all M, hence $|C_M(x)| = C_M(x)$ for all M. Therefore, because of (a),

$$\frac{1}{2\pi} \int_{-\pi}^{\pi} |C_M(x)|\, dx = 1 \qquad \text{for all } M.$$

Taking C to be 1 we see that (b) is true.

(c) Suppose $\epsilon > 0$ has been given. Then, for $\delta \leq |x| \leq \pi$ we have

$$0 \leq C_M(x) \leq \frac{1}{M} \frac{\sin^2(Mx/2)}{\sin^2(\delta/2)} \leq \frac{1}{M} \frac{1}{\sin^2(\delta/2)} \,. \tag{4.134}$$

If M is sufficiently large, then

$$\frac{1}{M} \frac{1}{\sin^2(\delta/2)} < \epsilon. \tag{4.135}$$

Combining (4.135) and (4.134) yields $0 \leq C_M(x) < \epsilon$, so $|C_M(x)| < \epsilon$ for M sufficiently large. Thus, (c) is true. ∎

Example 4.10:
The kernel I_M defined by

$$I_M(x) = \begin{cases} M & \text{for } |x| \leq \frac{L}{M} \\ 0 & \text{for } |x| > \frac{L}{M} \end{cases} \qquad (M = 1, 2, 3, \ldots)$$

is a summation kernel over $[-L, L]$.

We leave the verification of Example 4.10 to the reader as an exercise. The kernel I_M is called the *point impulse kernel.*

The following theorem illustrates the importance of summation kernels.

THEOREM 4.5:
Suppose that $\mathcal{P}_M$ is a summation kernel. Furthermore, suppose that f has period $2L$ and $\int_{-L}^{L} |f(x)| \, dx$ is finite (converges). Then, for each point x_0 of continuity of f,

$$\lim_{M \to \infty} \mathcal{P}_M * f(x_0) = f(x_0). \tag{4.136}$$

Moreover, if f is continuous, then

$$\lim_{M \to \infty} \left[\sup_{x \in \mathbf{R}} |\mathcal{P}_M * f(x) - f(x)| \right] = 0. \tag{4.137}$$

REMARK 4.7:   Equation (4.137) says that $\mathcal{P}_M * f$ converges *uniformly* to f over the whole real line, since sup is just another way of saying *maximum.*

Furthermore, since convolution is commutative, we have $\mathcal{P}_M * f = f * \mathcal{P}_M$. Therefore, Theorem 4.5 says that *a filtered Fourier series for a continuous function f converges uniformly to f over the whole real line.* ∎

PROOF OF THEOREM   Suppose that x_0 is a point of continuity of f, and that $\epsilon > 0$ has been given. We begin by observing that property (a) of Definition 4.4 yields

$$f(x_0) = f(x_0) \cdot 1 = f(x_0) \frac{1}{2L} \int_{-L}^{L} \mathcal{P}_M(u) \, du$$

$$= \frac{1}{2L} \int_{-L}^{L} f(x_0) \mathcal{P}_M(u) \, du. \qquad (4.138)$$

Consequently,

$$\mathcal{P}_M * f(x_0) - f(x_0) = \frac{1}{2L} \int_{-L}^{L} [f(x_0 - u) - f(x_0)] \mathcal{P}_M(u) \, du. \qquad (4.139)$$

Applying absolute values to (4.139) and bringing the absolute values inside the integral sign yields

$$|\mathcal{P}_M * f(x_0) - f(x_0)| \leq \frac{1}{2L} \int_{-L}^{L} |f(x_0 - u) - f(x_0)||\mathcal{P}_M(u)| \, du. \qquad (4.140)$$

The right hand side of (4.140) can be expressed as

$$\frac{1}{2L} \int_{-L}^{L} |f(x_0 - u) - f(x_0)||\mathcal{P}_M(u)| \, du = I + J \qquad (4.141)$$

where

$$I = \frac{1}{2L} \int_{|u|<\delta} |f(x_0 - u) - f(x_0)||\mathcal{P}_M(u)| \, du \qquad (4.141a)$$

and

$$J = \frac{1}{2L} \int_{\delta \leq |u| \leq L} |f(x_0 - u) - f(x_0)||\mathcal{P}_M(u)| \, du. \qquad (4.141b)$$

[*Note:* the integral $\int_{|u|<\delta}$ in (4.141a) is shorthand for $\int_{-\delta}^{\delta}$, while $\int_{\delta \le |u| \le L}$ is short-hand for the sum of two integrals $\int_{-L}^{-\delta} + \int_{\delta}^{L}$.]

Now, we shall first show that the integral I is small when δ is small. Since x_0 is a point of continuity for f we will have

$$|f(x_0 - u) - f(x_0)| < \epsilon \qquad \text{when} \qquad |(x_0 - u) - x_0| < \delta \qquad (4.142)$$

provided δ is sufficiently small. Choosing such a δ, and making use of the fact that $|(x_0 - u) - x_0| = |u|$, we obtain

$$|f(x_0 - u) - f(x_0)| < \epsilon \qquad \text{for} \qquad |u| < \delta. \qquad (4.143)$$

Using (4.143), the integral I in (4.141a) is bounded as follows:

$$I \le \frac{1}{2L} \int_{|u|<\delta} \epsilon |\mathcal{P}_M(u)| \, du \le \epsilon C \qquad (4.144)$$

where we used property (b) from Definition 4.4 to get the last inequality.

We now turn to the integral J in (4.141b), where δ is the *fixed* value chosen to get (4.144). By the triangle inequality for absolute values

$$|f(x_0 - u) - f(x_0)| \le |f(x_0 - u)| + |f(x_0)|. \qquad (4.145)$$

Hence, the integral J in (4.141b) is bounded in the following way:

$$J \le \frac{1}{2L} \int_{\delta \le |u| \le L} |f(x_0 - u)||\mathcal{P}_M(u)| \, du$$

$$+ \frac{1}{2L} \int_{\delta \le |u| \le L} |f(x_0)||\mathcal{P}_M(u)| \, du. \qquad (4.146)$$

Using property (c) from Definition 4.4, we can further bound the terms on the right side of Inequality (4.146). For M sufficiently large, we get

$$J \le \frac{1}{2L} \int_{\delta \le |u| \le L} |f(x_0 - u)|\epsilon \, du + \frac{1}{2L} \int_{\delta \le |u| \le L} |f(x_0)|\epsilon \, du. \qquad (4.147)$$

By enlarging the integration intervals to $[-L, L]$, we have

$$J \le \epsilon \frac{1}{2L} \int_{-L}^{L} |f(x_0 - u)| \, du + \epsilon \frac{1}{2L} \int_{-L}^{L} |f(x_0)| \, du. \qquad (4.148)$$

The first term on the right side of (4.148) equals

$$\epsilon \frac{1}{2L} \int_{-L}^{L} |f(v)| \, dv \qquad \left(= \frac{\epsilon}{2L} \|f\|_1 \right)$$

because of the periodicity of f. The second term on the right side of (4.148) just equals $\epsilon |f(x_0)|$. Therefore, (4.148) becomes

$$J \le \epsilon \left[\frac{1}{2L} \|f\|_1 + |f(x_0)| \right]. \tag{4.149}$$

Using inequalities (4.144) and (4.149) in (4.141) we get, for all M sufficiently large,

$$|\mathcal{P}_M * f(x_0) - f(x_0)| \le \epsilon \left[C + \frac{1}{2L} \|f\|_1 + |f(x_0)| \right]. \tag{4.150}$$

Since ϵ can be taken arbitrarily small, we have proved (4.136).

To complete the proof we must show that (4.137) holds. Here, we must use a result from advanced calculus. If the function f is continuous on the *closed* interval $[-L, L]$, then there exists a $\delta > 0$ such that

$$|f(x - u) - f(x)| < \epsilon \qquad \text{when} \qquad |(x - u) - x| < \delta \tag{4.151}$$

for *all* x in $[-L, L]$. This is known as the *uniform continuity* of f, it is proved in every text on advanced calculus (see, for instance, [Kr] or [Ru,2]). Formula (4.151) says that

$$|f(x - u) - f(x)| < \epsilon \qquad \text{when} \qquad |u| < \delta. \tag{4.152}$$

But, then, replacing x_0 by x in all of the calculations above from (4.138) through (4.150), we have for all x in $[-L, L]$

$$|\mathcal{P}_M * f(x) - f(x)| \le \epsilon \left[C + \frac{1}{2L} \|f\|_1 + |f(x)| \right] \tag{4.153}$$

provided M is chosen sufficiently large. Since f and $\mathcal{P}_M * f$ both have period $2L$, Formula (4.153) holds for all x in $\mathbf{R}$. If we define $\|f\|_{\text{sup}}$ to be the maximum of $|f(x)|$ over $\mathbf{R}$, then we have for all M sufficiently large

$$|\mathcal{P}_M * f(x) - f(x)| \le \epsilon \left[C + \frac{1}{2L} \|f\|_1 + \|f\|_{\text{sup}} \right]. \tag{4.154}$$

Since the right side of (4.154) is a constant, and (4.154) holds for all x in $\mathbf{R}$, we have for all M sufficiently large

$$\sup_{x \in \mathbf{R}} |\mathcal{P}_M * f(x) - f(x)| \leq \epsilon \left[C + \frac{1}{2L} \|f\|_1 + \|f\|_{\sup} \right]. \qquad (4.155)$$

Because ϵ can be taken arbitrarily small, we have proved (4.137). ■

REMARK 4.8:   Using the notation introduced in the proof above, we can rewrite (4.155) as $\lim_{M \to \infty} \|\mathcal{P}_M * f - f\|_{\sup} = 0$. It is said that $\mathcal{P}_M * f$ *converges to f in the Sup-Norm over* $\mathbf{R}$. ■

4.10   Further Analysis of Fourier Series Partial Sums

In this section we will show how Gibbs' phenomenon and other aspects of Fourier series partial sums can be analyzed by using the convolution form of partial sums. Throughout the discussion we will examine the partial sums of the following function:

$$f(x) = \begin{cases} 1 & \text{if } 0 < x < \pi \\ 0 & \text{if } -\pi < x < 0. \end{cases} \qquad (4.156)$$

We will use 8192 points and an interval of $[-\pi, \pi]$.

Let's examine the partial sums $S_{40}(x)$, $S_{80}(x)$, and $S_{160}(x)$ near $x = 0$ [see Figure 4.20(a)]. Let $x = x_M$ yield the first maximum of $S_M(x)$ to the right of $x = 0$. Using the *Trace* procedure of *FAS* we obtain the estimates shown in Table 4.1.

Table 4.1   Maximum Values of Some Partial Sums

40 harmonics	$x_{40} \approx 7.823 \times 10^{-2}$	$S_{40}(x_{40}) \approx 1.08959$
80 harmonics	$x_{80} \approx 3.912 \times 10^{-2}$	$S_{80}(x_{80}) \approx 1.0895$
160 harmonics	$x_{160} \approx 1.917 \times 10^{-2}$	$S_{160}(x_{160}) \approx 1.08955$

We know that $S_M(x_M) = f * D_M(x_M)$ where D_M is Dirichlet's kernel. Consequently, since $f(x) = 0$ on the interval $(-\pi, 0)$ and $f(x) = 1$ on the interval $(0, \pi)$, it follows that

$$S_M(x_M) = \frac{1}{2\pi} \int_0^\pi D_M(x_M - x) \, dx. \qquad (4.157)$$

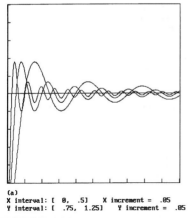
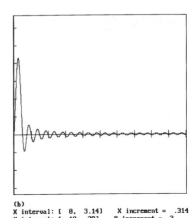

(a)
X interval: [0, .5]   X increment = .05
Y interval: [.75, 1.25]   Y increment = .05

(b)
X interval: [0, 3.14]   X increment = .314
Y interval: [-10, 20]   Y increment = 3

FIGURE 4.20
(a) Graphs of the partial sums $S_{40}(x)$, $S_{80}(x)$, and $S_{160}(x)$ for the function f in (4.156). (b) Graph of $D_{40}(x_{40} - x)/(2\pi)$ for $x_{40} = 0.07823$.

Formula (4.157) shows that $S_{40}(x_{40})$ equals the integral of $D_{40}(x_{40} - x)/(2\pi)$ over the x-interval $[0, \pi]$. By graphing $D_{40}(x_{40} - x)/(2\pi)$ over $[0, \pi]$ [see Figure 4.20(b)] and then selecting *Graphs* on the display menu, and selecting *Integrals* and *Integral*, we can approximate the integral of this function. The value we get is 1.089475. Notice how this value is approximately the same as the value of $S_{40}(x_{40})$ given in Table 4.1. We can also obtain approximations of $\int_0^\pi D_{80}(x_{80}-x)/(2\pi)\, dx$ and of $\int_0^\pi D_{160}(x_{160} - x)/(2\pi)\, dx$ in the same way. We get values of 1.08946 and 1.08935, which are approximately the same values as in Table 4.1.

The discussion above illustrates that Gibbs' phenomenon for the function f results from the fact that

$$S_M(x_M) = \int_0^\pi D_M(x_M - x)/(2\pi)\, dx \approx 1.089\ldots$$

for all sufficiently large M. More precisely, as $M \to \infty$, we have

$$x_M \to 0 \quad \text{and} \quad S_M(x_M) \to 1.089\ldots > 1. \tag{4.158}$$

Formula (4.158) shows that the overshooting of the value 1 *does not go away as* $M \to \infty$. (For a rigorous proof of (4.158), see [Wa, Chapter 2.5].)

The convolution expression for partial sums can be used to analyze other aspects of partial sums. For example, we can examine the origin of the oscillation about the value 1 exhibited by the partial sums for the function f given in Formula (4.156) [see Figure 4.20(a)]. Let $x = c_M$ yield the first intersection of $S_M(x)$ with $f(x) = 1$ to the right of $x = x_M$. Using the *Trace* procedure of *FAS* we obtain the estimates shown in the middle column of Table 4.2.

Table 4.2   Some Intersections of $S_M(x)$ with $f(x)$

40 harmonics	$c_{40} \approx 0.12195$	$S_{40}(c_{40})$ = 1
80 harmonics	$c_{80} \approx 6.059 \times 10^{-2}$	$S_{80}(c_{80})$ = 1
160 harmonics	$c_{160} \approx 0.03$	$S_{160}(c_{160})$ = 1

In Figure 4.21(a) we show a graph of $D_{40}(c_{40} - x)/(2\pi)$ over the interval $[0, \pi]$.

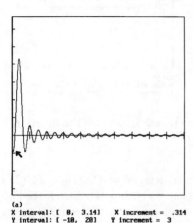
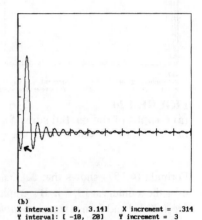

(a)
X interval: [0, 3.14]   X increment = .314
Y interval: [-10, 20]   Y increment = 3

(b)
X interval: [0, 3.14]   X increment = .314
Y interval: [-10, 20]   Y increment = 3

FIGURE 4.21
(a) Graph of $D_{40}(c_{40} - x)/(2\pi)$. (b) Graph of $D_{40}(t_{40} - x)/(2\pi)$. The arrows indicate regions of increasing negative area.

We have

$$S_{40}(c_{40}) = \int_0^{\pi} D_{40}(c_{40} - x)/(2\pi)\, dx$$

so $S_M(c_{40})$ is the integral of $D_{40}(c_{40} - x)/(2\pi)$ over $[0, 2\pi]$. Comparing Figures 4.20(b) and 4.21(a) we see that the main difference between them is the presence of the negative-valued region, marked by the arrow, in Figure 4.21(a). Consequently, the integral of $D_{40}(c_{40} - x)/(2\pi)$ will be noticeably smaller than the integral of $D_{40}(x_{40} - x)/(2\pi)$. Using *FAS* to approximate the integral of $D_{40}(c_{40} - x)/(2\pi)$ we get 1.0005. This value, as it should be, is approximately the same as the value of $S_{40}(c_{40})$, which is 1. The results for $M = 80$ and $M = 160$ are similar. For $M = 80$ we get a value of 1.002 and for $M = 160$ we get a value of 1.004.

Furthermore, if we let $x = t_M$ yield the first minimum of $S_M(x)$ to the right of $x = 0$, then we obtain the estimates shown in Table 4.3.

As before, we have $S_M(t_M) = \int_0^{\pi} D_M(t_M - x)/(2\pi)\, dx$. The graph of $D_{40}(t_{40} - x)/(2\pi)$ over $0 \le x \le \pi$ is shown in Figure 4.21(b). Notice that the main difference between the two graphs in Figure 4.21 is that there is more negative-valued area immediately to the right of the origin (marked by the arrow) in Figure 4.21(b). Consequently the integral over $[0, \pi]$ of $D_{40}(t_{40} - x)/(2\pi)$ is

Table 4.3   Some Minimum Values of Partial Sums

40 harmonics	$t_{40} \approx 0.156466$	$S_{40}(t_{40}) \approx 0.9512$
80 harmonics	$t_{80} \approx 7.823 \times 10^{-2}$	$S_{80}(t_{80}) \approx 0.95135$
160 harmonics	$t_{160} \approx 3.912 \times 10^{-2}$	$S_{160}(t_{160}) \approx 0.9514$

less than the integral of $D_{40}(c_{40} - x)/(2\pi)$ (which is 1). Using *FAS* to approximate the integral of $D_{40}(t_{40} - x)/(2\pi)$ over $[0, \pi]$, we get a value of 0.9513 which is approximately the same as the value of $S_{40}(t_{40})$ given in Table 4.3. Similar results are obtained for $M = 80$ and $M = 160$. For $M = 80$ we get a value of 0.9514 and for $M = 160$ we get a value of 0.9515.

Continuing this kind of analysis we can see that as x increases from $x = x_M$ onward towards $x = \pi$, the values of $S_M(x)$ will oscillate about the value $f(x) = 1$.

The discussion above shows how the convolution form for $S_M(x)$ helps us to understand some of the features of the graph of $S_M(x)$. Similar analyses can be carried out for filtered partial sums. We give some examples in the exercises.

References

For further discussion of applications of Fourier series to physical problems, see [Wa], [We], and [Str]. More discussion of filters can be found in [Ha], [Op-S], and [Ra-G].

Exercises

Section 1

4.1 Graph approximate solutions (using, say, 200 harmonics and 1024 points) to Problem (4.1) using the following functions, times, and diffusion constants.

(a) $u(x, 0) = 40(10x - x^2)$, $0 \le x \le 10$, $a^2 = 0.12$ (cast iron), $t = 5, 10, 20,$ and 40 sec.

(b) $u(x, 0) = 80 \exp(-(x - 8) \wedge 20)$, $0 \le x \le 16$, $a^2 = 0.005$ (concrete), $t = 30, 60, 120,$ and 300 sec.

(c)
$$u(x, 0) = \begin{cases} 100 & \text{for } 17 < x < 23 \\ 0 & \text{for } 0 < x < 17 \text{ and } 23 < x < 32 \end{cases}$$
$a^2 = 1.14$ (copper), $t = 0.5, 1.0, 30,$ and 60 sec.

(d)
$$u(x, 0) = \begin{cases} 100 & \text{for } 9 < x < 11 \text{ and } 29 < x < 32 \\ 0 & \text{for } 0 < x < 9 \text{ and } 11 < x < 29 \text{ and } 32 < x < 40 \end{cases}$$
$a^2 = 0.0038$ (brick), $t = 60, 120, 300,$ and 600 sec.

4.2 For each of the examples in Exercise 4.1, determine estimates for the maximum errors involved in using a 40 harmonic partial sum approximation to $u(x, t)$ (using the given values for t). If an error estimate is more than 10^{-5}, then find some number of harmonics which will yield such accuracy.

4.3 Generalize the estimate (4.12) as follows. First, show that

$$|b_n| \le \frac{2}{L} \|f\|_1, \qquad \left(\|f\|_1 = \int_0^L |f(x)| \, dx \right) \qquad (4.159)$$

then show that

$$\left| u(x, t) - \sum_{n=1}^{M} b_n e^{-(n\pi a/L)^2 t} \sin \frac{n\pi x}{L} \right| \le \frac{2}{L} \|f\|_1 \frac{e^{-(M+1)^2 (a\pi/L)^2 t}}{1 - e^{-(2M+3)(a\pi/L)^2 t}}.$$
$$(4.160)$$

4.4 Suppose $t \ge 0.01$ sec, $a^2 = 1.14$ (copper), and $L = 10$. How many harmonics M are needed to ensure that

$$\left| u(x, t) - \sum_{n=1}^{M} b_n e^{-(n\pi a/L)^2 t} \sin \frac{n\pi x}{L} \right| \le 10^{-5} \|f\|_1.$$

Hint: plot the function on the right side of Inequality (4.160) in Exercise 4.3 (except for the factor $\|f\|_1$), using x in place of M. Use the *Trace* procedure to determine when values of the graph are less than 10^{-5}.

4.5 Explain why as $t \to \infty$, we have

$$u(x, t) \approx b_1 e^{-(a\pi/L)^2 t} \sin \frac{\pi x}{L}.$$

That is, for t sufficiently large, $u(x, t)$ is approximated by the first term in its series expansion. Using *FAS*, illustrate this result for each of the functions in Exercise 4.1.

4.6 Using Fourier cosine series expansions, derive a filtered cosine series solution to the following heat conduction problem:

$$\frac{\partial u}{\partial t} = a^2 \frac{\partial^2 u}{\partial x^2} \qquad \text{(heat equation)}$$

$$\frac{\partial u}{\partial x}(0, t) = 0, \quad \frac{\partial u}{\partial x}(L, t) = 0 \qquad \text{(boundary conditions)}$$

$$u(x, 0) = g(x) \qquad \text{(initial condition).} \qquad (4.161)$$

REMARK 4.9: To see an animation of a solution of (4.161) evolving in time, view the Graphbook *TEMP_CH* which can be found on the computer disk accompanying this book.

4.7 Graph approximate solutions to Problem (4.161) in Exercise 4.6, using 40 harmonics and 1024 points, and using the same functions, times, and diffusion constants as in Exercise 4.1.

Section 2

4.8 Using *FAS,* graph solutions to (4.16) for $c = 100$, $L = 10$, $t = 0.01, 0.02, 0.03$, and 0.04, for the following initial conditions.

(a) $f(x) = 0.1 \exp[-((x-1)/0.4)^{10}]$, $g(x) = 0$

(b) $f(x) = 0.1 \sin(0.3\pi x)$, $g(x) = 0$

(c) $f(x) = 0.1 \sin(0.4\pi x)$, $g(x) = 0$

(d) $f(x) = 0$, $g(x) = \sin(0.3\pi x)$

(e) $f(x) = 0$, $g(x) = \sin(0.4\pi x)$

(f) $f(x) = x(x-10)(x+40)$, $g(x) = 3x(x-10)$

(g) $f(x) = 0.2 \exp[-(x-3)^{10}]$, $g(x) = x(10-x)$

4.9 Suppose c in Exercise 4.8 is changed to 200. Repeat the exercise and explain how (why) the new results differ from the previous ones.

4.10 Show that for $f(x) = 0$, $g(x) = \sin(n\pi x/L)$, a solution to (4.16) is

$$y(x,t) = \frac{L}{nc\pi} \sin(nc\pi t/L) \sin(n\pi x/L).$$

Describe what this solution looks like for $n = 1, 2, 3$, and 4.

4.11 Show that for $f(x) = K \sin(n\pi x/L)$, $g(x) = 0$, a solution to (4.16) is

$$y(x,t) = K \cos(nc\pi t/L) \sin(n\pi x/L).$$

Describe what this solution looks like for $n = 1, 2, 3$, and 4.

4.12 Use the following formula,

$$f(x) = 1000\text{sumk}\{[1 - (-1)^k] \cos(2k\pi t) \sin(k\pi x/50)/(k\pi) \wedge 3\} \backslash k{=}1,200$$

to plot exact partial sums of the Fourier sine series described in Example 4.2. You should use time values of $t = 0, 0.1, 0.2$, and 0.3 to produce graphs like the ones shown in Figure 4.3. Using 1024 points, find the Sup-Norm differences between these exact partial sums and the *FAS Sine Series, Cos* filtered, partial sums.

Section 3

4.13 Continue Example 4.5 for the times $t = 0.2$ and 0.25 sec. Notice how the basic form of the graph of $|S_{2000}(x,t)|^2$ has *stabilized.*

4.14 Use *FAS* to calculate $|S_{2000}(x,t)|^2$ for $t = 0.1, 0.2, 0.3, 0.4$, and 0.5 sec, given the following initial functions over the interval $[0, 64]$. Use 8192 points, and use the mass of the electron for m.

(a)

$$\psi(x,0) = \begin{cases} 1 & \text{if } 30.5 < x < 31 \text{ or } 33 < x < 33.5 \\ 0 & \text{if } x < 30.5 \text{ or } 31 < x < 33 \text{ or } x > 33.5 \end{cases}$$

$$= (-1.5 < x - 32 < -1) + (1 < x - 32 < 1.5)$$

(b) $\psi(x,0) = (-2.5 < x - 32 < -2) + (2 < x - 32 < 2.5)$

(c) $\psi(x,0) = (-3.5 < x - 32 < -3) + (3 < x - 32 < 3.5)$

4.15 Analyze the results of Exercise 4.14, using concepts of quantum mechanics. What effects do you observe as the distance between the rectangles in the initial functions in Exercise 4.14 is increased?

4.16 Calculate $|S_{2000}(x,t)|^2$ for $t = 0.1, 0.2$, and 0.3 sec, given the following initial functions (use 8192 points over the interval $[0, 64]$ and use m equal to the mass of an electron).

   (a) $\psi(x,0) = \exp[-((x-32)/2)^{10}]$

   (b) $\psi(x,0) = \exp[-((x-32)/0.5)^{10}]$

   (c) $\psi(x,0) = \exp(-\pi(x-32)^2)$

   (d) $\psi(x,0) = 2\exp(-4\pi(x-32)^2)$

4.17 Given that the mass of a neutron is many times greater than the mass of an electron (*it is part of this exercise for you to look up how many times greater it is*), calculate $|S_{2000}(x,t)|^2$ for the functions and times t in Examples 4.4 and 4.5 using the mass of the neutron for m. How does neutron diffraction compare with electron diffraction?

4.18 Compute $|S_{2000}(x,t)|^2$ for the functions given in Examples 4.4 and 4.5, for the times $t = 1.0, 2.0, 4.0$, and 8.0 sec (assume that m equals the mass of an electron). Give a physical interpretation for the increasingly erratic appearance of the wave function as t increases.

4.19 With regard to Exercise 4.17, for what time t_n will a neutron diffraction pattern be *identical* to an electron diffraction pattern at time t_e?

Section 4

4.20 For the following function compute Cesàro, dlVP, hanning, and Hamming filtered Fourier series using 40 harmonics and 1024 points:

$$f(x) = \begin{cases} -1 & \text{for } -1 < x < 0 \\ 1 & \text{for } 0 < x < 1. \end{cases}$$

Compare all four of these filtered partial sums within the x-y window: $0 \le x \le 1$ and $0.98 \le y \le 1.02$.

4.21 For the function $f(x) = 1 - |x - 1|$ on the interval $0 \le x \le 2$, use *FAS* to estimate how many harmonics are needed to approximate the function to within ± 0.01 for each of the filters: Cesàro, dlVP, hanning, and Hamming.

Section 5

4.22 *Lanczos Filter.* For each of the following functions, graph filtered Fourier series partial sums using 10, 20, and 30 harmonics with sinc (x) as the filter function (use $[-\pi, \pi]$ and 1024 points):

(a) $\begin{cases} 1 & \text{if } 0 < x < \pi \\ -1 & \text{if } -\pi < x < 0 \end{cases}$

(b) x^2

(c) x

(d) $\exp(-(x/2) \wedge 10)$

4.23 Suppose $\omega \approx \omega_3 = 30\pi$ in Example 4.7. Using *FAS,* draw approximate graphs of $y(x, t)$ in (4.78) for $\omega = 29.99\pi$ and the times $t = 1.01, 1.02, 1.03,$ and 1.04 sec. Compare your graphs with those for Exercise 4.11. Then repeat this comparison, using $\omega = 29.9999\pi$ and the same times t.

4.24 Generalizing Exercise 4.23, what type of motion would you expect if $\omega = 59.9999\pi$ or $\omega = 69.9999\pi$? Confirm your answer using *FAS.*

4.25 Explain the results of the previous two exercises using Formula (4.77).

4.26 Derive a series solution to (4.64) if the force function is $F(x, t) = \rho G(x) \cos \omega t$ and the initial conditions are $f(x) = 0$, $g(x) = 0$. You should obtain the following series solution (*assuming $\omega \neq \omega_n$ for all n*):

$$y(x, t) = \sum_{n=1}^{\infty} \left[\frac{\cos \omega t - \cos \omega_n t}{\omega_n^2 - \omega^2} \right] K_n \sin \frac{n\pi x}{L},$$

$$\left(K_n = \frac{2}{L} \int_0^L G(x) \sin \frac{n\pi x}{L} \, dx \right). \tag{4.162}$$

4.27 Using 200 harmonics, $L = 10$, $c = 100$, and $\omega = 29.99\pi$ graph approximations to the series in Equation (4.162) in Exercise 4.26 [using the function in (4.73) for G] for times $t = 1.01, 1.02, 1.03, 1.04$ and 1.05 sec.

4.28 Same problem as Exercise 4.27, but use $G(x) = 2 \exp(-[((x - 5)/4) \wedge 20])$.

4.29 Same problem as Exercise 4.27, but use $G(x) = 2 \exp(-[((x - 5)/4) \wedge 20])$ and $\omega = 69.99999\pi$ and $t = 1.001, 1.002, 1.003, 1.004,$ and 1.005.

Section 6

4.30 Check that (4.88) is correct.

4.31 Using $M = 8, 16,$ and 32 harmonics, and 1024 points, graph the kernels for each of the following filters (use $[-\pi, \pi]$ as the interval): (a) Cesàro , (b) dlVP, (c) hanning, (d) Hamming, (e) Gauss (damping constant 0.001), (f) Riesz (damping power 3.81).

4.32 (a) Graph the 20 harmonic Cesàro filtered Fourier series partial sum for the function

$$f(x) = \exp(-(x/2) \wedge 20)$$

over the interval $[-\pi, \pi]$ using 1024 points. (b) Graph the convolution over $[-\pi, \pi]$ of this function $f(x)$ with the *PSF* $\mathcal{P}_{20}$ for Cesàro filtering. Use 1024 points. (c) Check that the graphs found in (a) and (b) match, to a high degree of accuracy. (d) Repeat (a) to (c), but use 40 harmonics.

4.33 Repeat Exercise 4.32 using hanning, Hamming, and dlVP filtering.

4.34 For each of the functions given below, graph the convolution $f * \Delta_{1024}$, over the given interval, using 1024 points. Check that in each case $f * \Delta_{1024}$ is the same as f to a high degree of accuracy.

   (a) $f(x) = \exp(-x \wedge 2)$, interval $[-8, 8]$

   (b) $f(x) = \begin{cases} 1 & \text{for } |x| < 1 \\ 0 & \text{for } |x| > 1 \end{cases}$, interval $[-3, 3]$

   (c) $f(x) = x \exp(-x \wedge 2)$, interval $[-4, 4]$

4.35 Show that if f and g are functions of period $2L$, then $f * g = g * f$.

Section 7

4.36 Show that $u * v_j = v * u_j$ for any two sequences $\{u_j\}$ and $\{v_j\}$ of period N.

4.37 How many real multiplications are needed to calculate $\{u * v_j\}$ if $\{u_j\}$ and $\{v_j\}$ are real sequences? [*Hint:* use the methods in Sections 3.5 and 3.9 of Chapter 3.]

Section 8

4.38 Verify (4.129).

4.39 Combine (4.114a) and (4.129) to draw graphs of the hanning kernel for $M = 8$, 16, and 32 harmonics. Compare these graphs with the ones obtained as solutions to Exercise 4.31(c).

4.40 Show that Hamming's kernel H_M satisfies

$$H_M(x) = 0.54 D_M(x) + 0.23 D_M\left(x - \frac{\pi}{M}\right) + 0.23 D_M\left(x + \frac{\pi}{M}\right).$$

Use this formula in combination with (4.114a) to draw graphs of Hamming's kernel for $M = 8$, 16, and 32 harmonics. Compare these graphs with the ones obtained as solutions to Exercise 4.31(d).

4.41 Using (4.124), graph Cesàro's kernel for $M = 8$, 16, and 32 harmonics. Compare these graphs with the ones obtained as solutions to Exercise 4.31(a).

4.42 Using *FAS*, graph the hanning and Hamming filtered Fourier series, using 32 harmonics, of

$$f(x) = \begin{cases} 1 & \text{for } 0 < x < 1 \\ 0 & \text{for } -1 < x < 0. \end{cases}$$

Change the x-interval to $[0, 1]$ and the y-interval to $[0.98, 1.02]$. Explain the appearance of the graphs in terms of the *PSFs* for the two filters.

4.43 Repeat Exercise 4.42, but now use the dlVP kernel and the Riesz kernel with damping power 3.81.

Section 9

4.44 Verify that I_M in Example 4.10 is a summation kernel.

4.45 Using *FAS*, calculate $f * I_M$ for $M = 8$, 16, and 32, where

$$f(x) = \exp(-(x/0.3) \wedge 10).$$

Use an interval of $[-0.5, 0.5]$ and 1024 points. [*Hint:* use the formula $f(x) = M \operatorname{rect}(Mx)$ to plot I_M.]

4.46 Show that the dlVP kernel, using $2M$ harmonics, is a summation kernel. (*Remark:* the case of an odd number of harmonics is more difficult. The method for this case will be discussed in Section 5.10 of the next chapter.)

4.47 Show that Dirichlet's kernel is *not* a summation kernel.

Section 10

4.48 Using the definitions of x_M and t_M in the text, show that the following two approximations

$$x_M \approx \frac{\pi}{M + 1/2}, \quad t_M \approx \frac{2\pi}{M + 1/2}$$

are valid for large values of M.

4.49 Notice that in Tables 4.1 and 4.3 we have $x_{40} \approx t_{80}$ and $x_{80} \approx t_{160}$. Show that for large M, the approximation $x_M \approx t_{2M}$ is valid.

4.50 Show that a Gibbs' phenomenon occurs for the hanning filtered Fourier series partial sums of the function f in (4.156). Estimate the amount of overshooting at the origin. That is, to what limit does $H_M(x_M)$ tend to as $M \to \infty$ (where $H_M(x)$ is the hanning filtered partial sum using M harmonics, and $x = x_M$ yields the first maximum of $H_M(x)$ to the right of $x = 0$)?

4.51 Show that a Gibbs' phenomenon occurs for the dlVP filtered Fourier series partial sums (using even numbers of harmonics) of the function f in (4.156). Estimate the amount of overshooting at the origin. That is, to what limit does $V_{2M}(x_{2M})$ tend to as $2M \to \infty$ (where $V_{2M}(x)$ is the dlVP filtered partial sum using $2M$ harmonics, and $x = x_{2M}$ yields the first maximum of $V_{2M}(x)$ to the right of $x = 0$)?

4.52 Show that Gibbs' phenomenon does *not* occur for the Cesàro filtered Fourier series partial sums of the function f in (4.156).

5

Fourier Transforms

In this chapter, we will discuss the Fourier transform, which is one of the most important tools in pure and applied mathematics. We will describe its principal properties, including the concept of convolution, using *FAS* to provide many illustrations. Several applications to mathematical physics and signal processing will be described. And, we will examine two important topics in communication theory, Poisson summation and sampling theory.

5.1 Introduction

In this section we define the Fourier transform. Although it is possible to motivate the definition of the Fourier transform by way of Fourier series, it is logically simpler to just begin with the definition of the Fourier transform and then show its properties and applications.

Throughout this chapter we will use the following notation. Let $\|f\|_1$ be defined by

$$\|f\|_1 = \int_{-\infty}^{\infty} |f(x)| \, dx. \tag{5.1}$$

If the integral in (5.1) converges, then we will say that $\|f\|_1$ is finite and has the value to which the integral converges (if the integral diverges, then we will say that $\|f\|_1 = \infty$). Similarly, we define $\|f\|_2$ by

$$\|f\|_2 = \left[\int_{-\infty}^{\infty} |f(x)|^2 \, dx \right]^{\frac{1}{2}}. \tag{5.2}$$

Again, if the integral in (5.2) converges, then we will say that $\|f\|_2$ is finite and $\|f\|_2^2$ has the value to which the integral converges (if the integral diverges, then we will say that $\|f\|_2 = \infty$).

We can now give the definition of the Fourier transform.

DEFINITION 5.1:   *Given a function f for which $\| f \|_1$ is finite, the **Fourier***
***transform** of f is denoted by $\hat{f}$ and is defined as a function of u by*

$$\hat{f}(u) = \int_{-\infty}^{\infty} f(x)e^{-i2\pi ux} \, dx.$$

Here are some simple examples.

Example 5.1:

Suppose that the function rect is defined by

$$\text{rect}(x) = \begin{cases} 1 & \text{if } |x| < 0.5 \\ 0.5 & \text{if } |x| = 0.5 \\ 0 & \text{if } |x| > 0.5 \end{cases} \tag{5.3}$$

then

$$\widehat{\text{rect}}(u) = \int_{-0.5}^{0.5} e^{-i2\pi ux} \, dx = \frac{e^{-i2\pi ux}}{-i2\pi u}\bigg|_{x=-0.5}^{x=0.5}$$

$$= \frac{e^{-i\pi u} - e^{i\pi u}}{-i2\pi u} = \frac{\sin \pi u}{\pi u}.$$

The transform of rect(x) is therefore sinc (u). Using the notation $f(x) \xrightarrow{\mathcal{F}} \hat{f}(u)$
to denote the Fourier transform operation, we have

$$\text{rect}(x) \xrightarrow{\mathcal{F}} \text{sinc } (u). \tag{5.4}$$

See Figure 5.1.

Example 5.2:

Suppose that f is defined by

$$f(x) = \begin{cases} e^{-2\pi x} & \text{for } x > 0 \\ 0 & \text{for } x < 0. \end{cases}$$

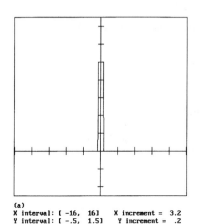

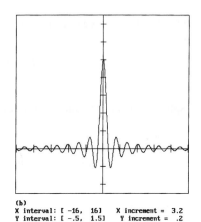

(a)
X interval: [-16, 16]   X increment = 3.2
Y interval: [-.5, 1.5]   Y increment = .2

(b)
X interval: [-16, 16]   X increment = 3.2
Y interval: [-.5, 1.5]   Y increment = .2

FIGURE 5.1
(a) Graph of the rect function. (b) Graph of its Fourier transform.

then

$$\hat{f}(u) = \int_0^\infty e^{-2\pi x} e^{-i2\pi ux} \, dx = \left. \frac{e^{-2\pi x(1+iu)}}{-2\pi(1+iu)} \right|_{x=0}^{x\to\infty}$$

$$= \frac{1}{2\pi(1+iu)}$$

since $e^{-2\pi x(1+iu)} \to 0$ as $x \to \infty$.

Similarly, as the reader may show, the function

$$g(x) = \begin{cases} 0 & \text{for } x > 0 \\ e^{2\pi x} & \text{for } x < 0 \end{cases}$$

has Fourier transform

$$\hat{g}(u) = \frac{1}{2\pi(1-iu)} \, .$$

Example 5.3:

This example will show that

$$e^{-2\pi|x|} \xrightarrow{\mathcal{F}} \frac{1}{\pi} \frac{1}{1+u^2} \, . \tag{5.5}$$

SOLUTION We need to evaluate the Fourier transform integral

$$\int_{-\infty}^{\infty} e^{-2\pi|x|} e^{-i2\pi ux} \, dx.$$

Splitting the integral into a sum of two integrals, we have

$$\int_{-\infty}^{\infty} e^{-2\pi|x|} e^{-i2\pi ux} \, dx = \int_{0}^{\infty} e^{-2\pi x} e^{-i2\pi ux} \, dx + \int_{-\infty}^{0} e^{2\pi x} e^{-i2\pi ux} \, dx$$

$$= \hat{f}(u) + \hat{g}(u)$$

where f and g are the functions defined in Example 5.2. Using the results of Example 5.2 for $\hat{f}$ and $\hat{g}$, we have

$$\int_{-\infty}^{\infty} e^{-2\pi|x|} e^{-i2\pi ux} \, dx = \frac{1}{2\pi(1+iu)} + \frac{1}{2\pi(1-iu)} = \frac{1}{\pi} \frac{1}{1+u^2}$$

and (5.5) is proved. See Figure 5.2. ∎

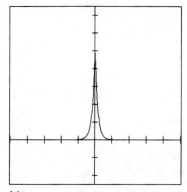

(a)
X interval: [-5, 5] X increment = 1
Y interval: [-.5, 1.5] Y increment = .2

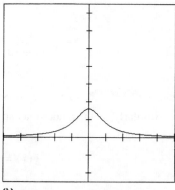

(b)
X interval: [-5, 5] X increment = 1
Y interval: [-.5, 1.5] Y increment = .2

FIGURE 5.2
(a) Graph of $e^{-2\pi|x|}$. (b) Graph of its Fourier transform.

Example 5.4:
Find the Fourier transform of

$$\Lambda(x) = \begin{cases} 1 - |x| & \text{for } |x| \le 1 \\ 0 & \text{for } |x| > 1. \end{cases}$$

SOLUTION   Since $\Lambda = 0$ outside of the interval $[-1, 1]$ and Λ is an even function, we have

$$\hat{\Lambda}(u) = \int_{-1}^{1} (1 - |x|) \cos 2\pi ux \, dx - i \int_{-1}^{1} (1 - |x|) \sin 2\pi ux \, dx$$

$$= 2 \int_{0}^{1} (1 - x) \cos 2\pi ux \, dx.$$

Performing an integration by parts with this last integral, we obtain

$$\hat{\Lambda}(u) = \frac{1}{\pi u} \int_{0}^{1} \sin 2\pi ux \, dx = \frac{1 - \cos 2\pi u}{2\pi^2 u^2}$$

$$= \frac{\sin^2 \pi u}{\pi^2 u^2} = \text{sinc}^2(u).$$

Thus, $\Lambda(x) \xrightarrow{\mathcal{F}} \text{sinc}^2(u)$.   ∎

Example 5.5:
  We will calculate the Fourier transform of $f(x) = e^{-\pi x^2}$. The definition of Fourier transform yields

$$\hat{f}(u) = \int_{-\infty}^{\infty} e^{-\pi x^2} e^{-i2\pi ux} \, dx. \tag{5.6}$$

If we differentiate both sides of (5.6) with respect to u, then by differentiating under the integral sign, we get

$$\frac{d\hat{f}}{du} = \int_{-\infty}^{\infty} \frac{\partial}{\partial u} \left[e^{-\pi x^2} e^{-i2\pi ux} \right] dx$$

$$= \int_{-\infty}^{\infty} -i2\pi x e^{-\pi x^2} e^{-i2\pi ux} \, dx$$

$$= \int_{-\infty}^{\infty} i \frac{d}{dx} \left[e^{-\pi x^2} \right] e^{-i2\pi ux} \, dx. \tag{5.7}$$

(*Note:* the method of differentiating under the integral sign is a common technique. For further discussion, see [Wa, Chapter 6.2].)

Integrating by parts in the last integral in (5.7), we obtain

$$\frac{d\hat{f}}{du} = ie^{-i2\pi ux}e^{-\pi x^2}\Big|_{x \to -\infty}^{x \to \infty} - \int_{-\infty}^{\infty} 2\pi u e^{-\pi x^2}e^{-i2\pi ux}\,dx$$

$$= -2\pi u \int_{-\infty}^{\infty} e^{-\pi x^2}e^{-i2\pi ux}\,dx.$$

Thus,

$$\frac{d\hat{f}}{du} = -2\pi u \hat{f}(u). \tag{5.8}$$

Solving (5.8) for the unknown function $\hat{f}$, we get

$$\hat{f}(u) = \hat{f}(0)e^{-\pi u^2}. \tag{5.9}$$

It remains to find $\hat{f}(0)$, where

$$\hat{f}(0) = \int_{-\infty}^{\infty} e^{-\pi x^2}\,dx. \tag{5.10}$$

Squaring both sides of (5.10), and replacing the dummy variable x by y in one of the integrals, we have

$$\hat{f}(0)^2 = \int_{-\infty}^{\infty} e^{-\pi x^2}\,dx \int_{-\infty}^{\infty} e^{-\pi y^2}\,dy$$

$$= \int_{-\infty}^{\infty}\int_{-\infty}^{\infty} e^{-\pi(x^2+y^2)}\,dx\,dy. \tag{5.11}$$

Changing to polar coordinates, we get

$$\int_{-\infty}^{\infty}\int_{-\infty}^{\infty} e^{-\pi(x^2+y^2)}\,dx\,dy = \int_{0}^{2\pi}\int_{0}^{\infty} e^{-\pi r^2} r\,dr\,d\theta$$

$$= \int_{0}^{2\pi} \left[\frac{-1}{2\pi}e^{-\pi r^2}\right]\Big|_{r=0}^{r \to \infty} d\theta = 1.$$

Thus, on returning to (5.11), we see that $\hat{f}(0)^2 = 1$. Since $\hat{f}(0) > 0$, as we can see from (5.10), we must have $\hat{f}(0) = 1$. Using this fact in (5.9) we have

$$e^{-\pi x^2} \xrightarrow{\mathcal{F}} e^{-\pi u^2}. \tag{5.12}$$

Formula (5.12) expresses the remarkable fact that $e^{-\pi x^2}$ Fourier transforms to itself (as a function of u).

5.2 Properties of Fourier Transforms

This section covers some important properties of Fourier transforms, for example, how they behave with respect to shifting, scaling, and differentiation.

We begin with the following theorem.

THEOREM 5.1:
The Fourier transform operation $f \xrightarrow{\mathcal{F}} \hat{f}$ has the following properties:
*(a) **Linearity:** For all constants a and b,*

$$af + bg \xrightarrow{\mathcal{F}} a\hat{f} + b\hat{g}.$$

*(b) **Scaling:** For each positive constant ρ,*

$$f\left(\frac{x}{\rho}\right) \xrightarrow{\mathcal{F}} \rho\hat{f}(\rho u) \quad \text{and} \quad f(\rho x) \xrightarrow{\mathcal{F}} \frac{1}{\rho}\hat{f}\left(\frac{u}{\rho}\right).$$

*(c) **Shifting:** For each real constant c,*

$$f(x - c) \xrightarrow{\mathcal{F}} \hat{f}(u)e^{-i2\pi cu}.$$

*(d) **Modulation:** For each real constant c,*

$$f(x)e^{i2\pi cx} \xrightarrow{\mathcal{F}} \hat{f}(u - c).$$

PROOF The proof of (a) is straightforward, so we leave the details to the reader. To prove (b), we make the change of variables $s = x/\rho$ in the following Fourier transform integral:

$$f\left(\frac{x}{\rho}\right) \xrightarrow{\mathcal{F}} \int_{-\infty}^{\infty} f\left(\frac{x}{\rho}\right) e^{-i2\pi ux}\, dx = \int_{-\infty}^{\infty} f(s) e^{-i2\pi u(\rho s)}\, d(\rho s)$$

$$= \rho \int_{-\infty}^{\infty} f(s) e^{-i2\pi(\rho u)s}\, ds = \rho \hat{f}(\rho u).$$

Thus, $f(x/\rho) \xrightarrow{\mathcal{F}} \rho \hat{f}(\rho u)$. Substituting $1/\rho$ in place of ρ, it follows that $f(\rho x) \xrightarrow{\mathcal{F}} (1/\rho) \hat{f}(u/\rho)$ and (b) is verified.

To prove (c) we make the change of variable $s = x - c$ in the following Fourier transform integral:

$$f(x - c) \xrightarrow{\mathcal{F}} \int_{-\infty}^{\infty} f(x - c) e^{-i2\pi ux}\, dx = \int_{-\infty}^{\infty} f(s) e^{-i2\pi u(s+c)}\, ds$$

$$= \int_{-\infty}^{\infty} f(s) e^{-i2\pi us}\, ds\, e^{-i2\pi cu} = \hat{f}(u) e^{-i2\pi cu}.$$

Thus, (c) holds.

To prove (d), we note that $e^{i2\pi cx} e^{-i2\pi ux} = e^{-i2\pi(u-c)x}$, hence

$$f(x) e^{i2\pi cx} \xrightarrow{\mathcal{F}} \int_{-\infty}^{\infty} f(x) e^{-i2\pi(u-c)x}\, dx = \hat{f}(u - c)$$

and (d) holds. ∎

Here are some examples of Theorem 5.1.

Example 5.6:

(a) Find the Fourier transform of $(1/\rho) e^{-\pi x^2/\rho^2}$ for each $\rho > 0$.

SOLUTION   Using scaling, linearity, and $e^{-\pi x^2} \xrightarrow{\mathcal{F}} e^{-\pi u^2}$ we have

$$\frac{1}{\rho} e^{-\pi x^2/\rho^2} \xrightarrow{\mathcal{F}} \frac{\rho}{\rho} e^{-\pi \rho^2 u^2} = e^{-\pi \rho^2 u^2}.$$

∎

(b) Find the Fourier transform of $\text{rect}(x - 4) + \text{rect}(x + 4)$.

SOLUTION   Using shifting and linearity, we have

$$\text{rect}(x-4)+\text{rect}(x+4) \xrightarrow{\mathcal{F}} \widehat{\text{rect}}(u)e^{-i2\pi 4u} + \widehat{\text{rect}}(u)e^{i2\pi 4u}.$$

Since $\widehat{\text{rect}}(u) = \text{sinc}\,(u)$ and $e^{-i2\pi 4u} + e^{i2\pi 4u} = 2\cos 8\pi u$, we have

$$\text{rect}(x-4)+\text{rect}(x+4) \xrightarrow{\mathcal{F}} 2\,\text{sinc}\,(u)\cos(8\pi u).$$

∎

(c) Find the Fourier transform of $e^{-2\pi |x|}\cos 6\pi x$.

SOLUTION   Rewriting $\cos 6\pi x$ as $(1/2)e^{i2\pi 3x} + (1/2)e^{-i2\pi 3x}$ and then using linearity and modulation, we get

$$e^{-2\pi |x|}\cos 6\pi x \xrightarrow{\mathcal{F}} \frac{1}{2}(e^{-2\pi |x|})^{\wedge}(u-3) + \frac{1}{2}(e^{-2\pi |x|})^{\wedge}(u+3).$$

Hence,

$$e^{-2\pi |x|}\cos 6\pi x \xrightarrow{\mathcal{F}} \frac{1}{2\pi}\frac{1}{1+(u-3)^2} + \frac{1}{2\pi}\frac{1}{1+(u+3)^2}.$$

∎

(d) Find the Fourier transform of $e^{-2\pi \rho |x|}$ for all $\rho > 0$.

SOLUTION   Using the scaling property, we have

$$e^{-2\pi \rho |x|} \xrightarrow{\mathcal{F}} \frac{1}{\rho}\left(\frac{1}{\pi}\frac{1}{1+(u/\rho)^2}\right).$$

Hence,

$$e^{-2\pi \rho |x|} \xrightarrow{\mathcal{F}} \frac{1}{\pi}\frac{\rho}{\rho^2+u^2}, \qquad (\rho > 0).$$

∎

Another useful property of Fourier transforms is the conversion of differentiation into multiplication by the transform variable. To be more precise, we have the following theorem.

THEOREM 5.2:
 (a) Suppose that $\|f\|_1$ and $\|f'\|_1$ are finite and that f' is continuous, then $f'(x) \xrightarrow{\mathcal{F}} i2\pi u \hat{f}(u)$. (b) Suppose that $\|f\|_1$ and $\|xf(x)\|_1$ are finite, then $\hat{f}$ is differentiable and

$$xf(x) \xrightarrow{\mathcal{F}} \frac{i}{2\pi} \hat{f}'(u).$$

PROOF   (a) Using integration by parts, we have

$$f'(x) \xrightarrow{\mathcal{F}} \int_{-\infty}^{\infty} f'(x)e^{-i2\pi ux}\, dx = \left. f(x)e^{-i2\pi ux} \right|_{x \to -\infty}^{x \to \infty}$$

$$+ i2\pi u \int_{-\infty}^{\infty} f(x)e^{-i2\pi ux}\, dx.$$

By the Fundamental Theorem of Calculus, we have $f(x) = \int_0^x f'(s)ds + f(0)$. Consequently, the limits $\lim_{x \to \infty} f(x)$ and $\lim_{x \to -\infty} f(x)$ exist and are finite. If either of these limits were not 0, then

$$\int_{-\infty}^{\infty} f(x)\, dx = \int_0^{\infty} f(x)\, dx + \int_{-\infty}^0 f(x)\, dx$$

would not converge (since at least one of the integrals on the right side would diverge). Therefore, both limits are 0 and

$$f'(x) \xrightarrow{\mathcal{F}} i2\pi u \int_{-\infty}^{\infty} f(x)e^{-i2\pi ux}\, dx = i2\pi u \hat{f}(u)$$

and (a) is proved.
    To prove (b) we differentiate under the integral sign as follows

$$\hat{f}'(u) = \frac{d}{du} \int_{-\infty}^{\infty} f(x)e^{-i2\pi ux}\, dx = \int_{-\infty}^{\infty} \frac{\partial}{\partial u}\left[f(x)e^{-i2\pi ux} \right] dx$$

$$= \int_{-\infty}^{\infty} -i2\pi x f(x)e^{-i2\pi ux}\, dx.$$

But, this shows that

$$-i2\pi x f(x) \xrightarrow{\mathcal{F}} \hat{f}'(u). \tag{5.13}$$

Multiplying both sides of (5.13) by $i/2\pi$ yields (b).   ∎

Example 5.7:
(a) Using Theorem 5.2(b), we have

$$xe^{-\pi x^2} \xrightarrow{\mathcal{F}} \frac{i}{2\pi} \frac{d}{du} \left[e^{-\pi u^2} \right]$$

hence

$$xe^{-\pi x^2} \xrightarrow{\mathcal{F}} -iue^{-\pi u^2}.$$

Thus, $xe^{-\pi x^2}$ transforms into $-i$ times itself (as a function of u).
(b)   Using the previous example, and Theorem 5.2(b),

$$x^2 e^{-\pi x^2} \xrightarrow{\mathcal{F}} \frac{i}{2\pi} \frac{d}{du} \left[-iue^{-\pi u^2} \right].$$

Hence,

$$x^2 e^{-\pi x^2} \xrightarrow{\mathcal{F}} \frac{1}{2\pi} e^{-\pi u^2} - u^2 e^{-\pi u^2}.$$

(c) From (b), it follows that

$$\left(x^2 - \frac{1}{4\pi} \right) e^{-\pi x^2} \xrightarrow{\mathcal{F}} -\left(u^2 - \frac{1}{4\pi} \right) e^{-\pi u^2}.$$

Thus, $[x^2 - 1/(4\pi)]e^{-\pi x^2}$ Fourier transforms into -1 times itself (as a function of u).

REMARK 5.1:   Bringing absolute values inside the integral sign and using

$$|e^{-i2\pi ux}| = 1$$

we obtain

$$|\hat{f}(u)| \le \|f\|_1 \tag{5.14}$$

which shows that $|\hat{f}|$ is bounded by the constant $\|f\|_1$.
   Using Theorem 5.2(a), we can say a little more than (5.14). If $\|f\|_1$ and $\|f'\|_1$ are both finite and f' is continuous, then by applying (5.14) to f' in place of f and using $\widehat{(f')}(u) = i2\pi u \hat{f}(u)$, we have

$$|i2\pi u \hat{f}(u)| \le \|f'\|_1. \tag{5.15}$$

In other words, for all $u \neq 0$

$$|\hat{f}(u)| \le \frac{\|f'\|_1}{2\pi |u|}. \tag{5.16}$$

From (5.16) it follows that

$$\lim_{|u| \to \infty} \hat{f}(u) = 0 \tag{5.17}$$

when $\| f \|_1$ and $\| f' \|_1$ are both finite and f' is continuous. Actually, (5.17) is known to be true whenever $\| f \|_1$ is finite (this is known as the *Riemann-Lebesgue Lemma,* but we will not need this fact in the sequel). ∎

Another important property of Fourier transforms is that they are continuous functions.

THEOREM 5.3:
If $\| f \|_1$ is finite, then the Fourier transform $\hat{f}$ is a continuous function.

PROOF   See [Wa, Chapter 6.2]. ∎

5.3   Inversion of Fourier Transforms

In this section, we shall state some inversion theorems for Fourier transforms. The proofs of these theorems are fairly technical and are omitted here, although references will be given.

We begin with the conceptually simplest inversion theorem.

THEOREM 5.4:
If f is continuous and $\| f \|_1$ and $\| \hat{f} \|_1$ are both finite, then

$$f(x) = \int_{-\infty}^{\infty} \hat{f}(u) e^{i2\pi u x} \, du$$

*for all x in **R**.*

PROOF   See [Wa, Chapter 6.4]. ∎

REMARK 5.2:   The integral in Theorem 5.4 is also called a Fourier transform. Sometimes we might call it an *inverse Fourier transform,* or we might call it a *Fourier transform with positive exponent* (the Fourier transform in Definition 5.1 being called a Fourier transform with *negative* exponent). If $\| g \|_1$ is finite, then we define $\tilde{g}$ by

$$\tilde{g}(x) = \int_{-\infty}^{\infty} g(u) e^{i2\pi ux} \, du.$$

Hence, the result in Theorem 5.4 can be rewritten as $\tilde{\hat{f}} = f$ or $\hat{f} \xrightarrow{\mathcal{F}^{-1}} f$. ∎

As an example of Theorem 5.4, consider the triangle function

$$\Lambda(x) = \begin{cases} 1 - |x| & \text{for } |x| \le 1 \\ 0 & \text{for } |x| > 1. \end{cases}$$

We have, by Example 5.4,

$$\hat{\Lambda}(u) = \operatorname{sinc}^2(u) = \frac{\sin^2(\pi u)}{(\pi u)^2}.$$

Both Λ and $\hat{\Lambda}$ are continuous. That $\|\Lambda\|_1$ is finite is obvious. The finiteness of $\|\hat{\Lambda}\|_1$ follows from comparing it to the integrals of $1/(\pi u)^2$ for $|u| \ge 1$. It follows from Theorem 5.4 that

$$\Lambda(x) = \int_{-\infty}^{\infty} \operatorname{sinc}^2(u) \, e^{i2\pi ux} \, du. \tag{5.18}$$

Let's examine Formula (5.18) more closely. The integral on the right side of (5.18) converges absolutely, since for $|u| > 0$

$$\left| \frac{\sin^2(\pi u)}{(\pi u)^2} e^{i2\pi ux} \right| \le \frac{1}{(\pi u)^2}. \tag{5.19}$$

This inequality can be used for estimating approximations to the inversion integral in (5.18). For example, if we only integrate from $-L$ to L, for some finite positive L, then we have

$$\left| \Lambda(x) - \int_{-L}^{L} \operatorname{sinc}^2 u \, e^{i2\pi ux} \, du \right| = \left| \int_{|u|>L} \operatorname{sinc}^2 u \, e^{i2\pi ux} \, du \right|$$

$$\le \int_{|u|>L} \frac{1}{(\pi u)^2} \, du = \frac{2}{\pi^2 L}.$$

For example, if we take $L = 32$, then $2/(\pi^2 L) \approx 0.006$. Hence, for *all* x,

$$\left| \Lambda(x) - \int_{-32}^{32} \operatorname{sinc}^2 u \, e^{i2\pi u x} \, du \right| \leq 0.006. \tag{5.20}$$

In Figure 5.3 we have shown a graph of the inverse Fourier transform of sinc^2 over the interval $[-32, 32]$, using 4096 points.

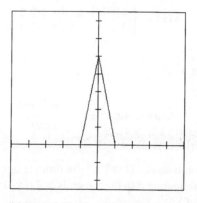

X interval: [-5, 5]   X increment = 1
Y interval: [-.5, 1.5]   Y increment = .2

FIGURE 5.3
Inverse Fourier transform of sinc^2 over $[-32, 32]$.

Theorem 5.4 can be used to derive new Fourier transforms. If we have $f \xrightarrow{\mathcal{F}} \hat{f}$ and $\| \hat{f} \|_1$ is finite, then we can Fourier transform $\hat{f}$. We leave it as an exercise for the reader to show, based on Theorem 5.4, that

$$\hat{f}(x) \xrightarrow{\mathcal{F}} f(-u).$$

In particular, since $\Lambda(x) \xrightarrow{\mathcal{F}} \operatorname{sinc}^2(u)$, it follows that $\operatorname{sinc}^2(x) \xrightarrow{\mathcal{F}} \Lambda(-u)$. Since Λ is an even function, we obtain

$$\operatorname{sinc}^2(x) \xrightarrow{\mathcal{F}} \Lambda(u).$$

Similarly, the reader may wish to verify that

$$\frac{1}{\pi} \frac{1}{1 + x^2} \xrightarrow{\mathcal{F}} e^{-2\pi |u|}.$$

Here is a second inversion theorem. It is less restrictive in its hypotheses than Theorem 5.4, in that it does not require f to be continuous nor $\|\hat{f}\|_1$ to be finite.

THEOREM 5.5:
If $\|f\|_1$ is finite, then

$$\lim_{L \to \infty} \int_{-L}^{L} \hat{f}(u) e^{i2\pi ux} \, du = \frac{1}{2} [f(x+) + f(x-)]$$

provided f is Lipschitz from the left and right at x. Moreover, if f is also continuous at x, then

$$\lim_{L \to \infty} \int_{-L}^{L} \hat{f}(u) e^{i2\pi ux} \, du = f(x).$$

PROOF   See [Wa, Chapter 6.5].   ∎

The reader might compare this theorem to the Fourier series convergence theorem (Theorem 1.5) in Chapter 1. Here is an example of how Theorem 5.5 works. The function

$$f(x) = \begin{cases} i\pi e^{-2\pi x} & \text{for } x > 0 \\ -i\pi e^{2\pi x} & \text{for } x < 0 \end{cases} \tag{5.21}$$

has Fourier transform $u/(1 + u^2)$. But, the inversion integral

$$\int_{-\infty}^{\infty} \frac{u}{1 + u^2} e^{i2\pi ux} \, du \tag{5.22}$$

does not converge for $x = 0$. Since, for $x = 0$, the inversion integral in (5.22) is defined by

$$\int_{-\infty}^{\infty} \frac{u}{1 + u^2} \, du = \int_{0}^{\infty} \frac{u}{1 + u^2} \, du + \int_{-\infty}^{0} \frac{u}{1 + u^2} \, du \tag{5.23}$$

and both integrals on the right side of (5.23) diverge. However, the integrals

$$\int_{-L}^{L} \frac{u}{1 + u^2} \, du \ (= 0)$$

converge to 0 as $L \to \infty$, which is precisely what Theorem 5.5 says should happen at $x = 0$ (in particular, $0 = (1/2)[f(0+) + f(0-)]$). In Figure 5.4 we show graphs obtained using *FAS* to approximate the following integral

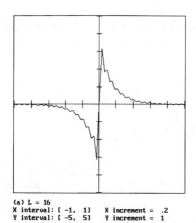 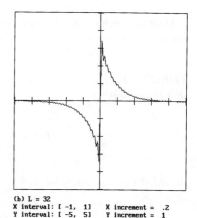

(a) L = 16
X interval: [-1, 1]   X increment = .2
Y interval: [-5, 5]   Y increment = 1

(b) L = 32
X interval: [-1, 1]   X increment = .2
Y interval: [-5, 5]   Y increment = 1

FIGURE 5.4
Calculation of (5.24) for $L = 16$ and $L = 32$.

$$\int_{-L}^{L} \frac{u}{1 + u^2} e^{i2\pi ux}\, du \tag{5.24}$$

for increasing values of L. As we can see from this figure, the convergence of the integral in (5.24) to $f(x)$ is very reminiscent of Fourier series convergence. (There is even a Gibbs' phenomenon!) We allude to this similarity in the following definition.

DEFINITION 5.2:   *If L is a positive real number, then $\mathcal{S}_L^f$ is defined by*

$$\mathcal{S}_L^f(x) = \int_{-L}^{L} \hat{f}(u) e^{i2\pi ux}\, du.$$

For example, Theorem 5.5 could be restated as

$$\mathcal{S}_L^f(x) \to \frac{1}{2}[\, f(x+) + f(x-)\,], \quad \text{as } L \to \infty \tag{5.25}$$

provided f is Lipschitz from the left and right at x.

Here is our final convergence theorem. It is analogous to the completeness of Fourier series.

THEOREM 5.6:
If $\|f\|_2 < \infty$, then $\lim_{L \to \infty} \|f - \mathcal{S}_L^f\|_2 = 0$. That is, $\mathcal{S}_L^f$ converges to f in the 2-Norm.

PROOF A proof can be found in [Ru, Chapter 9]. ∎

To understand Theorem 5.6 completely one must know how the Fourier transform $\hat{f}$ is defined when $\|f\|_2 < \infty$. There are many functions for which $\|f\|_2 < \infty$ but $\|f\|_1 = \infty$, in which case Definition 5.1 does not apply. There is a subtler definition of the Fourier transform $\hat{f}$ in this case. It is known that $\int_{-L}^{L} f(x)e^{-i2\pi ux}\,dx$ converges as $L \to \infty$, using the 2-Norm $\|\cdot\|_2$ to measure the magnitude of differences, to a function $g(u)$ for which $\|g\|_2 < \infty$. That is,

$$\lim_{L\to\infty} \left\| g(u) - \int_{-L}^{L} f(x)e^{-i2\pi ux}\,dx \right\|_2 = 0. \tag{5.26}$$

This function $g(u)$ is called the Fourier transform $\hat{f}(u)$. (See [Ru, Chapter 9] for more details.)

As an example of Theorem 5.6, suppose $f(x) = \text{rect}(x)$. Then, $\text{rect}(x) \xrightarrow{\mathcal{F}} \text{sinc }(u)$. If we use *FAS* to approximate the integral

$$S_L^{\text{rect}}(x) = \int_{-L}^{L} \text{sinc }(u)\,e^{i2\pi ux}\,du \tag{5.27}$$

(using, say, 1024 points), then the 2-Norms $\|\text{rect} - S_L^{\text{rect}}\|_2$ can be approximated. Here are some results for various values of L:

$$\|\text{rect} - S_8^{\text{rect}}\|_2 \approx 0.06, \qquad \|\text{rect} - S_{16}^{\text{rect}}\|_2 \approx 0.04,$$

$$\|\text{rect} - S_{32}^{\text{rect}}\|_2 \approx 0.03, \qquad \|\text{rect} - S_{64}^{\text{rect}}\|_2 \approx 0.02. \tag{5.28}$$

5.4 The Relation between Fourier Transforms and DFTs

In this section we will explain how the Fourier transform of a function can be approximated by a discrete Fourier transform (DFT). We will limit ourselves to a brief introduction to this topic; a more thorough discussion can be found in [Br-H, Chapter 6].

First, let's assume that the transform $\hat{f}$ can be approximated by an integral over the interval $[0, L]$ for some large L (this would be the case, for example, if $f(x) = 0$ for $x < 0$). This case is a bit easier to handle than the more general case, which would be to assume that $\hat{f}$ can be approximated by an integral over the interval $[-L/2, L/2]$ for some large L. We will discuss this latter case by reduction to the first case.

Assuming that $\hat{f}$ is approximated by an integral over $[0, L]$, we have

$$\hat{f}(u) \approx \int_0^L f(x)e^{-i2\pi ux}\,dx. \tag{5.29}$$

The integral is then approximated by a uniform, left-endpoint sum

$$\int_0^L f(x)e^{-i2\pi ux}\,dx \approx \sum_{j=0}^{N-1} f\left(j\frac{L}{N}\right)e^{-i2\pi ujL/N}\frac{L}{N}.$$

Thus,

$$\hat{f}(u) \approx \frac{L}{N}\sum_{j=0}^{N-1} f\left(j\frac{L}{N}\right)e^{-i2\pi ujL/N}. \tag{5.30}$$

To convert the right side of (5.30) into a DFT we replace u by k/L for $k = 0, \pm 1, \pm 2, \dots$

$$\hat{f}\left(\frac{k}{L}\right) \approx \frac{L}{N}\sum_{j=0}^{N-1} f\left(j\frac{L}{N}\right)e^{-i2\pi jk/N}. \tag{5.31}$$

The right side of (5.31) *is equal to* L/N *times the* N-*point DFT of the sequence*

$$\left\{f\left(j\frac{L}{N}\right)\right\}_{j=0}^{N-1}.$$

The computer, of course, is programmed to compute an N-point FFT of $\{f(jL/N)\}_{j=0}^{N-1}$, call it $\{F_k\}$. (To be more precise, *FAS* actually uses the more accurate *trapezoidal rule* approximation of the integral in (5.29); this involves a multiple of the DFT of $\{(1/2)[f(0) + f(L)], f(L/N), \dots, f[(N-1)L/N]\}$.) By periodicity, we know that F_{-k} satisfies

$$F_{-k} = F_{N-k} \tag{5.32}$$

and, hence

$$\hat{f}\left(\frac{k}{L}\right) \approx \frac{L}{N}F_k \quad \text{for } k = 0, 1, \dots, \tfrac{1}{2}N$$

$$\hat{f}\left(\frac{-k}{L}\right) \approx \frac{L}{N}F_{N-k} \quad \text{for } k = 1, \dots, \tfrac{1}{2}N - 1. \tag{5.33}$$

Because the right side of (5.31) is a multiple of a DFT, it has period N in k. Therefore, it is hopeless to expect that the approximation in (5.31) will be valid for *all* integers k (since $\hat{f}$ is typically *not* periodic). As we did for Fourier series, we shall assume that $|k| \leq N/8$. This assumption will serve to remove the grossest of aliasing errors.

Now, suppose that we wish to approximate $\hat{f}$ by an integral over $[-L/2, L/2]$. Then we will have

$$\hat{f}(u) \approx \int_{-\frac{1}{2}L}^{\frac{1}{2}L} f(x)e^{-i2\pi ux}\,dx. \tag{5.34}$$

We will show that $\{\hat{f}(k/L)\}$ can be approximated by a constant multiple of a DFT. We have

$$\hat{f}\left(\frac{k}{L}\right) \approx \int_{-\frac{1}{2}L}^{\frac{1}{2}L} f(x)e^{-i2\pi kx/L}\,dx. \tag{5.35}$$

First, we split the integral in (5.35) into two integrals:

$$\int_{-\frac{1}{2}L}^{\frac{1}{2}L} f(x)e^{-i2\pi kx/L}\,dx = \int_{-\frac{1}{2}L}^{0} f(x)e^{-i2\pi kx/L}\,dx$$

$$+ \int_{0}^{\frac{1}{2}L} f(x)e^{-i2\pi kx/L}\,dx. \tag{5.36}$$

Second, we substitute $t = x + L$ into the first integral on the right side of (5.36), and obtain

$$\int_{-\frac{1}{2}L}^{\frac{1}{2}L} f(x)e^{-i2\pi kx/L}\,dx = \int_{\frac{1}{2}L}^{L} f(t-L)e^{-i2\pi kt/L}\,dt$$

$$+ \int_{0}^{\frac{1}{2}L} f(x)e^{-i2\pi kx/L}\,dx$$

$$= \int_{\frac{1}{2}L}^{L} f(x-L)e^{-i2\pi kx/L}\,dx$$

$$+ \int_{0}^{\frac{1}{2}L} f(x)e^{-i2\pi kx/L}\,dx.$$

Thus, if we define g by

$$
g(x) = \begin{cases}
f(x) & \text{for } 0 \leq x < \frac{1}{2}L \\[2mm]
\frac{1}{2}f(\frac{1}{2}L) + \frac{1}{2}f(-\frac{1}{2}L) & \text{for } x = \frac{1}{2}L \\[2mm]
f(x - L) & \text{for } \frac{1}{2}L < x \leq L
\end{cases}
\tag{5.37}
$$

we have

$$
\int_{-\frac{1}{2}L}^{\frac{1}{2}L} f(x)e^{-i2\pi kx/L}\,dx = \int_0^L g(x)e^{-i2\pi kx/L}\,dx.
\tag{5.38}
$$

Using (5.38), we have from (5.35) that

$$
\hat{f}\left(\frac{k}{L}\right) \approx \int_0^L g(x)e^{-i2\pi kx/L}\,dx.
\tag{5.39}
$$

Hence, as we did above for (5.29), we can approximate the right side of (5.39) by L/N times an N-point DFT

$$
\hat{f}\left(\frac{k}{L}\right) \approx \frac{L}{N}\sum_{j=0}^{N-1} g\left(j\frac{L}{N}\right)e^{-i2\pi jk/N}.
\tag{5.40}
$$

In the discussion above, we mentioned the restriction: $|k| \leq N/8$. *FAS* actually computes the entire FFT of $\{g(jL/N)\}$ (or $\{f(jL/N)\}$ when working over the interval $[0, L]$) and connects these FFT values by line segments. An example should make this point clearer.

Example 5.8:
Compare the computer calculation of the Fourier transform of rect and its exact transform, sinc .

SOLUTION   In Figure 5.5 we compare the *FAS* computed transform of rect(x), computed over the interval $[-16, 16]$ using 1024 points, with its exact transform sinc (x). We see in Figure 5.5(a) that the graphs are a reasonably close match over the interval $[-4, 4]$ where $|k| \leq N/8$. In Figure 5.5(b), however, we see the difference between the two graphs over $[6, 16]$. The graph of the computed transform damps down more severely towards 0 as we approach the end of the interval.

To expand the range of the computed transform *and improve its accuracy,* we increase the number of points used *and* increase the size of the interval over which

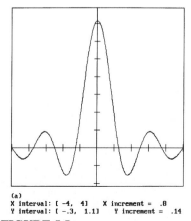

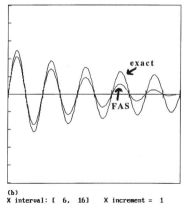

(a)
X interval: [-4, 4]   X increment = .8
Y interval: [-.3, 1.1]   Y increment = .14

(b)
X interval: [6, 16]   X increment = 1
Y interval: [-.1, .1]   Y increment = .02

FIGURE 5.5
Comparison of *FAS* calculated transform of rect **(using interval** $[-16, 16]$ **and 1024 points) and the exact transform** sinc.

we transform. If rect is transformed over $[-32, 32]$ using 4096 points, then the computed transform will be graphed over the interval $[-T, T]$ where

$$T = \frac{(\text{number of points})/2}{\text{length of } [-32, 32]} = \frac{2048}{64} = 32. \tag{5.41}$$

Also, the number of functional values of rect used per unit length is $4096/64 = 64$ values/unit-length, which is greater than $1024/32 = 32$ values/unit-length used in the previous computation. In Figure 5.6 we show the graph of the computed transform of rect and the graph of its exact transform, sinc. In particular, notice the improved match of the computed transform to sinc over the interval $[6, 16]$.

REMARK 5.3: In Formulas (5.31) and (5.40) the transform $\hat{f}$ is evaluated at integer multiples of $1/L$ where L is the length of the interval over which f was transformed. The quantity $1/L$ is called the *frequency resolution* for the computed transform. The frequency resolution is improved, made *finer,* by increasing the length of the interval over which the function is transformed.

On the other hand, in order to capture more rapid oscillation in a function, that function must be evaluated more frequently. The number of values per unit length is given by N/L. To increase N/L, which is called the *sampling frequency,* one must either increase N or decrease L. The frequency resolution $1/L$ and sampling frequency N/L can both be improved, but they are at odds with one another. To create twice as fine a frequency resolution one can double L, but then, in order to sample twice as frequently, one must increase N by a factor of 4. This is what we did in the previous example. ∎

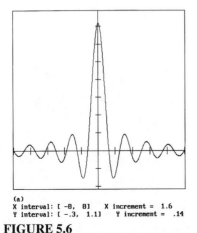
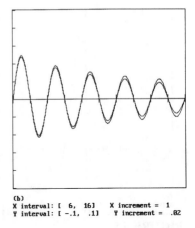

(a)
X interval: [-8, 8]   X increment = 1.6
Y interval: [-.3, 1.1]   Y increment = .14

(b)
X interval: [6, 16]   X increment = 1
Y interval: [-.1, .1]   Y increment = .02

FIGURE 5.6
Comparison of *FAS* **calculated transform of** rect **(using interval** $[-32, 32]$ **and 4096 points) and the exact transform** sinc.

When comparing graphs of a computed transform with the original function, it is a good idea to use the same number of points *and* the same interval for both these functions. This is possible, *using integer values for the interval,* for the following choices of numbers of points and intervals: 256 points with interval $[-8, 8]$, or 1024 points with interval $[-16, 16]$, or 4096 points with interval $[-32, 32]$. We will always restrict ourselves to these choices when it is desirable to compare a function with its computed transform.

5.5   Convolution — an Introduction

In this section we shall introduce the concept of *convolution,* which is one of the most important ideas in Fourier analysis. One of the great advantages of FFTs is that they provide a very fast way of computing approximations to convolutions.

To motivate the need for convolutions we will look at an example. We will show how convolution arises in describing the solution to the following problem. This problem is known as *Dirichlet's problem for the upper half plane.* It deals with the electrostatic potential $H(x, y)$ that is produced in the upper half-plane of $\mathbf{R}^2$ by a charge density $f(x)$ lying along the x-axis. Consequently, it is of some importance in the theory of electromagnetism.

Example 5.9:
Find a function $H(x, y)$ that satisfies

$$\frac{\partial^2 H}{\partial x^2} + \frac{\partial^2 H}{\partial y^2} = 0, \quad (-\infty < x < \infty, \ y > 0) \tag{5.42a}$$

$$H(x, 0) = f(x), \quad (-\infty < x < \infty). \tag{5.42b}$$

SOLUTION   We solve (5.42a) and (5.42b) by Fourier transforming the problem. Define $\hat{H}(u, y)$ by

$$\hat{H}(u, y) = \int_{-\infty}^{\infty} H(x, y)e^{-i2\pi ux} \, dx. \tag{5.43}$$

Or, by Fourier inversion,

$$H(x, y) = \int_{-\infty}^{\infty} \hat{H}(u, y)e^{i2\pi ux} \, du. \tag{5.44}$$

By differentiating the right side of (5.44) twice under the integral sign, we get

$$\frac{\partial^2 H}{\partial x^2} = \int_{-\infty}^{\infty} \hat{H}(u, y)\frac{\partial^2}{\partial x^2}\left[e^{i2\pi ux}\right] du$$

$$= \int_{-\infty}^{\infty} -(2\pi u)^2 \hat{H}(u, y)e^{i2\pi ux} \, du. \tag{5.45}$$

Similarly,

$$\frac{\partial^2 H}{\partial y^2} = \int_{-\infty}^{\infty} \frac{\partial^2 \hat{H}(u, y)}{\partial y^2}e^{i2\pi ux} \, du. \tag{5.46}$$

By combining (5.45) and (5.46) we obtain from (5.42a) that

$$0 = \frac{\partial^2 H}{\partial x^2} + \frac{\partial^2 H}{\partial y^2} = \int_{-\infty}^{\infty}\left[\frac{\partial^2 \hat{H}(u, y)}{\partial y^2} - (2\pi u)^2 \hat{H}(u, y)\right]e^{i2\pi ux} \, du. \tag{5.47}$$

Since the function in brackets in (5.47) is being inverse Fourier transformed to the 0-function, we conclude that it must also be the 0-function. That is,

$$\frac{\partial^2 \hat{H}(u, y)}{\partial y^2} - (2\pi u)^2 \hat{H}(u, y) = 0. \tag{5.48}$$

Now, by putting $y = 0$ into (5.43), we obtain from $H(x, 0) = f(x)$ in (5.42b):

$$\hat{H}(u, 0) = \int_{-\infty}^{\infty} f(x)e^{-i2\pi ux} \, dx = \hat{f}(u). \tag{5.49}$$

Formulas (5.48) and (5.49) result in the Fourier transformed problem of finding $\hat{H}(u, y)$, which satisfies

$$\frac{\partial^2 \hat{H}(u, y)}{\partial y^2} = (2\pi u)^2 \hat{H}(u, y), \qquad (-\infty < u < \infty, \ y > 0) \quad (5.50a)$$

$$\hat{H}(u, 0) = \hat{f}(u), \qquad (-\infty < u < \infty). \tag{5.50b}$$

To solve (5.50a) and (5.50b) we *fix* some value of u, then (5.50a) is an *ordinary* differential equation,

$$\frac{d^2 \hat{H}(u, y)}{dy^2} = (2\pi u)^2 \hat{H}(u, y)$$

which has solution

$$\hat{H}(u, y) = a \, e^{-2\pi |u|y} + b \, e^{2\pi |u|y} \tag{5.51}$$

where a and b depend on u, but not on x.

In order to recover $H(x, y)$ from Formula (5.44) we *reject the part of $\hat{H}$ in (5.51) which has an infinite integral over* **R**, namely, $b \, e^{2\pi |u|y}$. We do this by putting $b = 0$, so that

$$\hat{H}(u, y) = a \, e^{-2\pi |u|y}. \tag{5.52}$$

Comparing (5.52) with (5.50b) we see that

$$\hat{H}(u, y) = \hat{f}(u) \, e^{-2\pi |u|y}. \tag{5.53}$$

Now, to find $H(x, y)$ we substitute the expression for $\hat{H}$ in (5.53) into the integral in (5.44) obtaining

$$H(x, y) = \int_{-\infty}^{\infty} \hat{f}(u)e^{-2\pi |u|y} e^{i2\pi ux} \, du. \tag{5.54}$$

Formula (5.54) says that a solution $H(x, y)$ to (5.42a) and (5.42b) can be obtained by Fourier transforming f, multiplying by $e^{-2\pi |u|y}$, and then inverse Fourier transforming back to get $H(x, y)$.

We will now show how to describe this solution procedure in terms of convolution. Replacing $\hat{f}(u)$ in (5.54) by the integral which defines it, we get

$$H(x, y) = \int_{-\infty}^{\infty} \left[\int_{-\infty}^{\infty} f(s)e^{-i2\pi us} \, ds \right] e^{-2\pi |u| y} e^{i2\pi ux} \, du. \qquad (5.55)$$

Interchanging integrals and combining the exponentials yields

$$H(x, y) = \int_{-\infty}^{\infty} f(s) \left[\int_{-\infty}^{\infty} e^{-2\pi |u| y} e^{i2\pi u(x-s)} \, du \right] ds. \qquad (5.56)$$

Changing variables $(u = -v)$ and using Example 5.6(d) with y in place of ρ *and* $x - s$ *as the transform variable,* we obtain

$$\int_{-\infty}^{\infty} e^{-2\pi |u| y} e^{i2\pi u(x-s)} \, du = \int_{-\infty}^{\infty} e^{-2\pi |v| y} e^{-i2\pi v(x-s)} \, dv$$

$$= \frac{1}{\pi} \frac{y}{y^2 + (x-s)^2} \cdot \qquad (5.57)$$

Using this last result back in (5.56) we get

$$H(x, y) = \int_{-\infty}^{\infty} f(s) \left[\frac{1}{\pi} \frac{y}{y^2 + (x-s)^2} \right] ds. \qquad (5.58)$$

Formula (5.58) expresses our solution H to (5.42a) and (5.42b) as a *convolution* of $f(x)$ with the function $(1/\pi)[y/(y^2 + x^2)]$. This can be seen from the formal definition of convolution given in Definition 5.3 below. ∎

DEFINITION 5.3: *If f and g are two functions, with* $\|f\|_1$ *and* $\|g\|_1$ *finite, then the* **convolution** *of f and g is denoted by* $f * g$ *and is defined by*

$$f * g(x) = \int_{-\infty}^{\infty} f(s)g(x-s) \, ds.$$

Notice that the previous definition of convolution given in Formula (4.83d), Chapter 4, referred to convolution *over an interval* $[-L, L]$. *When no reference is given to any interval,* then we are referring to the convolution defined in Definition 5.3.

If we define $_yP$ by

$$_yP(x) = \frac{1}{\pi} \frac{y}{y^2 + x^2}, \qquad (y > 0) \tag{5.59}$$

then we can express (5.58) as

$$H(x, y) = f * {}_yP(x). \tag{5.60}$$

It is an easy exercise to show that

$$f * g = g * f, \qquad (commutativity). \tag{5.61}$$

Hence we also have

$$H(x, y) = {}_yP * f(x). \tag{5.62}$$

In the next section we shall generalize what we have done here and give further examples of how convolutions can be used to solve some important problems. We close this section by describing how *FAS* can be used to approximate convolutions.

FAS will compute as an approximation to $f * g$ a discrete version of

$$\int_{-L}^{L} f(s)g(x - s)\, ds. \tag{5.63}$$

(How this is done was explained in Section 4.7 of Chapter 4.) For suitably chosen (large enough) values of L, the integral in (5.63) will provide good approximations to $f * g(x)$. For instance, we can approximate $H(x, y) = f * {}_yP(x)$. For each value of y, the function $_yP(x)$ is a specific function of x. If $f(x) = \text{rect}(x)$, then we show the graph of $f * {}_yP$ for $y = 2.0$ in Figure 5.7(b). For $f(x) = \text{rect}(x + 1) + \text{rect}(x - 1)$, we show the graphs of $f * {}_yP$ for $y = 0.5, 1.0$, and 2.0 in Figure 5.8(b) to (d). It is interesting how the graph of $f * {}_yP$ for $y = 2.0$ has a similar form for both of these functions f. In both cases, the interval used was $[-L, L] = [-8, 8]$, and 1024 points were used. These graphs were obtained by following the procedure for graphing convolutions described in the user's manual for *FAS* (see Appendix A). For both of these examples it is possible to find exact formulas for $f * {}_yP$ and compare them with the approximate convolutions (see Exercise 5.19).

Example 5.10:
Suppose that $f(x) = \text{rect}(x)$ and $g(x) = \text{rect}(x - 1)$. Then

$$f * g(x) = g * f(x) = \int_{-\infty}^{\infty} \text{rect}(x - s)\,\text{rect}(s - 1)\, ds$$

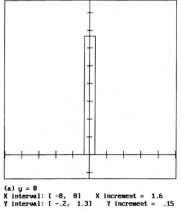

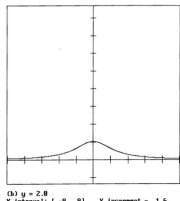

(a) y = 0
X interval: [-8, 8]   X increment = 1.6
Y interval: [-.2, 1.3]   Y increment = .15

(b) y = 2.0
X interval: [-8, 8]   X increment = 1.6
Y interval: [-.2, 1.3]   Y increment = .15

FIGURE 5.7
Graph of a solution $H(x, y)$ to (5.42a) and (5.42b) for $y = 0$, and $y = 2.0$.

$$= \int_{0.5}^{1.5} \text{rect}(x - s)\, ds.$$

By substituting $v = x - s$, $dv = -ds$, we obtain

$$f * g(x) = \int_{x-1.5}^{x-0.5} \text{rect}(v)\, dv.$$

We leave it as an exercise for the reader to finish the calculation, obtaining

$$f * g(x) = \begin{cases} 0 & \text{if } x < 0 \text{ or } x > 2 \\ x & \text{if } 0 \leq x \leq 1 \\ 2 - x & \text{if } 1 < x \leq 2 \end{cases}$$

$$= x(0 \leq x \leq 1) + (2 - x)(1 < x \leq 2). \tag{5.64}$$

We show graphs of this exact expression for $f * g$ and the *FAS* calculated version in Figure 5.9. The difference is hardly noticeable when using the interval $[-8, 8]$ and 1024 points. We have calculated the maximum difference between the two graphs, and it is about 0.0078.

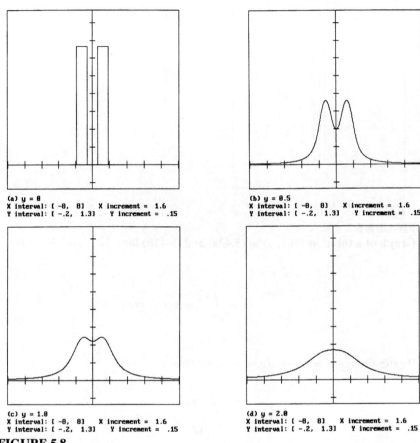

(a) y = 0
X interval: [-8, 8]   X increment = 1.6
Y interval: [-.2, 1.3]   Y increment = .15

(b) y = 0.5
X interval: [-8, 8]   X increment = 1.6
Y interval: [-.2, 1.3]   Y increment = .15

(c) y = 1.0
X interval: [-8, 8]   X increment = 1.6
Y interval: [-.2, 1.3]   Y increment = .15

(d) y = 2.0
X interval: [-8, 8]   X increment = 1.6
Y interval: [-.2, 1.3]   Y increment = .15

FIGURE 5.8
Graph of a solution $H(x, y)$ to (5.42a) and (5.42b) for $y = 0, 0.5, 1.0$, and 2.0.

5.6   The Convolution Theorem

In this section we shall discuss the *convolution theorem,* which generalizes the work done in the previous section. To motivate this theorem, look again at (5.54) and (5.60). The function $H(x, y) = f *_y P(x)$ is a convolution, and it was obtained by inverse Fourier transforming the product $\hat{f}(u)e^{-2\pi|u|y}$. In general, whenever a product is inverse Fourier transformed, a convolution results. More precisely, we have the following theorem.

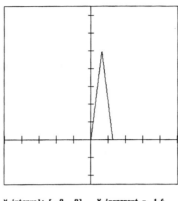

X interval: [-8, 8]   X increment = 1.6
Y interval: [-.5, 1.5]   Y increment = .2

FIGURE 5.9
Convolution of rect(x) **and** rect$(x - 1)$.

THEOREM 5.7: CONVOLUTION
The inverse Fourier transform of $\hat{f}\hat{g}$ is the convolution $f * g$.

PROOF   By Fourier inversion, we have

$$\mathcal{F}^{-1}(\hat{f}\hat{g}) = \int_{-\infty}^{\infty} \hat{f}(u)\hat{g}(u)e^{i2\pi ux}\, du$$

$$= \int_{-\infty}^{\infty} \left[\int_{-\infty}^{\infty} f(s)e^{-i2\pi us}\, ds\right] \hat{g}(u)e^{i2\pi ux}\, du.$$

Interchanging integrals and combining exponentials, we obtain

$$\mathcal{F}^{-1}(\hat{f}\hat{g})(x) = \int_{-\infty}^{\infty} f(s) \left[\int_{-\infty}^{\infty} \hat{g}(u)e^{i2\pi u(x-s)}\, du\right] ds. \qquad (5.65)$$

The integral in brackets in (5.65) is a Fourier inversion integral of $\hat{g}(u)$, with variable $x - s$ instead of x, so we will get $g(x - s)$ for that integral. Thus, we have shown that

$$\mathcal{F}^{-1}(\hat{f}\hat{g}) = f * g. \qquad (5.66)$$

By applying the Fourier transform $\mathcal{F}$ to both sides of (5.66) we get the following theorem, which is also called the convolution theorem.

THEOREM 5.8: CONVOLUTION
The Fourier transform of $f * g$ is $\hat{f}\hat{g}$.

There are many important applications of these convolution theorems. First, we will describe another application to a partial differential equation problem, similar to the one solved in Section 5.5.

Example 5.11:
Find a function $F(x, t)$ that satisfies the following heat equation problem (for $-\infty < x < \infty$, and $t > 0$):

$$\frac{\partial F}{\partial t} = a^2 \frac{\partial^2 F}{\partial x^2} \qquad \text{(heat equation, diffusion constant } a^2) \qquad (5.67a)$$

$$F(x, 0) = f(x) \qquad \text{(initial condition).} \qquad (5.67b)$$

SOLUTION   We proceed similarly to the solution of (5.42a) and (5.42b). Define $\hat{F}(u, t)$ by

$$\hat{F}(u, t) = \int_{-\infty}^{\infty} F(x, t) e^{-i2\pi ux} \, dx. \qquad (5.68)$$

Hence, by Fourier inversion,

$$F(x, t) = \int_{-\infty}^{\infty} \hat{F}(u, t) e^{i2\pi ux} \, du. \qquad (5.69)$$

By differentiating under the integral sign, we have

$$\frac{\partial F(x, t)}{\partial t} = \int_{-\infty}^{\infty} \frac{\partial \hat{F}(u, t)}{\partial t} e^{i2\pi ux} \, du$$

$$a^2 \frac{\partial^2 F(x, t)}{\partial x^2} = \int_{-\infty}^{\infty} \hat{F}(u, t) \left[-(2\pi a)^2 u^2 \right] e^{i2\pi ux} \, du. \qquad (5.70)$$

Substituting the integrals from (5.70) back into (5.67a) and rearranging terms into one integral we have

$$\int_{-\infty}^{\infty} \left[\frac{\partial \hat{F}(u, t)}{\partial t} + (2\pi a)^2 u^2 \hat{F}(u, t) \right] e^{i2\pi ux} \, du = 0. \qquad (5.71)$$

Since the Fourier inverse of the function in brackets in (5.71) is the 0-function, we conclude that the function in brackets is also the 0-function. Therefore,

$$\frac{\partial \hat{F}(u, t)}{\partial t} + (2\pi a)^2 u^2 \hat{F}(u, t) = 0. \tag{5.72}$$

Now, by putting $t = 0$ in (5.68), we conclude from (5.67b) that

$$\hat{F}(u, 0) = \hat{f}(u). \tag{5.73}$$

Hence, our transformed problem is

$$\frac{\partial \hat{F}(u, t)}{\partial t} = -(2\pi a)^2 u^2 \hat{F}(u, t) \tag{5.74a}$$

$$\hat{F}(u, 0) = \hat{f}(u). \tag{5.74b}$$

Fixing a value of u turns (5.74a) into an *ordinary* differential equation

$$\frac{d\hat{F}}{dt} = -(2\pi a)^2 u^2 \hat{F}$$

which has solution
$$\hat{F}(u, t) = A e^{-(2\pi a)^2 u^2 t}. \tag{5.75}$$

Putting $t = 0$ in (5.75) and using (5.74b) yields $A = \hat{f}(u)$. Hence,(5.69) becomes

$$F(x, t) = \int_{-\infty}^{\infty} \hat{f}(u) e^{-(2\pi a)^2 u^2 t} e^{i 2\pi u x} \, du. \tag{5.76}$$

Now, we can apply the convolution theorem (Theorem 5.7) to express $F(x, t)$ as a convolution. In (5.76) we have a Fourier inversion of a product of two transforms. These transforms are $f \xrightarrow{\mathcal{F}} \hat{f}$ and

$$\frac{1}{(4\pi a^2 t)^{\frac{1}{2}}} e^{-\pi x^2/(4\pi a^2 t)} \xrightarrow{\mathcal{F}} e^{-(2\pi a)^2 u^2 t}.$$

Thus, by Theorem 5.7, our solution $F(x, t)$ to (5.67a) and (5.67b) is

$$F(x, t) = f *_t H(x) \tag{5.77}$$

where

$$_t H(x) = \frac{1}{(4\pi a^2 t)^{\frac{1}{2}}} e^{-\pi x^2/(4\pi a^2 t)}. \tag{5.78}$$

■

It is easy to calculate computer approximations to (5.77) using *FAS*. For example, suppose that $f(x) = 20 \, \text{rect}(x/2)$ and $a^2 = 1$. We will use for our second function

$$\frac{1}{(4\pi t)^{\frac{1}{2}}} e^{-x^2/(4t)}$$

for $t = 0.5$, 1.0, and 2.0. If we use 1024 points and the interval $[-8, 8]$, then we obtain the graphs shown in Figure 5.10. These graphs can be interpreted as the evolution of temperature, from the initial temperature $f(x) = 20 \, \text{rect}(x/2)$, through a homogeneous insulated rod (or a thin wire).

Here is another application of convolution.

Example 5.12:

In probability theory, a *probability density function* (p.d.f.) is a non-negative function f satisfying $\int_{-\infty}^{\infty} f(x) \, dx = 1$. For example, $f(x) = (1/\sqrt{\pi})e^{-x^2}$ is a p.d.f. For a p.d.f. there is associated a *cumulative distribution function* (*distribution,* for short) defined by $F(x) = \int_{-\infty}^{x} f(s) \, ds$. Graph the distribution $F(x)$ for the p.d.f. $f(x) = (1/\sqrt{\pi})e^{-x^2}$.

SOLUTION   We leave it as an exercise for the reader to show that $F(x) = f * \mathcal{H}(x)$ where $\mathcal{H}$ is the *Heaviside function* defined by

$$\mathcal{H}(x) = \begin{cases} 0 & \text{if } x < 0 \\ 1 & \text{if } x > 0. \end{cases} \tag{5.79}$$

By doing a convolution of $f(x) = (1/\sqrt{\pi})e^{-x^2}$ with $\mathcal{H}(x)$, over $[-16, 16]$ using 1024 points, one obtains the graph shown in Figure 5.11(a). For reasons which we will explain below, there is an undesirable behavior at the ends of the interval $[-16, 16]$ (i.e., the graph obviously decreases near ± 16, but $F(x)$ *never* decreases, because its derivative is $(1/\sqrt{\pi})e^{-x^2}$, which is always positive). Nevertheless, if we change the x-interval to, say, $[-10, 10]$, then we get the graph shown in Figure 5.11(b), and this graph is a good approximation to the distribution $F(x)$ over the interval $[-10, 10]$. (See Exercise 5.28 for a treatment of the accuracy of this approximation.)   ■

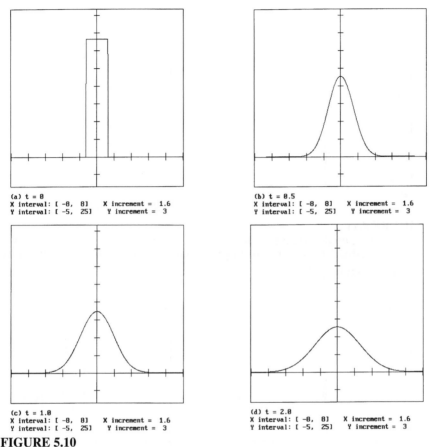

(a) t = 0
X interval: [-8, 8]   X increment = 1.6
Y interval: [-5, 25]   Y increment = 3

(b) t = 0.5
X interval: [-8, 8]   X increment = 1.6
Y interval: [-5, 25]   Y increment = 3

(c) t = 1.0
X interval: [-8, 8]   X increment = 1.6
Y interval: [-5, 25]   Y increment = 3

(d) t = 2.0
X interval: [-8, 8]   X increment = 1.6
Y interval: [-5, 25]   Y increment = 3

FIGURE 5.10
Graph of a solution $F(x, t)$ to (5.67a) and (5.67b) for $t = 0, 0.5, 1.0,$ and 2.0.

REMARK 5.4:   As we saw in Example 5.12, the computer approximation
of the convolution of $(1/\sqrt{\pi})e^{-x^2}$ with $\mathcal{H}(x)$ has some undesirable behavior at
the ends of the interval $[-16, 16]$. The reason for this is that, as we showed in
Section 4.7 of Chapter 4, an FFT algorithm is used for approximating the *periodic*
(or *cyclic*) convolution

$$f * g(x) = \frac{1}{2L} \int_{-L}^{L} f(s)g(x - s) \, ds \qquad (5.80)$$

where f and g are periodic functions with period $2L$ (in the example above,
$L = 16$). What is graphed in Figure 5.11(a) is $2L = 32$ times the *periodic*
convolution in (5.80), where f and g are the *periodic extensions* of the restrictions

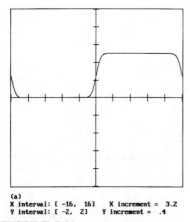

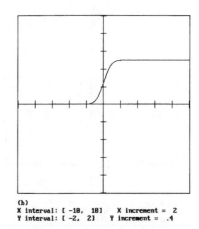

(a)
X interval: [-16, 16]   X increment = 3.2
Y interval: [-2,   2]   Y increment = .4

(b)
X interval: [-10, 10]   X increment = 2
Y interval: [-2,   2]   Y increment = .4

FIGURE 5.11
(a) *FAS* **computed convolution of** $(1/\sqrt{\pi})e^{-x^2}$ **and the Heaviside function over** $[-16, 16]$ **using** 1024 **points. (b) Approximate distribution for the probability density function (p.d.f.)** $(1/\sqrt{\pi})e^{-x^2}$.

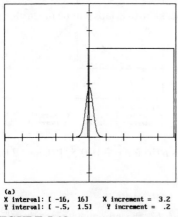

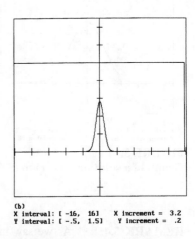

(a)
X interval: [-16, 16]   X increment = 3.2
Y interval: [-.5, 1.5]   Y increment = .2

(b)
X interval: [-16, 16]   X increment = 3.2
Y interval: [-.5, 1.5]   Y increment = .2

FIGURE 5.12
(a) **Graph of components of cyclic (periodic) convolution of** $(1/\sqrt{\pi})e^{-x^2}$ **and the Heaviside function over** $[-16, 16]$ **for** $x = 15.75$. **(b) Graph of the same components for the convolution over R for** $x = 15.75$.

to $[-16, 16]$ of $(1/\sqrt{\pi})e^{-x^2}$ and $\mathcal{H}(x)$. To see why the troublesome behavior near the endpoints ± 16 occurs, we show in Figure 5.12(a) what the components of (5.80) look like for $x = 15.75$. From Figure 5.12(a) it is clear that there will only be a *partial overlap* of $(1/\sqrt{\pi})e^{-x^2}$ and *not* the more complete overlap which the non-periodic convolution would have [as shown in Figure 5.12(b)]. However, as

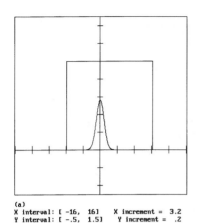
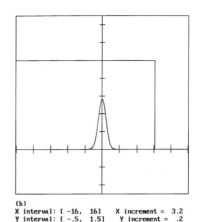

(a)
X interval: [-16, 16]   X increment = 3.2
Y interval: [-.5, 1.5]   Y increment = .2

(b)
X interval: [-16, 16]   X increment = 3.2
Y interval: [-.5, 1.5]   Y increment = .2

FIGURE 5.13
(a) Graph of components of cyclic (periodic) convolution of $(1/\sqrt{\pi})e^{-x^2}$ **and the Heaviside function over** $[-16, 16]$ **for** $x = 9.75$**. (b) Graph of the same components for the convolution over R for** $x = 9.75$**.**

we show in Figure 5.13, the periodic and non-periodic overlaps will be much closer when $x = 9.75$ (due to $(1/\sqrt{\pi})e^{-x^2}$ being so near 0 when $|x| \geq 6.25$). ∎

As a final application of the convolution theorem, we shall sketch the proof of Parseval's equalities.

THEOREM 5.9: PARSEVAL'S EQUALITIES

If $f \xrightarrow{\mathcal{F}} \hat{f}$, *then*

$$\int_{-\infty}^{\infty} |f(x)|^2 \, dx = \int_{-\infty}^{\infty} |\hat{f}(u)|^2 \, du. \qquad (5.81a)$$

Furthermore, if $g \xrightarrow{\mathcal{F}} \hat{g}$, *then*

$$\int_{-\infty}^{\infty} f(x)g^*(x) \, dx = \int_{-\infty}^{\infty} \hat{f}(u)\hat{g}^*(u) \, du. \qquad (5.81b)$$

PROOF   If we put g equal to f, then (5.81b) implies (5.81a). Therefore, it remains to prove (5.81b). The convolution theorem (Theorem 5.7) also holds for the inverse Fourier transform (by a simple change of variable). Therefore, we have

$$\int_{-\infty}^{\infty} f(x)g^*(x)e^{-i2\pi ux}\,dx = (\hat{f} * \widehat{g^*})(u)$$

$$= \int_{-\infty}^{\infty} \hat{f}(v)\widehat{g^*}(u-v)\,dv. \qquad (5.82)$$

Furthermore,

$$\widehat{g^*}(v) = \int_{-\infty}^{\infty} g^*(x)e^{-i2\pi vx}\,dx$$

$$= \left(\int_{-\infty}^{\infty} g(x)e^{i2\pi vx}\,dx\right)^* = \hat{g}^*(-v)$$

Hence we can rewrite (5.82) as

$$\int_{-\infty}^{\infty} f(x)g^*(x)e^{-i2\pi ux}\,dx = \int_{-\infty}^{\infty} \hat{f}(v)\hat{g}^*(v-u)\,dv. \qquad (5.83)$$

Putting $u = 0$ in (5.83) yields

$$\int_{-\infty}^{\infty} f(x)g^*(x)\,dx = \int_{-\infty}^{\infty} \hat{f}(v)\hat{g}^*(v)\,dv$$

and the theorem is proved.   ∎

5.7   An Application of Convolution in Quantum Mechanics

In this section we will discuss an example from quantum mechanics which makes use of the convolution theorem. Recall from the discussion at the beginning of section 4.3, that the potential-free version of the one-dimensional Schrödinger's equation can be expressed as

$$\frac{\partial \psi}{\partial t} = \frac{i\hbar}{2m}\frac{\partial^2 \psi}{\partial x^2}.$$

We now show how to use the convolution theorem to solve a problem involving this version of Schrödinger's equation.

Example 5.13:
Draw graphs that approximate $|\psi|^2$, where ψ satisfies

$$\frac{\partial \psi}{\partial t} = \frac{i\hbar}{2m} \frac{\partial^2 \psi}{\partial x^2} \qquad \text{(Schrödinger's equation)} \qquad (5.84a)$$

$$\psi(x, 0) = f(x) \qquad \text{(initial condition)} \qquad (5.84b)$$

for the function $f(x) = \text{rect}(x)$ and times $t = 0.1$ and 0.25 sec. Assume that m is the mass of an electron.

SOLUTION   We will make use of the solution found for Example 5.11. To make use of this solution, we let a^2 stand for the constant factor on the right side of Equation (5.84a). Thus, we can express the problem above as follows:

$$\frac{\partial \psi}{\partial t} = a^2 \frac{\partial^2 \psi}{\partial x^2}, \qquad \left(a^2 = \frac{i\hbar}{2m} \right) \qquad (5.85a)$$

$$\psi(x, 0) = f(x). \qquad (5.85b)$$

Therefore, using the Solution (5.76) found for (5.67a) and (5.67b), *but now replacing a^2 by $i\hbar/2m$*, we get

$$\psi(x, t) = \int_{-\infty}^{\infty} \hat{f}(u) e^{-(2\pi u)^2 (i\hbar/2m)t} e^{i2\pi ux} \, du \qquad (5.86)$$

Furthermore, instead of (5.77) and (5.78), we have

$$\psi(x, t) = f * {}_t F(x) \qquad (5.87)$$

where

$$_t F(x) = \frac{1}{[4\pi(i\hbar/2m)t]^{\frac{1}{2}}} e^{-\pi x^2/(4\pi t i\hbar/2m)}. \qquad (5.88)$$

We will now show how to approximate $|\psi(x, t)|^2$ using Formula (5.86). If we choose L sufficiently large, then we can approximate the integral $\int_{-\infty}^{\infty}$ by the integral $\int_{-L}^{L}$, obtaining

$$\psi(x,t) \approx \int_{-L}^{L} \hat{f}(u)e^{-(2\pi u)^2(i\hbar/2m)t}e^{i2\pi ux}\, du. \qquad (5.89)$$

Now, if $f(x) = \text{rect}(x)$, then $\hat{f}(u) = \text{sinc}(u)$. Substituting sinc (u) in place of $\hat{f}(u)$ in (5.89) and writing $e^{-(2\pi u)^2(i\hbar/2m)t}$ in terms of its real and imaginary parts, we get

$$\psi(x,t) \approx \int_{-L}^{L} \text{sinc}(u) \left\{ \cos\left(\frac{\hbar}{2m}(2\pi u)^2 t \right) - i\sin\left(\frac{\hbar}{2m}(2\pi u)^2 t \right) \right\} e^{i2\pi ux}\, du.$$
$$\qquad (5.90)$$

The integral in (5.90) is a filtered Fourier transform over $[-L, L]$, using a pos. exponent. The filter function has real part $\cos[(\hbar/2m)(2\pi u)^2 t]$ and imaginary part $-\sin[(\hbar/2m)(2\pi u)^2 t]$. Since (5.90) describes a Fourier transform operation, it follows that $|\psi(x,t)|^2$ is a *power spectrum*. Therefore, to graph $|\psi|^2$ we perform the following steps. First, we choose to do the *Fourier transform* procedure and choose *Complex* for the type of function. Second, we choose an interval $[-L, L]$ and choose the number of points. For this example, we shall use $L = 32$ and 4096 points. Third, we specify the real and imaginary parts of the function to be transformed. In this case, we enter $f(x) = \text{sinc}(x)$ and $f(x) = 0$, respectively. Fourth, we choose to do a *Power spectrum* with pos. exponent. And, we press y when asked if we want to apply a filter and choose to do a *Complex* filter. For the real part of the filter, we enter the following formula (*Note:* for an electron $\hbar/2m = 0.578$):

$$f(x) = \cos[0.578(2\pi x) \wedge 2t]\,\backslash t{=}0.1 \qquad (5.91)$$

and for the imaginary part of the filter we enter the following formula:

$$f(x) = -\sin[0.578(2\pi x) \wedge 2t]\,\backslash t{=}0.1. \qquad (5.92)$$

After transforming, a graph of $|\psi(x, 0.1)|^2$ is produced [see Figure 5.14(a)].

Repeating this procedure for $t = 0.25$, we obtain the graph of $|\psi(x, 0.25)|^2$, which is shown in Figure 5.14(b). ∎

Let's now examine a couple of theoretical points. First, we will prove the following theorem. This theorem says that if the initial state $\psi(x, 0) = f(x)$ generates a p.d.f. $|f(x)|^2$, then each subsequent state $\psi(x, t)$ also generates a p.d.f. $|\psi(x, t)|^2$.

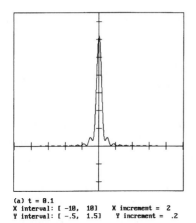

(a) t = 0.1
X interval: [-10, 10]   X increment = 2
Y interval: [-.5, 1.5]   Y increment = .2

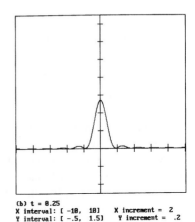

(b) t = 0.25
X interval: [-10, 10]   X increment = 2
Y interval: [-.5, 1.5]   Y increment = .2

FIGURE 5.14
Graphs of $|\psi(x, t)|^2$ for Example 5.13.

THEOREM 5.10:
Suppose ψ satisfies (5.84a) and (5.84b). Then, for all t-values,

$$\int_{-\infty}^{\infty} |\psi(x, t)|^2 \, dx = \int_{-\infty}^{\infty} |f(x)|^2 \, dx. \tag{5.93}$$

In particular, if $\int_{-\infty}^{\infty} |f(x)|^2 \, dx = 1$, then $\int_{-\infty}^{\infty} |\psi(x, t)|^2 \, dx = 1$ for all t-values.

PROOF   To prove (5.93) we use Parseval's Equality (5.81a) twice. Looking at (5.86), we see that it is the Fourier inversion integral corresponding to the following Fourier transform operation (for each *fixed* t-value):

$$\psi(x, t) \xrightarrow{\mathcal{F}} \hat{f}(u)e^{-(2\pi u)^2(i\hbar/2m)t}.$$

Hence, by Parseval's Equality (5.81a),

$$\int_{-\infty}^{\infty} |\psi(x, t)|^2 \, dx = \int_{-\infty}^{\infty} |\hat{f}(u)|^2 |e^{i(2\pi u)^2(-\hbar/2m)t}|^2 \, du$$

$$= \int_{-\infty}^{\infty} |\hat{f}(u)|^2 \, du. \tag{5.94}$$

We used the fact that $|e^{i\phi}|^2 = 1$ when ϕ is real valued to get the second equality above. By Parseval's equality again, we have $\int_{-\infty}^{\infty} |\hat{f}(u)|^2 \, du = \int_{-\infty}^{\infty} |f(x)|^2 \, dx$. Consequently, (5.93) holds for each t-value.   ∎

Second, we will look at the asymptotic behavior of $\psi(x, t)$. To be precise, we will show that (under certain assumptions) as $t \to \infty$,

$$|\psi(x, t)|^2 \approx \frac{1}{4\pi(\hbar/2m)t} \left| \hat{f}\left(\frac{x}{4\pi(\hbar/2m)t} \right) \right|^2 \qquad (5.95)$$

The approximation in (5.95) is an important one. It shows, for instance, that the complicated derivation of $|\psi(x, t)|^2$ given in Example 5.13 can be replaced by the simpler operation shown on the right side of (5.95). That is, for large enough t-values, we only need to calculate a scaled version of sinc^2. [For instance, $|\psi(x, 0.25)|^2 \approx (1/1.816)\,\mathrm{sinc}^2(x/1.816)$, and this approximation appears to be valid based on Figure 5.14(b).] It is worth noting that *if* $|f|^2$ *is a p.d.f., then both sides of (5.95) are p.d.f.s.* That $|\psi(x, t)|^2$ is a p.d.f. follows from Theorem 5.10. That $(1/[4\pi(\hbar/2m)t])|\hat{f}(x/(4\pi(\hbar/2m)t)|^2$ is a p.d.f. can be proved by Parseval's Equality (5.81a) and a change of variables (we leave this to the reader as an exercise). Formula (5.95) is analogous to Fraunhofer diffraction in optics (see Chapter 6, section 6.3). If m equals the mass of an electron, for instance, then Formula (5.95) says that electrons diffract in the same way as light waves do when the light undergoes Fraunhofer diffraction.

To prove (5.95) we rewrite (5.87) and (5.88), obtaining

$$\psi(x, t) = \frac{1}{[4\pi(i\hbar/2m)t]^{\frac{1}{2}}} \int_{-\infty}^{\infty} f(s) e^{-\pi(x-s)^2/(4\pi ti\hbar/2m)} \, ds. \qquad (5.96)$$

To simplify notation, we define $q(t)$ by

$$q(t) = 4\pi i(\hbar/2m)t \qquad (5.97)$$

Later, we shall make use of the fact that $1/q(t) \to 0$ as $t \to \infty$. If we expand the exponential inside the integral in (5.96), we get

$$e^{-\pi(x-s)^2/q(t)} = e^{-\pi x^2/q(t)} e^{-\pi s^2/q(t)} e^{2\pi xs/q(t)}. \qquad (5.98)$$

Now, suppose that for some positive number P,

$$f(s) = 0, \quad \text{for } |s| \geq \tfrac{1}{2} P \qquad (5.99)$$

which certainly is true for the function $f(s) = \mathrm{rect}(s)$ in Example 5.13. If we now let $t \to \infty$, we will have for all $|s| \leq \frac{1}{2} P$ *as soon as* t *is sufficiently large*

$$e^{-\pi s^2/q(t)} \approx e^0 = 1 \qquad (5.100)$$

hence

$$e^{-\pi(x-s)^2/q(t)} \approx e^{-\pi x^2/q(t)} e^{2\pi xs/q(t)}. \tag{5.101}$$

Using (5.101) and (5.99) back in (5.96), and also using the definition of $q(t)$, we have

$$\psi(x,t) = \frac{1}{q(t)^{\frac{1}{2}}} \int_{-\frac{1}{2}P}^{\frac{1}{2}P} f(s) e^{-\pi(x-s)^2/q(t)} \, ds$$

$$\approx \frac{1}{q(t)^{\frac{1}{2}}} \int_{-\frac{1}{2}P}^{\frac{1}{2}P} f(s) e^{-\pi x^2/q(t)} e^{2\pi xs/q(t)} \, ds.$$

Factoring $e^{-\pi x^2/q(t)}$ outside of this last integral yields

$$\psi(x,t) \approx \frac{e^{-\pi x^2/q(t)}}{q(t)^{\frac{1}{2}}} \int_{-\frac{1}{2}P}^{\frac{1}{2}P} f(s) e^{2\pi xs/q(t)} \, ds.$$

And, using (5.99) again, we have

$$\psi(x,t) \approx \frac{e^{-\pi x^2/q(t)}}{q(t)^{\frac{1}{2}}} \int_{-\infty}^{\infty} f(s) e^{2\pi xs/q(t)} \, ds.$$

Replacing $q(t)$ by $4\pi i(\hbar/2m)t$ inside the integral yields

$$\psi(x,t) \approx \frac{e^{-\pi x^2/q(t)}}{q(t)^{\frac{1}{2}}} \int_{-\infty}^{\infty} f(s) e^{-i2\pi s[x/(4\pi(\hbar/2m)t)]} \, ds.$$

Thus,

$$\psi(x,t) \approx \frac{e^{-\pi x^2/q(t)}}{q(t)^{\frac{1}{2}}} \hat{f}\left(\frac{x}{4\pi(\hbar/2m)t}\right).$$

And, since $|e^{i\phi}|^2 = 1$ for all real ϕ, we have

$$\left| e^{-\pi x^2/q(t)} \right|^2 = \left| e^{i\pi x^2/[4\pi(\hbar/2m)t]} \right|^2 = 1.$$

Therefore, we obtain

$$|\psi(x,t)|^2 \approx \frac{1}{|q(t)|} \left| \hat{f}\left(\frac{x}{4\pi(\hbar/2m)t}\right) \right|^2.$$

And replacing $|q(t)|$ by $4\pi(\hbar/2m)t$ yields

$$|\psi(x,t)|^2 \approx \frac{1}{4\pi(\hbar/2m)t} \left| \hat{f}\left(\frac{x}{4\pi(\hbar/2m)t} \right) \right|^2$$

for sufficiently large t. Thus, we have demonstrated (5.95). We obtained this asymptotic form for $|\psi(x,t)|^2$ using the assumption in (5.99). But, the argument above still applies if $f(s) \to 0$ as $|s| \to \infty$ rapidly enough (see, for example, Exercise 5.34). Actually, if we interpret the approximation in (5.95) as meaning a small difference in 2-Norm, then (5.95) holds for all functions f for which $\|f\|_2$ is finite, but we will omit the proof.

5.8   Filtering, Frequency Detection, and Removal of Noise

In this section, we introduce the concept of *filtering* a function before Fourier transforming it. This technique is widely used in signal processing. Let's begin with an example.

Example 5.14:

Suppose the function $f(x) = \cos(0.8\pi x)$ is Fourier transformed over the interval $[-32, 32]$ using 1024 points. The result is shown in Figure 5.15(b). Notice that the frequency 0.4 that characterizes $f(x)$ is clearly marked by a peak at the point 0.4. However, there is a very noticeable oscillation about the x-axis away from this peak. This phenomenon is known as *leakage*. Leakage occurs because the FFT method yields approximations of (see Section 5.4):

$$\left\{ \hat{f}\left(\frac{k}{L} \right) \right\}_{k=-\frac{1}{2}N}^{k=\frac{1}{2}N} = \left\{ \hat{f}\left(\frac{k}{64} \right) \right\}_{k=-512}^{k=512}.$$

The frequency 0.4, however, does not equal $k/64$ for any integer k. This failure to realize a precise frequency location results in leakage. That is the explanation on the frequency side. On the other hand, from the time (x-variable) side, if we look at the graph of $f(x)$ in Figure 5.15(a) we see that it does not fit very well in the window determined by $-32 \le x \le 32$. In fact, its periodic extension beyond $[-32, 32]$ *no longer represents the actual function $f(x)$*. See Figure 5.16(a). When this occurs, it is said that the *effect of the window becomes visible*. In fact, we can see that the pattern of the transform in Figure 5.15(b) looks a lot like the transform of rect $(x/64)$.

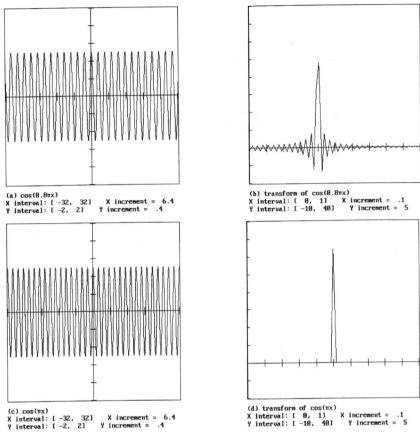

(a) cos(0.8πx)
X interval: [-32, 32]   X increment = 6.4
Y interval: [-2, 2]   Y increment = .4

(b) transform of cos(0.8πx)
X interval: [0, 1]   X increment = .1
Y interval: [-10, 40]   Y increment = 5

(c) cos(πx)
X interval: [-32, 32]   X increment = 6.4
Y interval: [-2, 2]   Y increment = .4

(d) transform of cos(πx)
X interval: [0, 1]   X increment = .1
Y interval: [-10, 40]   Y increment = 5

FIGURE 5.15
Fourier transforms of two cosine fuctions over the interval $[-32, 32]$ **using 1024 points.**

To see this last point more clearly, it helps to look at the function $g(x) = \cos \pi x$ [see Figure 5.15(c)] for which leakage does not occur. As we can see in Figure 5.15(d), the graph of $\hat{g}(u)$ has a peak centered on the frequency 0.5 and *no leakage*. For this example, $k/64 = 0.5$ has the solution $k = 32$ so the frequency 0.5 belongs to the set $\{k/64\}_{k=-512}^{k=512}$. On the other hand, as we show in Figure 5.16(b), the periodic extension of $g(x)$ beyond $[-32, 32]$ still represents the function $g(x)$. And, in this case, no window effects are visible.

Based on these considerations, a standard method of reducing leakage is to multiply the original function by a function that damps down to 0 at the edges of the x-interval *and whose transform does not exhibit the oscillations that the transform of* rect$(x/64)$ *does*. For example, if we transform the function $h(x) = \cos(0.8\pi x)[0.5 + 0.5\cos(\pi x/32)]$, then the transform $\hat{h}$ shown in Figure 5.17(b) results. We say that $\hat{h}(u)$ is a *hanning filtered Fourier transform* of

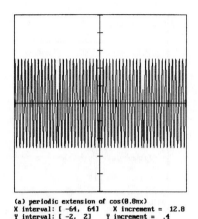

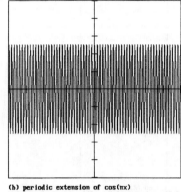

(a) periodic extension of cos(0.8πx)
X interval: [-64, 64]   X increment = 12.8
Y interval: [-2, 2]   Y increment = .4

(b) periodic extension of cos(πx)
X interval: [-64, 64]   X increment = 12.8
Y interval: [-2, 2]   Y increment = .4

FIGURE 5.16
Periodic extensions, period 64, **of two cosine functions (defined initially on**
$[-32, 32]$**).**

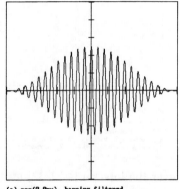

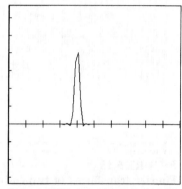

(a) cos(0.8πx), hanning filtered
X interval: [-32, 32]   X increment = 6.4
Y interval: [-2, 2]   Y increment = .4

(b) hanning filtered transform
X interval: [0, 1]   X increment = .1
Y interval: [-12, 24]   Y increment = 3.6

FIGURE 5.17
A hanning filtered Fourier transform.

$f(x)$. Comparing Figures 5.15(b) and 5.17(b) we can see that by hanning filtering
we have significantly reduced leakage.

A filtered Fourier transform is defined as follows.

DEFINITION 5.4:   *If* $\| f \|_1$ *is finite and F is a bounded, continuous function,*
*then the **filtered Fourier transform** of f (using the filter function F) is*

$$\int_{-\infty}^{\infty} F(s) f(s) e^{-i2\pi us} \, ds.$$

When we use *FAS* to graph filtered Fourier transforms, we employ the following approximation (if $[-L, L]$ is the chosen interval):

$$\int_{-\infty}^{\infty} F(s)f(s)e^{-i2\pi us}\, ds \approx \int_{-L}^{L} F(s)f(s)e^{-i2\pi us}\, ds. \tag{5.102}$$

When using the interval $[-L, L]$, some filters that are often used are

Filter name   *Filter function,* $F(s)$

Cesàro	$1 - |s/L|$
hanning	$0.5 + 0.5\cos(\pi s/L)$
Hamming	$0.54 + 0.46\cos(\pi s/L)$
Parzen	$1 - [N/(N+1)]|s/L|$
Welch	$1 - [N/(N+1)]^2(s/L)^2$

(*Note:* the parameter N in the Parzen and Welch filters denotes the number of points used by *FAS*.)

It is also possible to use *FAS* to design your own filter function. For instance, suppose you want to apply the filter $F(s) = e^{-|s|}$. You would then press *y* when *FAS* asks you if you want to apply a filter. Then, select *User* from the filter menu, and enter the formula $f(x) = \exp[-\mathrm{abs}\,(x)]$. A graph of the filter will be drawn. Then *FAS* will draw a graph of the product of this filter with your original function. This product function will then be Fourier transformed to obtain the filtered Fourier transform.

NOISE SUPPRESSION

One application of filtering is noise suppression in signal processing. In this case, the filter function multiplies the Fourier transform of the signal and then an inverse Fourier transform is performed. Let's look at an example.

Example 5.15:

Suppose a pulsed signal having a single fundamental frequency is transmitted, say a signal of the form

$$s(x) = (8\cos 100\pi x)e^{-10\pi x^2}. \tag{5.103}$$

for $-0.5 \le x \le 0.5$. See Figure 5.18(a). There are many situations where such signals are transmitted, such as sonar, or radar, or when a single bit of data is transmitted by a modem. The signal in (5.103) has a fundamental frequency of 50 and amplitude 8. Its Fourier transform is shown in Figure 5.18(b). Notice that the transform has two prominent spikes located at ± 50, corresponding to the fundamental frequency of 50 in the signal. Notice also that the transform is localized at ± 50 in the sense that it is essentially 0 except near the points ± 50.

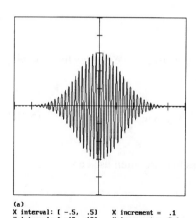

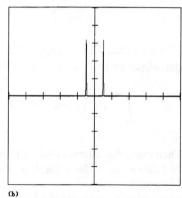

(a)
X interval: [-.5, .5]   X increment = .1
Y interval: [-13, 13]   Y increment = 2.6

(b)
X interval: [-512, 512]   X increment = 102.4
Y interval: [-2, 2]   Y increment = .4

FIGURE 5.18
(a) Transmitted signal. (b) Fourier transform of the signal, real and imaginary parts.

Now, when the signal $s(x)$ is received it may be corrupted by noise. Letting $n(x)$ stand for such noise, we can represent the received signal $f(x)$ as

$$f(x) = (\cos 100\pi x)e^{-10\pi x^2} + n(x). \tag{5.104}$$

See Figure 5.19(a) for a graph of such a noisy signal. The term $(\cos 100\pi x)e^{-10\pi x^2}$

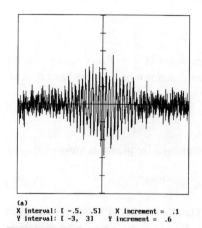

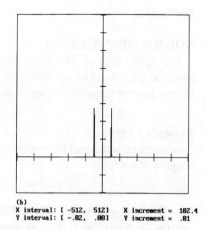

(a)
X interval: [-.5, .5]   X increment = .1
Y interval: [-3, 3]   Y increment = .6

(b)
X interval: [-512, 512]   X increment = 102.4
Y interval: [-.02, .08]   Y increment = .01

FIGURE 5.19
(a) Received, noisy signal. (b) Power spectrum of noisy signal; the spikes are located at ± 50.

represents the desired portion of the received signal; it has been obscured by the undesired noise $n(x)$. This amplitude of the desired signal is also 8 times smaller than the amplitude of the original signal (such diminution of amplitude is common in signal transmission). Its fundamental frequency, however, is still 50 (here we are assuming that neither frequency shifting nor phase shifting has occurred; those possibilities are treated in the exercises). The noise term $n(x)$ that we used in Formula (5.104) is a simulation of *white noise*. For further discussion of this type of noise, and a description of how we created the noise term in (5.104), see the subsection entitled WHITE NOISE below.

Now, looking at the signal shown in Figure 5.19(a), it is hard to imagine how one could be certain that the transmitted pulse has been received. If, however, we compute the *power spectrum* $|\hat{f}(u)|^2$ of the received signal, then we produce the graph shown in Figure 5.19(b). Notice that there are two spikes standing out clearly from the rest of the graph. These spikes are located at the positions ± 50, they tell us that *the received signal contains a fundamental frequency of 50*.

Having identified that there is a frequency of 50 present in the noisy signal we can try to reconstruct the original signal (this is not always necessary, sometimes it is only necessary to detect the presence of the fundamental frequency). To reconstruct the original signal, we shall perform a filtering operation on the Fourier transform of the noisy signal. This transform is shown in Figure 5.20(a). First, we choose to do a complex Fourier transform (choosing a pos. exponent so that we are doing

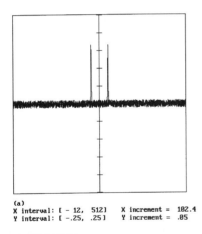

(a)
X interval: [- 12, 512]   X increment = 102.4
Y interval: [-.25, .25]   Y increment = .05

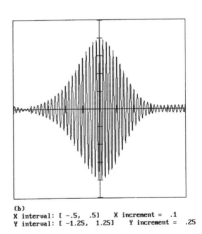

(b)
X interval: [-.5, .5]   X increment = .1
Y interval: [-1.25, 1.25]   Y increment = .25

FIGURE 5.20
(a) Transform of noisy signal, real and imaginary parts. (b) Reconstructed signal; compare with Figure 5.18(a).

an inverse Fourier transform). Second, we select the two graphs of the real and imaginary parts of the complex function shown in Figure 5.20(a) to be the real and imaginary parts of the complex function to be transformed. Finally, we choose to apply a real, user-created filter function, and enter the formula

$$f(x) = \text{rect}[(x - 50)/8] + \text{rect}[(x + 50)/8] \qquad (5.105)$$

for the filter function. This filter function equals 1 over two relatively small intervals centered at ±50 and is 0 elsewhere. Hence, multiplying it with the FFT shown in Figure 5.20(a) produces a product that matches the FFT near the two spikes, but equals 0 elsewhere. This product resembles the transform of the original signal. In Figure 5.20(b) we show the graph of the filtered inverse FFT. Considering how much noise was originally present, *it is an excellent reconstruction of the original signal* (divided by a factor of 8); *the noisy background has been almost entirely removed.*

The method described above, which we shall call *Fourier transform filtering,* is summarized as follows:

FOURIER TRANSFORM FILTERING PROCEDURE

Step 1. Measure noisy signal.

Step 2. Calculate power spectrum of noisy signal.

Step 3. From the power spectrum identify characteristic frequencies of the signal.

If signal reconstruction is desired, then the following additional steps may be needed:

Step 4. Compute FFT of noisy signal.

Step 5. Multiply FFT by a suitable filter.

Step 6. Compute inverse FFT.

In the exercises you will be asked to apply this method to several noisy signals (included with the disk accompanying this book) in order to determine the characteristic frequencies of the signal pulses that they contain and attempt to reconstruct these signal pulses.

WHITE NOISE

In Example 5.15 we used a noise term $n(x)$ to simulate random noise in a received signal. We will now describe what we mean by such noise. Perhaps the simplest way of defining a noise signal is to use some type of random number generator. Thus, it would be possible to define a noise term $n(x)$ by the formula

$$f(x) = b \operatorname{ran}(x) \tag{5.106}$$

where b is a positive constant and $\operatorname{ran}(x)$ is a random number generator [provided with *FAS;* see the User's Manual in Appendix A for further discussion of $\operatorname{ran}(x)$]. This will produce a graph containing random fluctuations varying between $\pm b$

with a mean of 0. There is, however, another way of generating noise which corresponds more closely to the kind of noise that is actually encountered in signal transmission. This is the noise that we have referred to as white noise. We will now explain what white noise is, and show how we constructed the simulation of white noise that we used in Example 5.15.

Let's assume that the noise $n(x)$ occurs over a fixed time interval of transmission, such as $-0.5 \leq x \leq 0.5$. Many different signals could be transmitted over time intervals of length 1 and in every case the random noise that occurs will have a different form. Consequently, we really need to examine an *ensemble* of noise functions, say $\{n_k(x)\}_{k=1}^{\infty}$. *White noise* is noise for which the average value of the power spectra from the ensemble is constant at each frequency. That is, the following limit holds:

$$\lim_{N \to \infty} \frac{1}{N} \sum_{k=1}^{N} |\hat{n}_k(u)|^2 = K \tag{5.107}$$

where K is a positive constant. In other words, on average, every frequency is represented with equal power. Clearly, this is not a physically realizable condition. All real signals will have the power damp down to zero as the frequency tends toward infinity. Hence, we shall assume that (5.107) holds only over some finite range of frequencies.

Another type of noise is *Gaussian noise*. For Gaussian noise, we replace (5.107) by

$$\lim_{N \to \infty} \frac{1}{N} \sum_{k=1}^{N} |\hat{n}_k(u)|^2 = K e^{-cu^2} \tag{5.108}$$

where K and c are positive constants. Gaussian noise is closer to a physically realizable condition. (For a good discussion of noise in random signals, see [Bas, Chapter 7].)

We will now describe how to use *FAS* to create a simulation of white noise. First, since a noise function $n(x)$ is real valued, its transform $\hat{n}(u)$ must satisfy $\hat{n}(-u) = \hat{n}^*(u)$. That is, if Re $\hat{n}$ is the real part of $\hat{n}$ and Im $\hat{n}$ is the imaginary part of $\hat{n}$, then

$$\text{Re}\,\hat{n}(-u) = \text{Re}\,\hat{n}(u), \qquad \text{Im}\,\hat{n}(-u) = -\text{Im}\,\hat{n}(u). \tag{5.109}$$

In other words, the real part of $\hat{n}$ is an even function and the imaginary part of $\hat{n}$ is an odd function. The equations in (5.109) tell us that we are only free to define $\hat{n}(u)$ for $u \geq 0$ since the values of $\hat{n}(-u)$ are then determined by these equations. Now, with this fact in mind, the following five-step procedure can be used in *FAS* to create a simulation of white noise:

Step 1. Enter the function $f(x) = a\,\text{ran}(x)$ for some positive constant a over the interval $[0, L]$. This produces the real part of the transform of the noise for positive frequencies.

Step 2. Enter the function $f(x) = b \operatorname{ran}(x)$ for some positive constant b over the interval $[0, L]$. This produces the imaginary part of the transform of the noise for positive frequencies.

Step 3. Enter the formula $f(x) = g1[\operatorname{abs}(x)]$ over the interval $[-L, L]$. This produces an even function, which is the real part of the transform of the noise function.

Step 4. Enter the formula $f(x) = \operatorname{sign}(x)g2[\operatorname{abs}(x)]$ over the interval $[-L, L]$. This produces an odd function, which is the imaginary part of the transform of the noise function.

Step 5. Perform a complex, inverse Fourier transform of the real and imaginary parts of the transform created in steps 3 and 4. Press n when asked if you want to apply a filter. The resulting inverse transformed function is the noise function $n(x)$. It will be defined over the interval $[-M/(4L), M/(4L)]$, where M equals the number of points used.

For example, to produce the noise function $n(x)$ used in Example 5.15, we chose $M = 1024$ points, $L = 512$, and $b = c = 0.01$. This produced a noise function defined over $[-0.5, 0.5]$.

We can check that (5.107) does hold (at least approximately) for the procedure described above. In Figure 5.21 we show graphs of the power spectrum averages

$$\frac{1}{N} \sum_{k=1}^{N} \left\{ [.01 \operatorname{ran}(x)]^2 + [.01 \operatorname{ran}(x)]^2 \right\} \tag{5.110}$$

for $N = 1000$ and $N = 30,000$. As you can see, the average appears to be approximately constant for the larger value of N. In fact, if rounding error is neglected, the average in (5.110) will converge as $N \to \infty$ to the constant $2(0.01)^2V$, where V is the variance of the random number generator $\operatorname{ran}(x)$. Thus, the procedure described above will produce a noise function $n(x)$ that simulates a member of the white noise ensemble $\{n_k(x)\}_{k=1}^{\infty}$ of random functions. We say "simulates" because of the following two limitations of the procedure: (1) the frequencies used were restricted to the interval $[-L, L]$; (2) only a discrete, finite set of frequencies was used. In regard to the first limitation, we mentioned previously that arbitrarily large frequencies of constant power are not physically realizable. (There are some who would say, in fact, that we are simulating *pink noise*, i.e., noise in the lower frequencies [the "red" end of the spectrum] are of equal power, while noise in the higher frequencies is nonexistent. We shall, however, stick to the more commonly used expression, white noise.) In regard to the second limitation, this limitation occurs with all of the examples throughout this text, since we have consistently been using finite sets of data to model continuous phenomena.

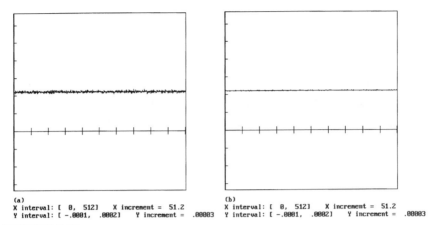

(a)
X interval: [0, 512]   X increment = 51.2
Y interval: [-.0001, .0002]   Y increment = .00003

(b)
X interval: [0, 512]   X increment = 51.2
Y interval: [-.0001, .0002]   Y increment = .00003

FIGURE 5.21
(a) Average of 1000 **random power spectrums. (b) Average of 30,000 random power spectrums.**

To model Gaussian noise, step 5 in the procedure above is modified slightly. A filtering operation is applied in this step, using a real valued Gaussian filter function $f(x) = \exp(-cx \wedge 2)$ for some positive constant c.

FURTHER EXAMPLES OF NOISE SUPPRESSION

We close this section with two more examples of noise suppression and frequency detection. First, we show how the Fourier transform filtering procedure can be used to recover a succession of pulses that represents a sequence of bits in a binary signal. Second, we illustrate a relatively new method of frequency detection called stochastic resonance.

Example 5.16:
Suppose that we wish to transmit the following sequence of eight bits: 1 0 1 1 0 1 0 1. Assuming that the time interval of transmission is $-4 \leq x \leq 4$, then this sequence could be represented by the following signal:

$$s(x) = (\cos 20\pi x)[\, e^{-10\pi(x+3.5)^2} + e^{-10\pi(x+1.5)^2} + e^{-10\pi(x+0.5)^2}$$

$$+ e^{-10\pi(x-1.5)^2} + e^{-10\pi(x-3.5)^2} \,]. \ (5.111)$$

This is a train of pulses all having fundamental frequency 10. If the reader graphs this signal it will be apparent that *the bit 1 is represented by the presence of a pulse between two integer values of time and the bit 0 is represented by the absence of such a pulse.* Now, suppose that the signal in (5.111) is received with white noise added to it. See Figure 5.22(a). After applying the Fourier transform filtering

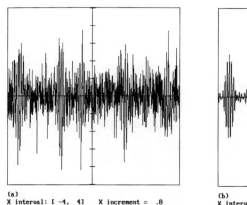 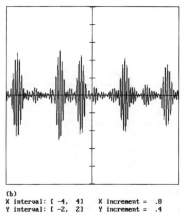

(a)
X interval: [-4, 4]   X increment = .8
Y interval: [-2, 2]   Y increment = .4

(b)
X interval: [-4, 4]   X increment = .8
Y interval: [-2, 2]   Y increment = .4

FIGURE 5.22
(a) Sequence of pulses obscured by noise. (b) Noise removed by Fourier transform filtering; the pulse sequence 1 0 1 1 0 1 0 1 is now apparent.

procedure, we obtain the signal shown in Figure 5.22(b). Clearly, we can read off from this signal the sequence of bits 1 0 1 1 0 1 0 1, which was the originally transmitted sequence.

The topic of noise removal is a fundamental one in signal processing. In addition to the classic method of Fourier transform filtering illustrated in the previous examples, many other interesting methods have been discovered. We close this section with a description of one of these methods, called *stochastic resonance*.

Example 5.17: STOCHASTIC RESONANCE
To illustrate this method of noise removal, consider the following noisy signal:

$$g(x) = \cos(50\pi x) + 10\mathrm{ran}(x). \tag{5.112}$$

This signal consists of a random noise term, $10\,\mathrm{ran}(x)$, of amplitude 10 added to a single-frequency term, $\cos(50\pi x)$, of amplitude 1 and frequency 25. The noise has a ten times greater amplitude than the cosine term, hence the cosine is completely obscured by the noise. Moreover, as we can see from Figure 5.23(a), the power spectrum of g does not reveal any easily discernible pair of spikes that would identify the frequency of the cosine term (hence, the Fourier transform filtering procedure cannot be used here). The method of stochastic resonance consists in multiplying the signal g by a filter function of the form

$$F(x) = [\,|g(x)| \geq c\,] \tag{5.113}$$

where c is a parameter that is greater than or equal to 0. The function $[\,|g(x)| \geq c\,]$

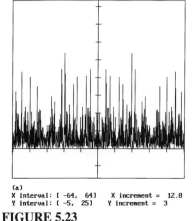 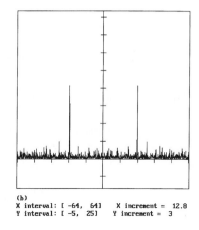

(a)
X interval: [-64, 64]   X increment = 12.8
Y interval: [-5, 25]   Y increment = 3

(b)
X interval: [-64, 64]   X increment = 12.8
Y interval: [-5, 25]   Y increment = 3

FIGURE 5.23
(a) **Power spectrum of noisy signal.** (b) **Power spectrum of filtered signal; the two spikes are located at** $x = \pm 25$.

equals 1 if $|g(x)| \geq c$ and equals 0 if $|g(x)| < c$. Thus, the filtered function is

$$[\,|g(x)| \geq c\,]\,g(x) = \begin{cases} g(x) & \text{if } |g(x)| \geq c \\ \\ 0 & \text{if } |g(x)| < c. \end{cases} \tag{5.114}$$

In other words, the filtered function corresponds to registering the signal $g(x)$ only when its intensity $|g(x)|$ is greater than or equal to the threshold value c. It is worth noting that many natural systems have this property of only responding when the stimulus is above a certain threshold. In *FAS* Formula (5.114) is implemented by entering the user-created filter function $f(x) = [\,\text{abs}(g1(x)) \geq c\,]$ where $g1$ stands for the graph of the signal g (the first graph displayed on screen).

The phenomenon of *stochastic resonance* occurs when there is a range of values of c, located somewhere between $c = 0$ and $c = \|g\|_{\sup}$ (the maximum value of g), which will enhance the detection of the underlying frequency of the obscured signal. In Figure 5.23(b) we show the power spectrum $|\hat{F}(u)|^2$ of the filtered function for the signal g defined in (5.112), using a value of $c = 9.5$. As you can see, the filtered power spectrum shows two prominent spikes. These spikes are located at $x = \pm 25$, hence we have identified the frequency 25 that was previously obscured by the noise. See [Mo-W] for an excellent, nontechnical description of stochastic resonance. This article also describes how some animals' nervous systems make use of stochastic resonance.

5.9   Poisson Summation

In this section, we will explore the relationship between Fourier series and Fourier transforms. This relationship is governed by a process known as *Poisson summation*.

Suppose first that f is a function that is 0 outside of the interval $[-L, L]$. Then the Fourier transform of f reduces to

$$\hat{f}(u) = \int_{-L}^{L} f(x)e^{-i2\pi ux}\,dx. \tag{5.115}$$

Substituting $n/2L$ for u in (5.115) and multiplying by $1/2L$ we get

$$\frac{1}{2L}\hat{f}\left(\frac{n}{2L}\right) = \frac{1}{2L}\int_{-L}^{L} f(x)e^{-in\pi x/L}\,dx. \tag{5.116}$$

Formula (5.116) says that *the Fourier series coefficients of $f(x)$ over the interval $[-L, L]$ are given by $\{(1/2L)\hat{f}(n/2L)\}_{n=-\infty}^{\infty}$ where $\hat{f}$ is the Fourier transform of f.* Or, expressed another way,

$$f \sim \sum_{n=-\infty}^{\infty} \frac{1}{2L}\hat{f}\left(\frac{n}{2L}\right) e^{in\pi x/L} \tag{5.117}$$

is the Fourier series expansion, period $2L$, of a function f that is 0 outside the interval $[-L, L]$.

Formula (5.117) can be generalized beyond the case of a function that is 0 outside of a finite interval. One version of this is the following theorem.

THEOREM 5.11: POISSON SUMMATION

Suppose that $\|f\|_1$ is finite. If $f_{\mathbf{P}}$ is defined by

$$f_{\mathbf{P}}(x) = \sum_{n=-\infty}^{\infty} f(x - 2nL)$$

then $f_{\mathbf{P}}$ is a periodic function, period $2L$, and has Fourier series

$$f_{\mathbf{P}} \sim \sum_{n=-\infty}^{\infty} \frac{1}{2L}\hat{f}\left(\frac{n}{2L}\right) e^{in\pi x/L}. \tag{5.118}$$

REMARK 5.5: (a) The function $f_\mathbf{P}$ is called the *periodization* of f. (b) If f is 0 outside of $[-L, L]$, then $f_\mathbf{P}$ is just the periodic extension of f and (5.118) reduces to (5.117). ∎

PROOF OF THEOREM Clearly, $f_\mathbf{P}$ has period $2L$. To prove that the Fourier series expansion in (5.118) is correct, let c_n stand for the nth Fourier series coefficient of $f_\mathbf{P}$, then

$$c_n = \frac{1}{2L} \int_{-L}^{L} f_\mathbf{P}(x) e^{-in\pi x/L}\, dx$$

$$= \frac{1}{2L} \int_{-L}^{L} \left[\sum_{k=-\infty}^{\infty} f(x - 2kL) e^{-in\pi x/L} \right] dx$$

$$= \sum_{k=-\infty}^{\infty} \left[\frac{1}{2L} \int_{-L}^{L} f(x - 2kL) e^{-in\pi x/L}\, dx \right].$$

Now, substituting $x + 2kL$ in place of x, we get

$$c_n = \sum_{k=-\infty}^{\infty} \frac{1}{2L} \int_{2kL-L}^{2kL+L} f(x) e^{-in\pi x/L}\, dx$$

$$= \frac{1}{2L} \int_{-\infty}^{\infty} f(x) e^{-i2\pi(n/2L)x}\, dx = \frac{1}{2L} \hat{f}\left(\frac{n}{2L}\right).$$

Thus, the Fourier series coefficient c_n of $f_\mathbf{P}$ equals $(1/2L)\hat{f}(n/2L)$. This says that (5.118) holds and our theorem is proved. ∎

There is an important corollary to this theorem, which we will also call *Poisson summation*.

THEOREM 5.12: POISSON SUMMATION
Suppose that f is continuous and $\|f\|_1$ is finite. If

$$\sum_{n=-\infty}^{\infty} f(x - 2nL)$$

converges uniformly for $|x| \le L$ and

$$\sum_{n=-\infty}^{\infty} \frac{1}{2L} \left| \hat{f} \left(\frac{n}{2L} \right) \right|$$

converges, then

$$\sum_{n=-\infty}^{\infty} f(x - 2nL) = \sum_{n=-\infty}^{\infty} \frac{1}{2L} \hat{f} \left(\frac{n}{2L} \right) e^{in\pi x/L}. \qquad (5.119)$$

PROOF   By Theorem 5.11, $f_{\mathbf{p}}(x) = \sum_{n=-\infty}^{\infty} f(x - 2nL)$ has the right side of (5.119) as its Fourier series. Since the sum defining $f_{\mathbf{p}}(x)$ converges uniformly and each term is a continuous function, a well-known theorem of advanced calculus says that $f_{\mathbf{p}}(x)$ is a continuous function. Also, since

$$\sum_{n=-\infty}^{\infty} \frac{1}{2L} \left| \hat{f} \left(\frac{n}{2L} \right) \right|$$

converges, the Weierstrass M-test says that the Fourier series

$$\sum_{n=-\infty}^{\infty} \frac{1}{2L} \hat{f} \left(\frac{n}{2L} \right) e^{in\pi x/L}$$

converges uniformly to a continuous function. Moreover, we know that this continuous function has the same Fourier coefficients as $f_{\mathbf{p}}(x)$. These two functions must therefore be the same. Thus, (5.119) holds.   ∎

We close this section with a simple example of Poisson summation. In the next section we will examine an important application.

Example 5.18:

Since $\rho/(\rho^2 + x^2) \xrightarrow{\mathcal{F}} \pi e^{-2\pi\rho|u|}$ for each $\rho > 0$, we have by Theorem 5.12

$$\sum_{n=-\infty}^{\infty} \frac{\rho}{\rho^2 + (x - n)^2} = \sum_{n=-\infty}^{\infty} \pi e^{-2\pi\rho|n|} e^{i2n\pi x}. \qquad (5.120)$$

If, for example, we put $x = 0$ and $\rho = 1$ in (5.120) we obtain

$$\sum_{n=-\infty}^{\infty} \frac{1}{1+n^2} = \sum_{n=-\infty}^{\infty} \pi e^{-2\pi|n|}. \tag{5.121}$$

It is interesting that by using geometric series one can determine the precise sums of the right-hand sides of (5.120) and (5.121). We leave it as an exercise for the reader to show that

$$\sum_{n=-\infty}^{\infty} \frac{\rho}{\rho^2 + (x-n)^2} = \frac{\pi(1 - e^{-4\pi\rho})}{1 + e^{-4\pi\rho} - 2e^{-2\pi\rho}\cos 2\pi x} \tag{5.122}$$

$$\sum_{n=-\infty}^{\infty} \frac{1}{1+n^2} = \frac{\pi(1 + e^{-2\pi})}{1 - e^{-2\pi}}. \tag{5.123}$$

5.10   Summation Kernels Arising from Poisson Summation

In this section we will show how a summation kernel (see Section 4.9, Chapter 4) can be analyzed using Poisson summation. As an example, we will show that the hanning kernel is a summation kernel.

For simplicity of notation we shall begin by assuming that we are dealing with Fourier series with period 1.

Using the functional notation for filter coefficients introduced in Section 4.5 of Chapter 4, let's suppose that the point spread function *(PSF)* $\mathcal{P}_M$ is defined by

$$\mathcal{P}_M(x) = \sum_{n=-M}^{M} F\left(\frac{n}{M}\right) e^{i2\pi nx} \tag{5.124}$$

where F is an even, continuous function on $[-1, 1]$. Extending F to be 0 outside of the interval $[-1, 1]$, we can rewrite (5.124) as

$$\mathcal{P}_M(x) = \sum_{n=-\infty}^{\infty} F\left(\frac{n}{M}\right) e^{i2\pi nx}. \tag{5.125}$$

Notice that F is still an even function.

We will now show that Poisson summation can be applied to (5.125) so that

$$\mathcal{P}_M(x) = \sum_{n=-\infty}^{\infty} M\hat{F}[M(x-n)].$$ (5.126)

Moreover, we can relax the requirement that F be 0 outside of $[-1, 1]$. Here is the theorem.

THEOREM 5.13:
Suppose that F is even and continuous and that F and its Fourier transform $\hat{F}$ satisfy

$$|F(x)| \le A\,(1+|x|)^{-1-\alpha}$$ (5.127)

$$|\hat{F}(u)| \le B\,(1+|u|)^{-1-\beta}$$ (5.128)

*for some positive constants A, α, B, and β. Then, for all x in **R**,*

$$\sum_{n=-\infty}^{\infty} F\left(\frac{n}{M}\right) e^{i2\pi nx} = \sum_{n=-\infty}^{\infty} M\hat{F}[M(x-n)].$$ (5.129)

REMARK 5.6:   Notice that when F is continuous and is 0 outside of $[-1, 1]$, then Inequality (5.127) is definitely satisfied for some A and α.   ∎

PROOF OF THEOREM   By Fourier inversion, we have

$$\hat{\hat{F}}(x) = F(-x) = F(x),$$

so F is the Fourier transform of $\hat{F}$. Inequality (5.127) ensures that $\sum_{n=-\infty}^{\infty} |F(n/M)|$ converges. In fact, by (5.127), we have

$$\sum_{n=-\infty}^{\infty} \left|F\left(\frac{n}{M}\right)\right| \le \sum_{n=-\infty}^{\infty} \frac{A}{(1+|n/M|)^{1+\alpha}}$$

and the sum on the right side converges *for each fixed M* (by comparison with

$\sum |n|^{-1-\alpha})$. Inequality (5.128) also ensures that

$$\sum_{n=-\infty}^{\infty} M \hat{F} [M(x-n)]$$

converges uniformly for all x in $[-1/2, 1/2]$. In fact, for $|n| \geq 1$,

$$M \left| \hat{F} [M(x-n)] \right| \leq \frac{BM}{(1+M|x-n|)^{1+\beta}} \cdot \tag{5.130}$$

Since $|n| \geq 1$, and x is in $[-1/2, 1/2]$, it follows that

$$|x-n| \geq \frac{1}{2}|n| \tag{5.131}$$

by a simple consideration of distance on the real line. Thus, for $|n| \geq 1$

$$M \left| \hat{F} [M(x-n)] \right| \leq \frac{BM}{(1+\frac{M}{2}|n|)^{1+\beta}} \cdot \tag{5.132}$$

Since

$$\sum_{|n| \geq 1} \frac{BM}{(1+\frac{M}{2}|n|)^{1+\beta}} = \sum_{n=1}^{\infty} 2 \frac{BM}{(1+\frac{M}{2}n)^{1+\beta}} \tag{5.133}$$

converges (by comparison with $\sum n^{-1-\beta}$), it follows from the Weierstrass M-test that

$$\sum_{n=-\infty}^{\infty} M \hat{F} [M(x-n)]$$

converges uniformly for all x in $[-1/2, 1/2]$.

We have now verified all the conditions for Poisson summation to be applied to the series $\sum_{n=-\infty}^{\infty} F(n/M) e^{i2\pi nx}$ which defines $\mathcal{P}_M(x)$. Since we showed above that $\hat{\hat{F}} = F$, we have by change of scale

$$M \hat{F} (Mx) \xrightarrow{\mathcal{F}} F \left(\frac{u}{M} \right).$$

Consequently, Theorem 5.12 yields (for $2L = 1$)

$$\sum_{n=-\infty}^{\infty} M\hat{F}[M(x-n)] = \sum_{n=-\infty}^{\infty} F\left(\frac{n}{M}\right) e^{i2\pi nx}$$

which is the same as (5.129). ∎

Based on Theorem 5.13 we can prove the following theorem, which tells us when the *PSF* $\mathcal{P}_M$ in (5.125) will be a summation kernel.

THEOREM 5.14:
Suppose that F is even and continuous, that $F(0) = 1$, and that (5.127) and (5.128) are valid. Then, the kernel $\mathcal{P}_M$ defined by

$$\mathcal{P}_M(x) = \sum_{n=-\infty}^{\infty} F\left(\frac{n}{M}\right) e^{i2\pi nx}$$

is a summation kernel over $[-1/2, 1/2]$.

PROOF   We need to check the three properties (a) to (c) in Definition 4.4 from section 4.9 of Chapter 4 (for $L = 1/2$).

(a) By integrating term by term

$$\int_{-\frac{1}{2}}^{\frac{1}{2}} \mathcal{P}_M(x)\, dx = \sum_{n=-\infty}^{\infty} F\left(\frac{n}{M}\right) \int_{-\frac{1}{2}}^{\frac{1}{2}} e^{i2\pi nx}\, dx$$

$$= F\left(\frac{0}{M}\right) \cdot 1 = 1$$

so (a) holds.

(b) We know from Theorem 5.13 that (5.129) is true. Making use of (5.129), and also (5.132) and (5.133), we have

$$|\mathcal{P}_M(x)| \le \sum_{n=-\infty}^{\infty} \left| M\hat{F}[M(x-n)] \right|$$

$$= \left| M\hat{F}(Mx) \right| + \sum_{|n|\ge 1}^{\infty} \left| M\hat{F}[M(x-n)] \right|$$

$$\le \left| M\hat{F}(Mx) \right| + \sum_{n=1}^{\infty} 2\frac{BM}{(1+\frac{M}{2}n)^{1+\beta}} .   \tag{5.134}$$

Since the numerical series

$$\sum_{n=1}^{\infty} 2 \frac{BM}{(1 + \frac{M}{2}n)^{1+\beta}}$$

converges, let's denote its sum by the constant S. Then, by (5.128) applied to $|M\hat{F}(Mx)|$ we can rewrite (5.134) as

$$|\mathcal{P}_M(x)| \leq B \frac{M}{(1 + M|x|)^{1+\beta}} + S.$$

Consequently,

$$\int_{-\frac{1}{2}}^{\frac{1}{2}} |\mathcal{P}_M(x)| \, dx \leq B \int_{-\frac{1}{2}}^{\frac{1}{2}} \frac{M}{(1 + M|x|)^{1+\beta}} \, dx + \int_{-\frac{1}{2}}^{\frac{1}{2}} S \, dx$$

$$= B \int_{-M/2}^{M/2} \frac{1}{(1 + |t|)^{1+\beta}} \, dt + S$$

$$\leq B \int_{-\infty}^{\infty} \frac{1}{(1 + |t|)^{1+\beta}} \, dt + S = \frac{2B}{\beta} + S.$$

Putting C equal $2B/\beta + S$ we see that (b) holds.

(c) We now assume that some $\delta > 0$ has been given and $\delta \leq |x| \leq 1/2$. Applying (5.134), and using (5.128) to bound $|M\hat{F}(Mx)|$, we obtain

$$|\mathcal{P}_M(x)| \leq B \frac{M}{(1 + M|x|)^{1+\beta}} + \sum_{n=1}^{\infty} 2 \frac{BM}{(1 + \frac{M}{2}n)^{1+\beta}}$$

$$\leq B \frac{M}{(M\delta)^{1+\beta}} + \sum_{n=1}^{\infty} \frac{BM2^{2+\beta}}{(Mn)^{1+\beta}}$$

$$= \frac{1}{M^\beta} \left[\frac{B}{\delta^{1+\beta}} + \sum_{n=1}^{\infty} \frac{B2^{2+\beta}}{n^{1+\beta}} \right]. \tag{5.135}$$

Since the series

$$\sum_{n=1}^{\infty} \frac{B2^{2+\beta}}{n^{1+\beta}}$$

converges, the quantity in brackets in (5.135) is a finite constant; let's call it D. Then (5.135) becomes

$$|\mathcal{P}_M(x)| \leq \frac{D}{M^\beta}, \qquad (\delta \leq |x| \leq \frac{1}{2}). \qquad (5.136)$$

Letting $M \to \infty$ in (5.136) we see that the right side of the inequality tends to 0. Hence, for any given $\epsilon > 0$, if M is taken sufficiently large we will have $D/M^\beta < \epsilon$ and then

$$|\mathcal{P}_M(x)| < \epsilon, \qquad (\delta \leq |x| \leq \frac{1}{2})$$

so (c) is true. ∎

Using Theorem 5.14, we can easily establish the following theorem, which can be used for demonstrating that many of the *PSFs* discussed in Chapter 4 are summation kernels.

THEOREM 5.15:

Suppose that F is even and continuous, that $F(0) = 1$, and that (5.127) and (5.128) are valid. Then, the kernel $\mathcal{P}_M$ defined by

$$\mathcal{P}_M(x) = \sum_{n=-\infty}^{\infty} F\left(\frac{n}{M}\right) e^{i\pi nx/L}$$

is a summation kernel over $[-L, L]$.

PROOF   If we make the change of variable, $x = 2Lt$, then

$$\mathcal{P}_M(2Lt) = \sum_{n=-\infty}^{\infty} F\left(\frac{n}{M}\right) e^{i2\pi nt}$$

is a summation kernel over $[-1/2, 1/2]$ (because of Theorem 5.14). By making the reverse change of variable, $t = x/(2L)$, it follows that $\mathcal{P}_M(x)$ is a summation kernel over $[-L, L]$. ∎

As an application of this theory we will now show that the hanning kernel is a summation kernel.

Example 5.19:
The hanning kernel, defined by

$$\mathcal{P}_M(x) = \sum_{|n| \leq M} \left[0.5 + 0.5 \cos\left(\frac{\pi n}{M}\right)\right] e^{inx}, \tag{5.137}$$

is a summation kernel over $[-\pi, \pi]$.

SOLUTION For the kernel $\mathcal{P}_M$ defined by (5.137) we have that

$$\mathcal{P}_M(x) = \sum_{n=-\infty}^{\infty} F\left(\frac{n}{M}\right) e^{inx}$$

where the function F is defined by

$$F(x) = \begin{cases} 0.5 + 0.5 \cos(\pi x) & \text{for } |x| \leq 1 \\ 0 & \text{for } |x| > 1. \end{cases} \tag{5.138}$$

It is easy to see that this function F is continuous and even, and that $F(0) = 1$. Since F is 0 outside of $[-1, 1]$ and continuous, we also have (5.127). In particular, if we take $A = 4$ and $\alpha = 1$ it is easy to check that (5.127) holds (just draw graphs of $F(x)$ and $4[1 + |x|]^{-2}$). We also have that

$$F(x) = 0.5 \operatorname{rect}(x/2) + 0.25 \operatorname{rect}(x/2)\left[e^{i\pi x} + e^{-i\pi x}\right] \tag{5.139}$$

so, by scaling and modulation properties,

$$\hat{F}(u) = \operatorname{sinc}(2u) + 0.5 \operatorname{sinc}\left[2\left(u - \frac{1}{2}\right)\right] + 0.5 \operatorname{sinc}\left[2\left(u + \frac{1}{2}\right)\right]. \tag{5.140}$$

Replacing sinc v by $(\sin \pi v)/(\pi v)$ and simplifying yields

$$\hat{F}(u) = \frac{-\sin(2\pi u)}{(2\pi u)(4u^2 - 1)} = \frac{-\operatorname{sinc}(2u)}{4u^2 - 1}. \tag{5.141}$$

Since $|\operatorname{sinc} v| \leq 1$ for all v in $\mathbf{R}$, it follows from (5.141) that (5.128) holds when $\beta = 1$ and B is chosen sufficiently large (since $\lim_{|u| \to \infty} (1 + |u|)^2/(4u^2 - 1) = 1/4$).

Because (5.127) and (5.128) are valid, it follows from Theorem 5.15 that the hanning kernel is a summation kernel. ∎

Example 5.20:
The kernel $\mathcal{P}_M$ defined by

$$\mathcal{P}_M(x) = \sum_{n=-\infty}^{\infty} e^{-\pi(n/M)^2} e^{i2\pi nx} \tag{5.142}$$

is a summation kernel over $[-1/2, 1/2]$. This is because the function $F(x) = e^{-\pi x^2}$ satisfies all the requirements of Theorem 5.14.

REMARK 5.7:   The summation kernel defined in (5.142) is related to the *theta function kernel,* which plays an important role in the theory of heat conduction. See [Wa, Chapter 4.7].   ∎

5.11   The Sampling Theorem

This section describes a very important concept in signal processing, known as the *sampling theorem.* The theory of Poisson summation and the theory of sampling are closely related. We begin with the definition of those functions to which the sampling theory applies.

DEFINITION 5.5:   *A function f is called **band limited** if its Fourier transform $\hat{f}$ is 0 outside of a finite interval $[-L, L]$.*

The sampling theorem says that a band limited function can be recovered from its *samples*

$$\left\{ f\left(\frac{n}{2L}\right) \right\}_{n=-\infty}^{n=\infty}$$

provided $\hat{f}$ is 0 outside of the interval $[-L, L]$. This theorem has many important applications. For instance, the construction of compact disc players uses sampling theory (see [Mo]). In this case, the frequencies that human ears are sensitive to lie in a finite range (20 to 20, 000 Hz) and, consequently, the recorded music can be sampled effectively. In fact, there is software available now which allows a PC to be used for analyzing and synthesizing sampled music. Another application of sampling theory is to the sending of multiple telephone messages along a *single* cable. By sampling the messages (at points separated in time by $1/2L$ units) gaps are created within which other sampled messages can be sent. Using a fiber optic cable, upwards of 25,000 simultaneous messages can be transmitted. Besides illustrating the wonder of fiber optics, this result also illustrates the power of the sampling theorem.

Here is the sampling theorem.

THEOREM 5.16:
Suppose that f is band limited. If $\hat{f}$ is 0 outside of $[-L, L]$, then

$$f(x) = \sum_{n=-\infty}^{\infty} f\left(\frac{n}{2L}\right) \text{sinc}\,(2Lx - n). \tag{5.143}$$

PROOF We will prove a very general version of the sampling theorem which yields (5.143) as a special case. Since $\hat{f}$ is 0 outside of $[-L, L]$ we can periodically extend it. In fact, if we use $\hat{f}_{\mathbf{P}}$ to denote the periodic extension of $\hat{f}$ with period $2L$, then

$$\hat{f}_{\mathbf{P}}(u) = \sum_{n=-\infty}^{\infty} \hat{f}(u - 2nL). \tag{5.144}$$

By Poisson summation, we know that the Fourier series for $\hat{f}_{\mathbf{P}}$ is [use $\hat{f}$ in place of f in (5.118)]

$$\hat{f}_{\mathbf{P}} \sim \sum_{n=-\infty}^{\infty} \frac{1}{2L} \hat{\hat{f}}\left(\frac{n}{2L}\right) e^{in\pi u/L}.$$

Since $\hat{\hat{f}}(x) = f(-x)$, we have

$$\hat{f}_{\mathbf{P}}(u) \sim \sum_{n=-\infty}^{\infty} \frac{1}{2L} f\left(\frac{-n}{2L}\right) e^{in\pi u/L}. \tag{5.145}$$

Now, suppose we multiply (5.145) by a function $W(u)$, known in sampling theory as a *window function*, which satisfies

$$W(u) = \begin{cases} 1 & \text{when } \hat{f}(u) \neq 0 \\ 0 & \text{when } |u| > L \end{cases} \tag{5.146}$$

and is bounded for all other u-values. For such a window function, we have

$$\hat{f}_{\mathbf{P}}(u)W(u) = \hat{f}(u)W(u) = \hat{f}(u) \tag{5.147}$$

because $\hat{f}(u) = 0$ for $|u| > L$.

For example, if we take

$$W(u) = \begin{cases} 1 & \text{when } |u| \leq L \\ 0 & \text{when } |u| > L \end{cases} \tag{5.148}$$

then (5.146) and (5.147) hold.

Multiplying both sides of (5.145) by $W(u)$, and using (5.147), we get

$$\hat{f}(u) \sim \sum_{n=-\infty}^{\infty} \frac{1}{2L} f\left(\frac{-n}{2L}\right) W(u) e^{in\pi u/L}. \tag{5.149}$$

Multiplying both sides of (5.149) by $e^{i2\pi ux}$ and integrating with respect to u from $-L$ to L, we obtain the following equality:

$$\int_{-L}^{L} \hat{f}(u) e^{i2\pi ux}\, du = \sum_{n=-\infty}^{\infty} f\left(\frac{-n}{2L}\right) \frac{1}{2L} \int_{-L}^{L} W(u) e^{i2\pi(x+n/2L)u}\, du. \tag{5.150}$$

We will explain why there is equality in (5.150) in Remark 5.8 below. But first, let's just assume that the equality is valid.

If we define $\mathbf{S}(x)$ by

$$\mathbf{S}(x) = \frac{1}{2L} \int_{-L}^{L} W(u) e^{i2\pi ux}\, du \tag{5.151}$$

then (5.150) becomes

$$\int_{-L}^{L} \hat{f}(u) e^{i2\pi ux}\, du = \sum_{n=-\infty}^{\infty} f\left(\frac{-n}{2L}\right) \mathbf{S}\left(x + \frac{n}{2L}\right). \tag{5.152}$$

Since $\hat{f}$ is 0 outside of $[-L, L]$, we have

$$\int_{-L}^{L} \hat{f}(u) e^{i2\pi ux}\, du = \int_{-\infty}^{\infty} \hat{f}(u) e^{i2\pi ux}\, du = f(x) \tag{5.153}$$

by Fourier inversion. Consequently, (5.152) becomes

$$f(x) = \sum_{n=-\infty}^{\infty} f\left(\frac{-n}{2L}\right) \mathbf{S}\left(x + \frac{n}{2L}\right). \tag{5.154}$$

Substituting $-n$ in place of n, and noting that the sum in (5.154) is over *all* integers, we obtain

$$f(x) = \sum_{n=-\infty}^{\infty} f\left(\frac{n}{2L}\right) \mathbf{S}\left(x - \frac{n}{2L}\right). \tag{5.155}$$

Formula (5.155) *describes a very general sampling theorem.* The data specific to the function f is the set of sample values

$$\left\{ f\left(\frac{n}{2L}\right) \right\}_{n=-\infty}^{\infty} .$$

The *reconstruction kernel* **S** is defined by Formula (5.151).

If we take for W the function in (5.148), then we have

$$\mathbf{S}(x) = \frac{1}{2L} \int_{-L}^{L} e^{i2\pi u x} \, du = \operatorname{sinc}(2Lx)$$

and (5.155) reduces to (5.143). ∎

REMARK 5.8: In the proof above we did not explain why (5.150) holds. We will now prove the following more precise version of that equality:

$$\int_{-L}^{L} \hat{f}(u) e^{i2\pi u x} \, du$$

$$= \lim_{M \to \infty} \sum_{n=-M}^{M} f\left(\frac{-n}{2L}\right) \frac{1}{2L} \int_{-L}^{L} W(u) e^{i2\pi(x+n/2L)u} \, du. \quad (5.156)$$

To prove (5.156) we use $\hat{f}(u) = \hat{f}(u) W(u)$ from (5.147), and we have

$$\left| \int_{-L}^{L} \hat{f}(u) \, e^{i2\pi u x} \, du - \sum_{n=-M}^{M} f\left(\frac{-n}{2L}\right) \frac{1}{2L} \int_{-L}^{L} W(u) e^{i2\pi(x+n/2L)u} \, du \right|$$

$$= \left| \int_{-L}^{L} \left[\hat{f}(u) - \sum_{n=-M}^{M} \frac{1}{2L} f\left(\frac{-n}{2L}\right) e^{in\pi u/L} \right] W(u) e^{i2\pi u x} \, du \right|$$

$$\leq \int_{-L}^{L} \left| \hat{f}(u) - \sum_{n=-M}^{M} \frac{1}{2L} f\left(\frac{-n}{2L}\right) e^{in\pi u/L} \right| |W(u)| \, du. \quad (5.157)$$

Now, using the symbol $S_M^{\hat{f}}(u)$ in place of the M-harmonic Fourier series partial sum for $\hat{f}$, we obtain from (5.157),

$$\left| \int_{-L}^{L} \hat{f}(u) e^{i2\pi ux} \, du - \sum_{n=-M}^{M} f\left(\frac{-n}{2L}\right) \frac{1}{2L} \int_{-L}^{L} W(u) e^{i2\pi(x+n/2L)u} \, du \right|$$

$$\leq \int_{-L}^{L} \left| \hat{f}(u) - S_M^{\hat{f}}(u) \right| |W(u)| \, du \leq \left\| \hat{f} - S_M^{\hat{f}} \right\|_2 \| W \|_2 \quad (5.158)$$

the last inequality being the well-known *Schwarz inequality* (see Appendix C) where the 2-Norms are taken over the interval $[-L, L]$.

If we now use the completeness relation from Theorem 4.2 in Chapter 4, we have (since $\hat{f}$ is bounded by $\| f \|_1$ its 2-Norm over $[-L, L]$ is finite):

$$\lim_{M \to \infty} \left\| \hat{f} - S_M^{\hat{f}} \right\|_2 = 0. \quad (5.159)$$

Comparing (5.159) with the last inequality in (5.158) we see that (5.156) holds.

Formula (5.156) justifies using (5.150) in the proof of the sampling theorem. Moreover, using the definition of **S** in (5.151), and Formula (5.153), in Inequality (5.158), we have also shown that for *all* x in **R**

$$\left| f(x) - \sum_{n=-M}^{M} f\left(\frac{n}{2L}\right) \mathbf{S}\left(x - \frac{n}{2L}\right) \right| \leq \left\| \hat{f} - S_M^{\hat{f}} \right\|_2 \| W \|_2 \quad (5.160)$$

Inequality (5.160) provides a useful estimate for the magnitude of the difference between f and a *partial sum* of the reconstruction series for f given in (5.155). It also shows that the sampling series converges uniformly over the whole real line. ∎

REMARK 5.9:   For a band limited function f, a frequently used value of L is the *smallest possible* value for which $\hat{f}(u) = 0$ when $|u| \geq L$. In this case, one must use the window defined in (5.148), and the sampling series must be the one in (5.143). The sampling rate, $2L$ samples/unit-length, is called the *Nyquist rate* when this smallest value of L is used. ∎

Example 5.21:
 Suppose $f(x) = \text{sinc}^2(2x - 1) + 3 \text{sinc}^2(2x + 1)$. Find the Nyquist sampling rate for f. Approximate $f(x)$ to within ± 0.001 for all x-values over the interval $[-4, 4]$, using a partial sum of the Nyquist sampling series for f.

SOLUTION   Since $\text{sinc}\,^2 x \xrightarrow{\mathcal{F}} \Lambda(u)$, the linearity, scaling, and shifting properties imply that

$$\hat{f}(u) = \frac{1}{2}\Lambda\left(\frac{1}{2}u\right)e^{-i\pi u} + \frac{3}{2}\Lambda\left(\frac{1}{2}u\right)e^{i\pi u}$$

$$= \left[\frac{1}{2}e^{-i\pi u} + \frac{3}{2}e^{i\pi u}\right]\Lambda\left(\frac{1}{2}u\right)$$

Since $\Lambda(u/2) = 0$ outside of the interval $[-2, 2]$, it follows that $\hat{f}(u) = 0$ outside of $[-2, 2]$. Moreover, $[-2, 2]$ is the smallest interval centered at 0 outside of which $\hat{f}(u) = 0$ (this is because $(1/2)e^{-i\pi u} + (3/2)e^{i\pi u} = 2$ at $u = 2$, hence continuity implies that $\hat{f}(u) \neq 0$ for u slightly less than 2). Therefore, the Nyquist sampling rate for f is 4.

A partial sum of the Nyquist sampling series for f is

$$S_M(x) = \sum_{k=-M}^{M} f\left(\frac{k}{4}\right)\text{sinc}\,(4x - k).$$

This sum can be graphed by *FAS* using the following formula:

$$f(x) = \text{sumk}\{(\,\text{sinc}\,(0.5k - 1) \wedge 2 + 3\,\text{sinc}\,(0.5k + 1) \wedge 2)\,\text{sinc}\,(4x - k)\}.$$

where sumk is the function in *FAS* for summing over k. A little experimentation (using 512 points, 1024 points, and 2048 points) shows that $M = 20$ yields a graph whose Sup-Norm difference from f over $[-4, 4]$ is approximately 2.8×10^{-4}. Thus, we can safely conclude that $S_{20}(x)$ approximates $f(x)$ to within ± 0.001 for all x-values in the interval $[-4, 4]$. (*Note:* although the sampling theorem applies to *all* x-values in the interval $[-4, 4]$, while *FAS* only performs its calculations for a finite set of values, we can still guage the validity of our approximations by successively doubling the number of points and seeing if the Sup-Norm differences remain relatively constant.)   ▌

As we noted above, the sampling theorem is frequently applied to the recording of musical signals for compact discs and in telephone transmission of voice signals. The basic scheme for both of these applications consists of the following three steps:

Step 1. Record samples of the signal and convert to digital data. This is called analog to digital conversion (or A/D conversion).

Step 2. Transmit digital data. For telephone signals this is done along copper wire or fiber optic cable. For CD players, this step involves the production and distribution of the compact discs.

Step 3. Reconstruction of the analog signal from the digital data. This step involves using a digital to analog converter (or D/A converter), an electronic device that produces a partial sum of the sampling series as an output. The D/A converter has the reconstruction kernel built into its circuitry; it only needs the digital data obtained from the original analog signal in order to produce its output.

A few observations need to be made about this procedure. First, in Step 1, when the digitization of the samples $\{f(k/2L)\}$ is performed, only a finite number of bits can be used. That is, each value $\{f(k/2L)\}$ is rounded off to the closest value that can be represented with this finite number of bits. The error produced by this rounding is called *quantization error.* We will not discuss quantization error any further. We shall assume that it is small enough to be ignored and simply refer to $\{f(k/2L)\}$ as the transmitted data. We shall also not discuss any details of Step 2, the transmission of the digitized data. There are many important issues that are of concern (e.g., error correction, optimal compression of the data, and elimination of noise), but, for reasons of space, we shall leave these issues aside.

Second, in regard to Step 3, the reconstruction of the analog signal, the result of Example 5.21 and other examples like it should be examined carefully. If, in Example 5.21, we view the interval $-4 \leq x \leq 4$ as the time interval of transmission (and reception) of the sampled values of the signal f, then using $M = 20$ would involve several samples $[f(k/4)$ for $k = \pm 20, \ldots, \pm 17]$ that are *not part of the transmitted values.* If we can only use samples over the interval $[-4, 4]$, then the largest value of M that we can choose is $M = 16$. When we use $M = 16$, then the Sup-Norm difference between f and the reconstruction S_{16} over the interval $[-4, 4]$ is approximately 2.4×10^{-3}. Although this is not as small a value as required in the example, it still produces an acceptable approximation of the signal f. See Figure 5.24.

Third, when a sampling series partial sum

$$S_M(x) = \sum_{k=-M}^{M} f\left(\frac{k}{2L}\right) \mathbf{S}\left(x - \frac{k}{2L}\right) \tag{5.161}$$

is used for the D/A conversion, it is important to consider the response time of the reconstruction kernel $\mathbf{S}$. In Figure 5.25 we show three reconstruction kernels (which we shall describe further in Section 5.13). The kernel $\mathbf{S}_3$ shown in Figure 5.25(c) appears to be essentially 0 for $|x| > 0.5$. Therefore, we could ignore those terms $f(k/2L)\mathbf{S}_3(x - k/2L)$ in (5.161) for which $|x - k/2L| > 0.5$. The time length of 0.5 is called the *response time* for the kernel. The kernel $\mathbf{S}_1(x) = \text{sinc}(4x)$

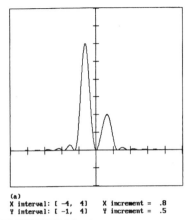

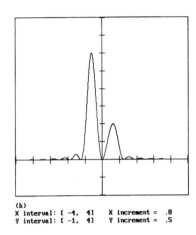

(a)
X interval: [−4, 4]   X increment = .8
Y interval: [−1, 4]   Y increment = .5

(b)
X interval: [−4, 4]   X increment = .8
Y interval: [−1, 4]   Y increment = .5

FIGURE 5.24
(a) Graph of the signal $f(x) = \operatorname{sinc}^2(2x - 1) + 3\operatorname{sinc}^2(2x + 1)$. (b) Graph of the Nyquist sampling series partial sum S_{16}. The ratio $\|f - S_{16}\|_2/\|f\|_2$ is about 9.76×10^{-4} (this gives quantitative support to the visual similarity between f and S_{16}).

shown in Figure 5.25(a) has a much longer response time. A D/A converter that uses this sinc function kernel would have to wait for most (if not all) of the data before it could begin producing output. On the other hand, a D/A converter that uses the kernel S_3 would have to wait for a much shorter set of data to be received (about 1/16 of the total data) before it begins producing output. We will examine these points in more detail in Section 5.13.

5.12 Aliasing

The sampling theorem proved in the previous section applies to *band limited* functions only. Not all functions are band limited, however. In this section we will discuss what happens if f is not band limited. It is still possible to obtain an approximate sampling reconstruction, if some error is allowed to exist between the original function f and its sampling series.

To begin, we will call a function *almost band limited* if there are positive constants A and α for which

$$|\hat{f}(u)| \le A\,[1 + |u|]^{-1-\alpha} \tag{5.162}$$

holds for all u in $\mathbf{R}$.

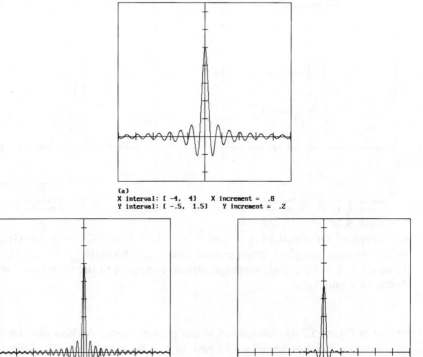

(a)
X interval: [-4, 4]   X increment = .8
Y interval: [-.5, 1.5]   Y increment = .2

(b)
X interval: [-4, 4]   X increment = .8
Y interval: [-.5, 1.5]   Y increment = .2

(c)
X interval: [-4, 4]   X increment = .8
Y interval: [-.5, 1.5]   Y increment = .2

FIGURE 5.25
(a) **The kernel** $S_1(x) = \text{sinc}\,(4x)$. (b) **The kernel** $S_2(x) = \text{sinc}\,(8x)$. (c) **The kernel** $S_3(x) = 0.75\,\text{sinc}\,(6x)\,\text{sinc}\,(2x)$.

Now, suppose that f is almost band limited but, *not* band limited. If we look again at (5.144), we see that $\hat{f}_{\mathbf{P}}$ is no longer the periodic extension of $\hat{f}$. For any given L, it is the *periodization* of $\hat{f}$ having period $2L$ that we defined in Remark 5.5(a). Nevertheless, by the Poisson summation theorem (Theorem 5.11), we still have Formula (5.145). If we use the window function

$$W(u) = \begin{cases} 1 & \text{if } |u| \leq L \\ 0 & \text{if } |u| > L \end{cases} \tag{5.163}$$

then, multiplying (5.145) by $W(u)$ and integrating over $-L \leq u \leq L$, we obtain a new form of (5.150). Namely,

$$\int_{-L}^{L} \hat{f}_{\mathbf{P}}(u)e^{i2\pi ux}\, du$$

$$= \sum_{n=-\infty}^{\infty} f\left(\frac{-n}{2L}\right) \frac{1}{2L} \int_{-L}^{L} W(u)e^{i2\pi(x+n/2L)u}\, du. \qquad (5.164)$$

Since the window W is the same one as in (5.148), we obtain

$$\int_{-L}^{L} \hat{f}_{\mathbf{P}}(u)e^{i2\pi ux}\, du = \sum_{n=-\infty}^{\infty} f\left(\frac{n}{2L}\right) \text{sinc}\,(2Lx - n). \qquad (5.165)$$

We now define the *alias* of f, denoting it by $\mathcal{A}_f$, to be

$$\mathcal{A}_f(x) = \int_{-L}^{L} \hat{f}_{\mathbf{P}}(u)e^{i2\pi ux}\, du \qquad (5.166)$$

where $\hat{f}_{\mathbf{P}}$ is defined by

$$\hat{f}_{\mathbf{P}}(u) = \sum_{n=-\infty}^{\infty} \hat{f}(u - 2nL). \qquad (5.167)$$

[*Note:* the Condition (5.162) implies that the series in (5.167) converges, as shown in the proof of Theorem 5.13.]

Thus, we have from (5.166) and (5.165)

$$\mathcal{A}_f(x) = \sum_{n=-\infty}^{\infty} f\left(\frac{n}{2L}\right) \text{sinc}\,(2Lx - n). \qquad (5.168)$$

Formula (5.168) shows that *the sampling series for f reconstructs the alias for f and not f itself.*

We will now obtain an estimate of the magnitude of the difference between f and its alias $\mathcal{A}_f$. We have

$$\left| f(x) - \mathcal{A}_f(x) \right| = \left| \int_{-\infty}^{\infty} \hat{f}(u)e^{i2\pi ux}\, du - \int_{-L}^{L} \hat{f}_{\mathbf{P}}(u)e^{i2\pi ux}\, du \right|. \qquad (5.169)$$

Using the definition of $\hat{f}_{\mathbf{P}}$ in (5.167), we get

$$\int_{-L}^{L} \hat{f}_{\mathbf{P}}(u) e^{i2\pi ux} \, du = \int_{-L}^{L} \sum_{n=-\infty}^{\infty} \hat{f}(u - 2nL) e^{i2\pi ux} \, du$$

$$= \sum_{n=-\infty}^{\infty} \int_{-L}^{L} \hat{f}(u - 2nL) e^{i2\pi ux} \, du.$$

Substituting $u + 2nL$ in place of u in the integral in the last sum above yields

$$\int_{-L}^{L} \hat{f}_{\mathbf{P}}(u) e^{i2\pi ux} \, du = \sum_{n=-\infty}^{\infty} \int_{-L+2nL}^{L+2nL} \hat{f}(u) e^{i2\pi ux} e^{i4\pi nLx} \, du. \qquad (5.170)$$

We can also rewrite $\int_{-\infty}^{\infty} \hat{f}(u) e^{i2\pi ux} \, du$ as a similar sum

$$\int_{-\infty}^{\infty} \hat{f}(u) e^{i2\pi ux} \, du = \sum_{n=-\infty}^{\infty} \int_{-L+2nL}^{L+2nL} \hat{f}(u) e^{i2\pi ux} \, du. \qquad (5.171)$$

Substituting from (5.170) and (5.171) into the right-hand side of (5.169) *and noting that the $n = 0$ term cancels*, we get (for the primed sums the $n = 0$ term is omitted):

$$\left| f(x) - A_f(x) \right| = \left| \sum_{n=-\infty}^{\infty}{}' \int_{-L+2nL}^{L+2nL} \hat{f}(u) e^{i2\pi ux} \left[1 - e^{i4\pi nLx} \right] du \right|$$

$$\leq \sum_{n=-\infty}^{\infty}{}' \int_{-L+2nL}^{L+2nL} |\hat{f}(u)| \left| 1 - e^{i4\pi nLx} \right| du$$

$$\leq \sum_{n=-\infty}^{\infty}{}' 2 \int_{-L+2nL}^{L+2nL} |\hat{f}(u)| \, du$$

$$= 2 \int_{\{u : |u| > L\}} |\hat{f}(u)| \, du. \qquad (5.172)$$

Thus, we have for *all* x in $\mathbf{R}$

$$\left| f(x) - A_f(x) \right| \leq 2 \int_{\{u : |u| > L\}} |\hat{f}(u)| \, du. \qquad (5.173)$$

Furthermore, using (5.162) we can bound the integral in (5.173), obtaining

$$\left| f(x) - \mathcal{A}_f(x) \right| \leq \frac{4A}{\alpha} \frac{1}{(L+1)^\alpha}. \tag{5.174}$$

Since $(4A/\alpha)(L+1)^{-\alpha} \to 0$ as $L \to \infty$, Inequality (5.174) shows that *for an almost band limited function f we can always replace f with its alias $\mathcal{A}_f$ and have negligible error when L is taken sufficiently large.* Or, put another way, *when the sampling rate of 2L samples per unit length is large enough, there will be negligible error between f and its sampling series.* This last result holds because (5.168) and (5.174) give

$$\left| f(x) - \sum_{n=-\infty}^{\infty} f\left(\frac{n}{2L}\right) \operatorname{sinc}(2Lx - n) \right| \leq \frac{4A}{\alpha} \frac{1}{(L+1)^\alpha}. \tag{5.175}$$

There are, of course, practical limitations. It is not always possible to sample at theoretically desired sampling rates.

If we use other sampling series, based on different kernels, we can obtain results that are similar to those described above. For instance, let's require that the window W satisfy the following three conditions:

$$|W(u)| \leq 1 \quad \text{for all } u\text{-values} \tag{5.176a}$$

$$W(u) = 0 \quad \text{if } |u| > L \tag{5.176b}$$

$$W(u) = 1 \quad \text{if } |u| \leq L/2. \tag{5.176c}$$

There are many examples of such windows (e.g., see Example 5.22 below). We define the alias function $\mathcal{A}_f$ for such a window, by

$$\mathcal{A}_f(x) = \int_{-L}^{L} \hat{f}_{\mathbf{P}}(u) W(u) e^{i2\pi ux} \, du. \tag{5.177}$$

It then follows that

$$\mathcal{A}_f(x) = \sum_{n=-\infty}^{\infty} f\left(\frac{n}{2L}\right) \mathbf{S}\left(x - \frac{n}{2L}\right) \tag{5.178}$$

where

$$\mathbf{S}(x) = \frac{1}{2L} \int_{-L}^{L} W(u) e^{i2\pi ux} \, du \tag{5.179}$$

is the kernel that we defined in Section 5.11. We leave it as an exercise for the reader to show that

$$|f(x) - \mathcal{A}_f(x)| \le \sum_{n=-\infty}^{\infty} \int_{-L+2nL}^{L+2nL} |\hat{f}(u) - \hat{f}(u)W(u + 2nL)e^{i4\pi nLx}|\, du$$

and that

$$|f(x) - \mathcal{A}_f(x)| \le \int_{-L}^{L} |\hat{f}(u)|\,|1 - W(u)|\, du + 2 \int_{\{u:|u|>L\}} |\hat{f}(u)|\, du. \quad (5.180)$$

Because of (5.176a) and (5.176c) it follows that

$$\int_{-L}^{L} |\hat{f}(u)|\,|1 - W(u)|\, du \le 2 \int_{\{u:L/2<|u|\le L\}} |\hat{f}(u)|\, du.$$

Hence, we can convert (5.180) into

$$|f(x) - \mathcal{A}_f(x)| \le 2 \int_{\{u:|u|>L/2\}} |\hat{f}(u)|\, du. \quad (5.181)$$

Using (5.162) we can bound the integral in (5.181), obtaining

$$|f(x) - \mathcal{A}_f(x)| \le \frac{4A}{\alpha} \frac{1}{(L/2 + 1)^{\alpha}}. \quad (5.182)$$

Replacing $\mathcal{A}_f(x)$ by the sampling series in (5.178), we also have

$$\left| f(x) - \sum_{n=-\infty}^{\infty} f\left(\frac{n}{2L}\right) \mathbf{S}\left(x - \frac{n}{2L}\right) \right| \le \frac{4A}{\alpha} \frac{1}{(L/2 + 1)^{\alpha}}. \quad (5.183)$$

As $L \to \infty$, the quantity $(4A/\alpha)(L/2 + 1)^{-\alpha} \to 0$. Therefore, we have found that *for a window W that satisfies (5.176a) to (5.176c) we can approximate an almost band limited function f by its sampling series, with kernel S derived from the window W, and have negligible error when the sampling rate is sufficiently large.* The sampling series converges to the alias function $\mathcal{A}_f$, but, as shown by (5.182), the difference between f and its alias can be made arbitrarily small when the sampling rate is chosen sufficiently large (there may, of course, be practical difficulties involved with sampling signals at very high rates).

Let's look at a couple of examples that illustrate the points we have just discussed.

Example 5.22:
Let the window W be defined by

$$W(u) = \begin{cases} 1 & \text{if } |u| \le L/2 \\ 2 - 2|u|/L & \text{if } L/2 < |u| \le L \\ 0 & \text{if } |u| > L. \end{cases} \qquad (5.184)$$

Show that the kernel **S** for this window is

$$S(x) = 0.75 \operatorname{sinc}(1.5Lx) \operatorname{sinc}(0.5Lx). \qquad (5.185)$$

SOLUTION   We have

$$S(x) = \frac{1}{2L} \int_{-L}^{L} W(u) e^{i2\pi ux} \, du = \frac{1}{L} \int_{0}^{L} W(u) \cos 2\pi ux \, du.$$

The second equality holds because W is an even function. Splitting the second integral above into a sum of two integrals we obtain

$$S(x) = \frac{1}{L} \int_{0}^{L/2} \cos 2\pi ux \, du + \frac{1}{L} \int_{L/2}^{L} (2 - 2u/L) \cos 2\pi ux \, du$$

$$= \frac{2}{L^2} \frac{1}{(2\pi x)^2} [\cos \pi Lx - \cos 2\pi Lx].$$

Using the trigonometric identity $\cos(\theta - \phi) - \cos(\theta + \phi) = 2 \sin \theta \sin \phi$ for $\theta = 1.5\pi Lx$ and $\phi = 0.5\pi Lx$, we obtain

$$S(x) = \frac{\sin(1.5\pi Lx) \sin(0.5\pi Lx)}{(L\pi x)^2}. \qquad (5.186)$$

Using the definition of the sinc function, Formula (5.186) can be rewritten as (5.185).   ∎

Example 5.23:
Let f_1 and f_2 be defined by

$$f_1(x) = \begin{cases} e^{-0.1\pi|x|} \sin \pi x & \text{if } |x| \le 4 \\ 0 & \text{if } |x| > 4 \end{cases}, \quad f_2(x) = \begin{cases} e^{-0.1\pi|x|} \sin^2 \pi x & \text{if } |x| \le 4 \\ 0 & \text{if } |x| > 4. \end{cases}$$

Approximate each of these two functions over the interval $[-4, 4]$ using two different sampling series. For one of the series, use the kernel $S_1(x) = \operatorname{sinc}(2Lx)$ and sampling rates of $2L = 4, 8, 16$, and 32. For the other series, use the same sampling rates, but use a different kernel defined by $S_2(x) = 0.75$ sinc $(1.5Lx)$ sinc $(0.5Lx)$. Using Sup-Norms, estimate the magnitude of the difference between the functions and their sampling series.

SOLUTION   Since f_1 and f_2 are both 0 outside of $[-4, 4]$, each sampling series equals one of its partial sums $S_M(x)$. For $2L = 4, 8, 16$, and 32 we will use $M = 16, 32, 64$, and 128, respectively. After graphing these partial sums and comparing with f_1 and f_2 for both 1024 points and 2048 points (as a check), we obtained the results shown in Tables 5.1 and 5.2 below. Based on these results, it cannot be said which kernel is quantitatively superior for reconstruction purposes.

■

Table 5.1   Comparison of Two Sampling Series for the Function f_1

2L, M	Sup-Norm diff., kernel S_1	Sup-Norm diff., kernel S_2
4, 16	2.15×10^{-2}	1.70×10^{-2}
8, 32	1.06×10^{-2}	8.53×10^{-3}
16, 64	5.28×10^{-3}	4.29×10^{-3}
32, 128	2.64×10^{-3}	2.16×10^{-3}

Table 5.2   Comparison of Two Sampling Series for the Function f_2

2L, M	Sup-Norm diff., kernel S_1	Sup-Norm diff., kernel S_2
4, 16	8.60×10^{-3}	2.33×10^{-2}
8, 32	1.78×10^{-3}	2.58×10^{-3}
16, 64	4.26×10^{-4}	6.11×10^{-4}
32, 128	1.05×10^{-4}	1.49×10^{-4}

5.13   Comparison of Three Kernels

In this section we shall compare three different reconstruction kernels and examine the notion of *response time* for a kernel. We shall produce sampling series reconstructions of the following signal,

$$f(x) = \begin{cases} \text{sinc}^2(2x-6) + \text{sinc}^2(2x) + \text{sinc}^2(2x+6) & \text{for } |x| \le 4 \\ 0 & \text{for } |x| > 4 \end{cases}$$

$$(5.187)$$

which we assume is transmitted for $-4 \le x \le 4$. See Figure 5.26(a). Since

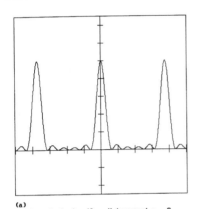

(a)
X interval: [-4, 4] X increment = .8
Y interval: [-.5, 1.5] Y increment = .2

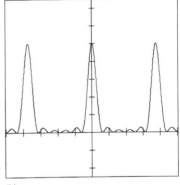

(b)
X interval: [-4, 4] X increment = .8
Y interval: [-.5, 1.5] Y increment = .2

FIGURE 5.26
(a) The signal $f(x)$, see Formula (5.187). (b) Reconstruction of f using the sinc kernel $S_1(x) = \text{sinc}(4x)$.

$f(x) = 0$ for $|x| > 4$, the sampling series partial sum

$$S_M(x) = \sum_{k=-M}^{M} f\left(\frac{k}{2L}\right) S\left(x - \frac{k}{2L}\right) \qquad (5.188)$$

equals the full series for the alias function

$$A_f(x) = \sum_{k=-\infty}^{\infty} f\left(\frac{k}{2L}\right) S\left(x - \frac{k}{2L}\right) \qquad (5.189)$$

provided $M \ge 8L$. Consequently, to simplify things, we shall always choose the sampling rate $2L$ to be a positive integer and set M equal to $8L$.

We will now describe the three kernels that we shall compare. First, since $\text{sinc}^2(2x-6) + \text{sinc}^2(2x) + \text{sinc}^2(2x-6)$ has a Nyquist rate of 4, we define the kernel S_1 by

$$S_1(x) = \text{sinc}(4x). \qquad (5.190)$$

Hence $2L = 4$ and $M = 16$ for this kernel. Second, we shall double the sampling rate for S_1 to get a sampling rate of $2L = 8$, and we shall use a sinc function kernel again. That is, we define the kernel $\mathbf{S}_2$ by

$$\mathbf{S}_2(x) = \text{sinc}\,(8x). \tag{5.191}$$

For this kernel, $M = 32$. Third, for the kernel $\mathbf{S}_3$ we shall use the kernel defined in Formula (5.185) with a sampling rate of $2L = 8$. That is, we define $\mathbf{S}_3$ by

$$\mathbf{S}_3(x) = 0.75\,\text{sinc}\,(6x)\,\text{sinc}\,(2x). \tag{5.192}$$

For this kernel, $M = 32$. These three kernels are graphed in Figure 5.25.

If we use the first kernel $\mathbf{S}_1$ in place of $\mathbf{S}$ in (5.188) and put $M = 16$ and $2L = 4$, then the sampling series

$$S_{16}^1(x) = \sum_{k=-16}^{16} f\left(\frac{k}{4}\right) \text{sinc}\,(4x - k) \tag{5.193}$$

produces what appears visually to be an acceptable approximation of f [see Figure 5.26(b)]. As a measure of accuracy we shall use the *relative 2-Norm error* $\|f - S_{16}^1\|_2/\|f\|_2$. If we think of S_{16}^1 as $f + (S_{16}^1 - f)$, then $S_{16}^1 - f$ can be thought of as *noise* added to the signal f to produce S_{16}^1. With this interpretation, $\|f - S_{16}^1\|_2/\|f\|_2$ is the reciprocal of the signal-to-noise ratio. Hence a high signal-to-noise ratio corresponds to a small value of $\|f - S_{16}^1\|_2/\|f\|_2$. In this case, we find that $\|f - S_{16}^1\|_2/\|f\|_2 \approx 3.8 \times 10^{-3}$ which indicates an acceptable reconstruction (roughly 99.6% accuracy).

If we double the sampling rate to $2L = 8$ and use the kernel $\mathbf{S}_2$, then the sampling series

$$S_{32}^2(x) = \sum_{k=-32}^{32} f\left(\frac{k}{8}\right) \text{sinc}\,(8x - k) \tag{5.194}$$

provides a better reconstruction. In this case, $\|f - S_{32}^2\|_2/\|f\|_2 \approx 3.2 \times 10^{-4}$. With the sampling rate of $2L = 8$ we can also use the kernel $\mathbf{S}_3$, in which case $M = 32$ and the sampling series is

$$S_{32}^3(x) = \sum_{k=-32}^{32} f\left(\frac{k}{8}\right) 0.75\,\text{sinc}\,(6x - 0.75k)\,\text{sinc}\,(2x - 0.25k). \tag{5.195}$$

For this reconstruction we obtain $\|f - S_{32}^3\|_2/\|f\|_2 \approx 5.0 \times 10^{-4}$. Both S_{32}^2 and S_{32}^3 have graphs that are visually indistinguishable from the original signal shown in Figure 5.26(a).

These results show that the second kernel $\mathbf{S}_2$ seems to provide the most accurate reconstruction of the given function f. There is, however, an important advantage that the third kernel S_3 has over the other two kernels. The kernel S_3 has a much shorter *response time* than either S_1 or S_2 (a brief discussion of response time was given at the end of Section 5.11). We will now take into account the response times of our three kernels. If we use a time span of length τ, then the sampling series used above should be modified as follows:

$$S_{16}^1(x) = \sum_{k=-16}^{16} f\left(\frac{k}{4}\right) \text{sinc}(4xk) \, [\, |x - k/4| \le \tau \,] \tag{5.196a}$$

$$S_{32}^2(x) = \sum_{k=-32}^{32} f\left(\frac{k}{8}\right) \text{sinc}(8xk) \, [\, |x - k/8| \le \tau \,] \tag{5.196b}$$

$$S_{32}^3(x) = \sum_{k=-32}^{32} f\left(\frac{k}{8}\right) 0.75 \text{ sinc}(6x - 0.75k)$$

$$\times \text{sinc}(2x - 0.25k) \, [\, |x - k/8| \le \tau \,]. \tag{5.196c}$$

In these formulas, the factors $[\, |x - k/4| \le \tau \,]$ and $[\, |x - k/8| \le \tau \,]$ are functions that equal 1 if the inequalities in brackets are true and equal 0 if the inequalities in brackets are false. Thus in (5.196a) only those terms are used for which k satisfies $|x - k/4| \le \tau$, while in (5.196b) and (5.196c) only those terms are used for which k satisfies $|x - k/8| \le \tau$. Using the sampling series in (5.196a) to (5.196c), we obtained the error measurements shown in Table 5.3. The relative 2-Norm error measures the total energy of all the errors (relative to the total energy of the signal) over all computed points. As shown by the data in the table (which remain the same for 1024 computed points, 2048 computed points, and 4096 computed points), the kernel S_3 produces smaller errors for much shorter time spans than either S_1 or S_2. If we use a *rule of thumb* that a reconstruction will be acceptable if the relative 2-Norm error is less than 10^{-2} (i.e., 99% accuracy), then S_3 has a response time of less than 0.5. The response time of S_2 is between 1.5 and 2.0, and the response time of S_1 is between 3.5 and 4.0. The graphs of the reconstructions from each kernel for a time span of $\tau = 0.5$ are shown in Figure 5.27. It is clear from these graphs that only the kernel S_3 provides an accurate reconstruction of f when such a short time span is used. As we pointed out at the end of Section 5.11, the kernel S_3 can begin reconstructing the signal f after receiving only 1/16 of the samples. The other kernels require more samples before they can begin reconstructing f. The kernel S_2 requires at least 3/16 of the samples, and the kernel S_1 requires about 1/2 of the samples. Of course, we obtained these results for the particular function f given in (5.187), but similar values can be obtained for other functions (see Exercises 5.72 to 5.74).

Examining response time on a theoretical basis is beyond the scope of this

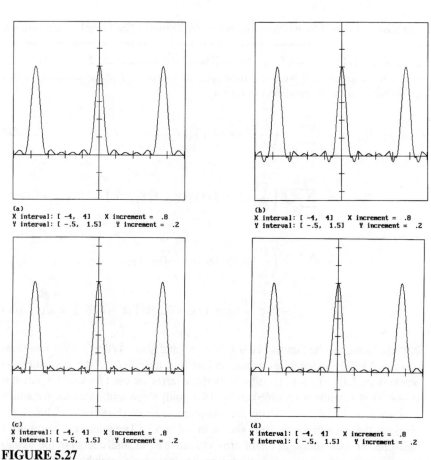

(a)
X interval: [-4, 4]   X increment = .8
Y interval: [-.5, 1.5]   Y increment = .2

(b)
X interval: [-4, 4]   X increment = .8
Y interval: [-.5, 1.5]   Y increment = .2

(c)
X interval: [-4, 4]   X increment = .8
Y interval: [-.5, 1.5]   Y increment = .2

(d)
X interval: [-4, 4]   X increment = .8
Y interval: [-.5, 1.5]   Y increment = .2

FIGURE 5.27
Graphs of sampling series reconstructions with time span $\tau = 0.5$. (a) Original signal, (b) reconstruction using kernel S_1, (c) using kernel S_2, (d) using kernel S_3.

text. For further information about various kernels in sampling theory, consult the references given at the end of this chapter.

5.14   Sine and Cosine Transforms

Besides Fourier transforms, there are two other transforms commonly used in Fourier analysis, the Fourier sine and Fourier cosine transforms. Since the theory of these transforms is so closely related to that of Fourier transforms, we shall be very brief. Fourier sine and Fourier cosine transforms bear the same relationship

Table 5.3 Relative 2-Norm Errors for Three Kernels

Time span, τ	Kernel S_1	Kernel S_2	Kernel S_3
4.0	8.13×10^{-3}	3.54×10^{-3}	4.84×10^{-4}
3.5	1.30×10^{-2}	4.84×10^{-3}	4.90×10^{-4}
3.0	1.76×10^{-2}	6.86×10^{-3}	5.00×10^{-4}
2.5	1.71×10^{-2}	6.80×10^{-3}	5.28×10^{-4}
2.0	2.09×10^{-2}	8.54×10^{-3}	5.65×10^{-4}
1.5	3.77×10^{-2}	1.58×10^{-2}	5.90×10^{-4}
1.0	4.10×10^{-2}	1.89×10^{-2}	1.08×10^{-3}
0.5	9.21×10^{-2}	4.13×10^{-2}	4.08×10^{-3}

to the Fourier transform as Fourier sine and cosine series have to Fourier series. In particular, if f is an odd function over $\mathbf{R}$, then

$$\hat{f}(u) = \int_{-\infty}^{\infty} f(x) \cos 2\pi ux \, dx - i \int_{-\infty}^{\infty} f(x) \sin 2\pi ux \, dx$$

$$= -2i \int_{0}^{\infty} f(x) \sin 2\pi ux \, dx. \tag{5.197}$$

Similarly, if f is even, then

$$\hat{f}(u) = 2 \int_{0}^{\infty} f(x) \cos 2\pi ux \, dx. \tag{5.198}$$

Based on (5.197) and (5.198) we make the following definition.

DEFINITION 5.6:   *If $\int_{0}^{\infty} |f(x)| \, dx$ is finite, then the **Fourier sine transform** of f is denoted by $\hat{f}_S$ and is defined by*

$$\hat{f}_S(u) = 2 \int_{0}^{\infty} f(x) \sin 2\pi ux \, dx.$$

*The **Fourier cosine transform** of f is denoted by $\hat{f}_C$ and is defined by*

$$\hat{f}_C(u) = 2 \int_{0}^{\infty} f(x) \cos 2\pi ux \, dx.$$

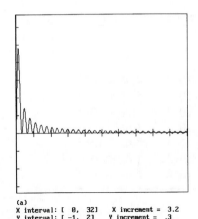
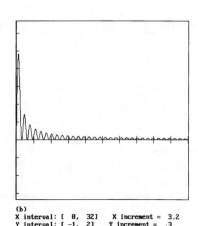

(a)
X interval: [0, 32]   X increment = 3.2
Y interval: [-1, 2]   Y increment = .3

(b)
X interval: [0, 32]   X increment = 3.2
Y interval: [-1, 2]   Y increment = .3

FIGURE 5.28
Graphs of Fourier sine transforms. (a) Graph of the *FAS* calculated transform of the function in (5.199). (b) Graph of the exact transform of that function.

Example 5.24:

Calculate the Fourier cosine and Fourier sine transforms of

$$f(x) = \begin{cases} 1 & \text{if } 0 \le x < 1 \\ 0 & \text{if } x > 1 \end{cases} \tag{5.199}$$

and compare them with the graphs generated by *FAS* using 1024 points and the interval [0, 16].

SOLUTION   For the Fourier sine transform $\hat{f}_S$ we have

$$\hat{f}_S(u) = 2 \int_0^1 \sin 2\pi u x \, dx = \frac{1 - \cos 2\pi u}{\pi u} = \frac{2 \sin^2 \pi u}{\pi u}$$

$$= 2 \operatorname{sinc}(u) \sin(\pi u). \tag{5.200}$$

On the other hand, using *FAS*, we choose *Sine* from the *Transform* menu, an interval of [0, 16] and 1024 points, and then enter the function in (5.199) using the formula $f(x) = (0 < x < 1)$. The graph produced by *FAS* is shown in Figure 5.28(a). For comparison, we have graphed the exact transform given in (5.200) in Figure 5.28(b).

For the Fourier cosine transform $\hat{f}_C$ we have

$$\hat{f}_C(u) = 2 \int_0^1 \cos 2\pi u x \, dx = \frac{\sin 2\pi u}{\pi u}$$

$$= 2 \operatorname{sinc}(2u). \tag{5.201}$$

Using *FAS* (choosing *Cosine* from the *Transform* menu, an interval of [0, 16] and 1024 points) we can approximate $\hat{f}_C$. The approximate graph is shown in Figure 5.29(a), while the exact transform shown in (5.201) is graphed in Figure 5.29(b).

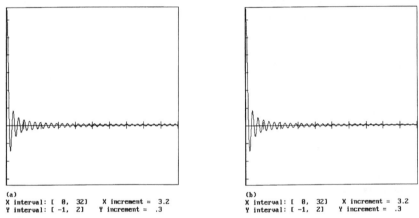

(a)
X interval: [0, 32]   X increment = 3.2
Y interval: [-1, 2]   Y increment = .3

(b)
X interval: [0, 32]   X increment = 3.2
Y interval: [-1, 2]   Y increment = .3

FIGURE 5.29
Graphs of Fourier cosine transforms. (a) Graph of the *FAS* calculated transform of the function in (5.199). (b) Graph of the exact transform of that function.

As the reader can see, for both the Fourier sine and Fourier cosine transforms, there is a good match (at the level of resolution of the graphs shown) between the exact transforms and the computer approximations. ∎

As the reader might expect, *FAS* uses a *fast sine transform* to approximate a Fourier sine transform and a *fast cosine transform* to approximate a Fourier cosine transform. Since the theory is essentially the same as we described in Section 5.4 for Fourier transforms, we will omit any further discussion of the methods used.

There are also inversion theorems for Fourier sine and Fourier cosine transforms that are similar to those described for the Fourier transform. In fact, these inversion theorems are just corollaries of the ones for the Fourier transform. The key facts are contained in Formulas (5.197) and (5.198). If f_o is the *odd extension* of f, then the Fourier transform of f_o, denoted by $\hat{f}_o$, satisfies $\hat{f}_o(u) = -i\,\hat{f}_S(u)$. Consequently, since $\hat{f}_o(u)$ is an odd function of u, inversion of the Fourier transform $\hat{f}_o$ yields

$$f_o(x) = \int_{-\infty}^{\infty} \hat{f}_o(u)e^{i2\pi ux}\, du = 2i \int_{0}^{\infty} \hat{f}_o(u) \sin 2\pi ux\, dx$$

$$= 2 \int_{0}^{\infty} \hat{f}_S(u) \sin 2\pi ux\, dx.$$

Therefore, for $x > 0$, $f(x)$ can be obtained by performing a Fourier sine transform of $\hat{f}_S(u)$. Similarly, $f(x)$ can be obtained by performing a Fourier cosine transform of $\hat{f}_C(u)$. [*Note:* just like for the Fourier transform, sometimes $(1/2)[\, f(x+) + f(x-)\,]$ is obtained by inversion, rather than $f(x)$.] We have shown that the Fourier sine and Fourier cosine transforms *are their own inverses.*

One area of science that makes extensive use of Fourier cosine transforms is the field of *Fourier transform spectroscopy.* This type of spectroscopy is the principal type used in chemistry. It provides important information about chemical structure and chemical reactions. See [Be] for a definitive introduction to Fourier transform spectroscopy.

References

For further discussion of Fourier transforms, see [Sn], [Br,2], and [Wa]. More discussion of filtering and noise can be found in [Ha], [Op-S], [Ra-G], and [Bas]. For more on sampling theory, see [St], [Je], [Fa], and [Hi]. See also [Dau], [Walt], and [Mey] for further information on kernels and sampling theory.

Exercises

Section 1

5.1 Using Definition 5.1, find the Fourier transforms of the following functions.

(a) $3 \operatorname{rect}(2x)$

(b) $\operatorname{rect}(x - 4) + \operatorname{rect}(x + 4)$

(c) $e^{-|x|}$

(d) $\operatorname{rect}(x/8) \cos 2\pi x$ [*Hint:* $\cos 2\pi x = (1/2)e^{i2\pi x} + (1/2)e^{-i2\pi x}$.]

(e) $xe^{-\pi x^2}$ [*Hint:* integrate by parts.]

5.2 Use *FAS* to draw the graphs of the Fourier transforms (negative exponent) of the following functions. [Consult the User's Manual for details of the Fourier transform procedure.]

(a) $xe^{-\pi x^2}$

(b) $\exp(-[(x-2)/2]^{20})$

(c) $f(x) = \begin{cases} \text{sqr}\,(1-x^2) & \text{if } |x| < 1 \\ 0 & \text{if } |x| < 1 \end{cases}$

(d) $\Lambda(x) = \begin{cases} 1 - |x| & \text{if } |x| \le 1 \\ 0 & \text{if } |x| > 1 \end{cases}$

5.3 Compute the transforms of the following functions using *FAS* and compare them to the exact transforms using Definition 5.1. (For all graphs use 1024 points and interval $[-16, 16]$.)

(a) $f(x) = \begin{cases} 1 - 3|x| & \text{if } |x| < 1/3 \\ 0 & \text{if } |x| > 1/3 \end{cases}$

(b) $f(x) = \text{rect}\,(3x)$

(c) $f(x) = xe^{-2|x|}$

(d) $\frac{1}{2}\,\text{rect}\,(x-2) - \frac{1}{2}\,\text{rect}\,(x+2)$

Section 2

5.4 Repeat Exercise 5.1, but now use the properties discussed in Section 5.2.

5.5 Suppose that f, f', and f'' are all continuous and $\|f\|_1$, $\|f'\|_1$, and $\|f''\|_1$ are all finite. Show that

$$|\hat{f}(u)| \le \frac{\|f''\|_1}{4\pi^2 u^2}, \qquad (u \ne 0).$$

5.6 Under the same hypotheses as in Exercise 5.5, show that for some positive constant A (dependent on f),

$$|\hat{f}(u)| \le A\,[1 + |u|]^{-2}, \qquad \text{all } u \in \mathbf{R}.$$

5.7 Find the exact Fourier transforms of the following functions (check your answers with *FAS*):

(a) $x^3 e^{-\pi x^2}$

(b) $xe^{-2\pi |x|}$

(c) $\begin{cases} \sin 8\pi x & \text{if } |x| < 3 \\ 0 & \text{if } |x| > 3 \end{cases}$

(d) $e^{-|x|} \cos 6\pi x$

(e) $e^{-|x-1|} + e^{-|x+1|}$

Section 3

5.8 Check the estimates in (5.28). (See the User's Manual for details on how to calculate 2-Norm differences in *FAS*.)

5.9 Repeat the estimates in (5.28), but use the function defined in (5.21) in place of rect.

5.10 Examine S_8^{rect}, S_{16}^{rect}, S_{32}^{rect}, and S_{64}^{rect} near the point $x = 0.5$ in order to observe *Gibbs' phenomenon*.

5.11 Repeat Exercise 5.10, but use the function defined in (5.21) and examine the graphs near $x = 0$.

5.12 If $e^{-2\pi|x|}$ is used instead of Λ, what number should replace 0.006 in (5.20)? (Again, use 32 for L.)

Section 4

5.13 Compare the computer calculation of Λ and its exact transform sinc^2.

5.14 Why is there less error involved in Exercise 5.13 than in Example 5.8?

5.15 Compare the computer calculation of the transform of $e^{-\pi x^2}$ with its exact transform $e^{-\pi u^2}$.

Section 5

5.16 Check (5.61). Also, show that $f * (g * h) = (f * g) * h$, (*associativity*).

5.17 Verify formula (5.64).

5.18 Using *FAS*, compute $f * f$, $f * f * f$, and $f * f * f * f$ for $f(x) = \text{rect}(x)$.

5.19 (a) Let $f(x) = \text{rect}(x)$. Show that

$$f *{}_y P(x) = \frac{1}{\pi}\left[\text{Tan}^{-1}\left(\frac{x+0.5}{y}\right) - \text{Tan}^{-1}\left(\frac{x-0.5}{y}\right)\right]. \qquad (5.202)$$

(b) Compare the exact expression for $f *{}_y P$, given in (5.202), with the *FAS*-calculated convolution of f with ${}_y P$, for $y = 0.5$, 1.0, and 2.0, by calculating their Sup-Norm differences over $[-10, 10]$ using 1024 points. (c) Let $f(x) = \text{rect}(x-1) + \text{rect}(x+1)$. Find an exact formula for $f *{}_y P$. (d) Compare the exact expression for $f *{}_y P$, found in (c), with the *FAS*-calculated convolution of f with ${}_y P$, for $y = 0.5$, 1.0, and 2.0, by calculating their Sup-Norm differences over $[-10, 10]$ using 1024 points.

5.20 Using *FAS*, compute $f *{}_y P(x)$ for $f(x) = e^{-|x|}$ and $y = 0.01, 0.05, 0.25, 1.25, 6.0$.

5.21 Repeat Exercise 5.20, but use $f(x) = \Lambda(x)$, and $f(x) = 1/(1 + x^2)$.

5.22 Using *FAS*, illustrate for the functions given in Exercises 5.20 and 5.21 that

$$\lim_{y\to 0+} f * {}_y P(x) = f(x).$$

Prove that this limit holds for the function $f(x) = \text{rect}(x)$, using formula (5.202) in Exercise 5.19. (A general proof of this limit can be found in [Wa, Chapter 7.1].)

5.23 Use Fourier transforms to solve the following system of equations for $W(x, t)$ [given $f(x)$]:

$$\frac{\partial^2 W}{\partial x^2} = \frac{\partial^2 W}{\partial t^2}, \qquad W(x, 0) = f(x), \qquad \frac{\partial W}{\partial t}(x, 0) = 0.$$

Graph this solution for $f(x) = e^{-0.5|x|}$ and $t = 0.1, 0.2, 0.4, 0.8$, and 1.6. Use the interval $[-5, 5]$ and 1024 points.

5.24 Use Fourier transforms to solve the following system of equations for W (given g):

$$\frac{\partial^2 W}{\partial x^2} = \frac{\partial^2 W}{\partial t^2}, \qquad W(x, 0) = 0, \qquad \frac{\partial W}{\partial t}(x, 0) = g(x).$$

Graph this solution for $g(x) = e^{-0.5|x|}$ and $t = 0.1, 0.2, 0.4, 0.8$, and 1.6. Use the interval $[-16, 16]$ and 1024 points.

Section 6

5.25 Using *FAS*, graph approximations to $f * {}_t H(x)$ for the following functions and values of a^2. Use $t = 0.1, 0.5, 1.0, 2.0$. Use the interval $[-10, 10]$ and 1024 points.

(a) $40 \, \text{rect}(x), \quad a^2 = 0.5$

(b) $20 \, \text{rect}(x - 2) + 20 \, \text{rect}(x + 2), \quad a^2 = 1$

(c) $20 \, \text{rect}(x/2), \quad a^2 = 4$

(d) $40/(1 + x^2), \quad a^2 = 1$

(e) $e^{-|x|}, \quad a^2 = 2$

(f) $\exp(-(x/3) \wedge 10), \quad a^2 = 2$

5.26 Using *FAS*, check that $\lim_{t\to 0+} f * {}_t H(x) = f(x)$ for each of the functions in Exercise 5.25(d) to (f); check also that the limit holds for points of continuity of the functions in (a) to (c). (A proof of the limit can be found in [Wa, Chapter 7.2].)

5.27 Using the convolution theorem, prove that ${}_t H * {}_\tau H = {}_{(t+\tau)} H$ for each $t > 0$ and $\tau > 0$. Also, show that ${}_y P * {}_\Upsilon P = {}_{(y+\Upsilon)} P$ holds for each $y > 0$ and $\Upsilon > 0$. (See (5.59) for the definition of ${}_y P$.) Using *FAS*, check these identities (do not use huge values of t, τ, y, or Υ).

5.28 This exercise compares two ways of computing the distribution $F(x)$ for the p.d.f. $f(x) = (1/\sqrt{\pi})e^{-x^2}$ [see Example 5.12]. (a) First, check that $f(x)$ is a p.d.f., i.e., show that

$$\int_{-\infty}^{\infty} (1/\sqrt{\pi})e^{-x^2}\, dx = 1.$$

(b) Define the *error function*, $\mathrm{erf}(x)$, by

$$\mathrm{erf}(x) = \frac{2}{\sqrt{\pi}} \int_0^x e^{-s^2}\, ds.$$

Show that $y = \mathrm{erf}(x)$ satisfies the initial-value problem

$$y' = \frac{2}{\sqrt{\pi}} e^{-x^2}, \quad y(0) = 0. \tag{5.203}$$

(c) Plot an approximate solution of (5.203) over $[0, 10.24]$ using 1024 points. [*Hint:* enter the formula $f(x) = (2/\mathrm{sqr}(\pi)) \exp(-x \wedge 2) \backslash \mathrm{Diff} \backslash y0 = 0.$] (d) Consult a table (see, e.g., [A-S, page 310]) to verify that the solution of (5.203) plotted by *FAS* is accurate to at least 10 decimal places for all computed values. (e) Show that the distribution function $F(x)$ satisfies

$$F(x) = 0.5 + 0.5\mathrm{erf}(x) \tag{5.204}$$

(f) Use (5.204) to plot a graph of $F(x)$ over $[-10.24, 10.24]$. [*Hint:* $\mathrm{erf}(x)$ is an odd function, so you can enter the formula $f(x) = 0.5 + 0.5\mathrm{sign}(x)g1(\mathrm{abs}(x))$ where $g1$ stands for the function graphed in part (c). Conclude that your graph of $F(x)$ is accurate to 10 decimal places for all computed values. (g) Recompute $F(x)$ over the interval $[-16, 16]$ and compare it with the convolution approximation described in Example 5.12. In particular, find their Sup-Norm differences over $[-10, 10]$.

5.29 (a) Use the definition of the error function in Exercise 5.28 to prove that $f * {}_t H(x)$, where $f(x) = 20\,\mathrm{rect}(x/2)$ and $a^2 = 1$, satisfies

$$f * {}_t H(x) = 10\left[\mathrm{erf}\left(\frac{x+1}{\sqrt{4t}}\right) - \mathrm{erf}\left(\frac{x-1}{\sqrt{4t}}\right)\right]. \tag{5.205}$$

(b) Use formula (5.205), and a method similar to the one described in Exercise 5.28, to produce a graph of $f * {}_t H$ for $a^2 = 1$ and $t = 0.5, 1.0,$ and 1.5, using the interval $[-8, 8]$ and 1024 points. (c) Compute the Sup-Norm differences between the approximation of $f * {}_t H$ from part (b) and the *FAS* convolution approximation discussed in the text.

5.30 Using *FAS*, graph the distributions for the following p.d.f.s over the interval $[-10, 10]$.

(a) $(1/\sqrt{\pi})e^{-x^2}$

(b) $\mathrm{rect}(x)$

(c) $\Lambda(x) = (1 - \mathrm{abs}(x))(\mathrm{abs}(x) \leq 1)$

(d) $(1/\pi)(1/(1+x^2))$

(e) $\text{rect}(x-1)$

(f) $\Lambda(x+1) = (1 - \text{abs}(x+1))(\text{abs}(x+1) \le 1)$

Section 7

5.31 Using for $f(x)$ the function $2\,\text{rect}\,(4x)$, and using 4096 points and an interval of $[-32, 32]$, check Formula (5.95) on *FAS* for times $t = 0.1, 0.15$, and 0.2 (assume that m = mass of an electron, so that $\hbar/2m \approx 0.578$).

5.32 Using the convolution theorem, show that $_tF * {}_\tau F = {}_{(t+\tau)}F$ for all $t > 0$ and $\tau > 0$.

5.33 Graph approximations to $|\psi|^2$ for $t = 0.25, 0.5, 1.0$, and 2.0, $\hbar/2m = 0.578$, and $f(x) = e^{-\pi x^2/2}$. Check Formula (5.95) for $t = 1.0$ and 2.0.

5.34 Show that if $f(x) = e^{-\pi x^2}$, then

$$f * {}_tF(x) = \frac{1}{[1 + 4\pi i(\hbar/2m)t]^{\frac{1}{2}}}\, e^{-\pi x^2/[1+4\pi i(\hbar/2m)t]}. \tag{5.206}$$

Using (5.206), show that (5.95) holds for sufficiently large t when $f(x) = e^{-\pi x^2}$.

Section 8

5.35 Can you give a physical interpretation of why the filtered Fourier transform method of noise suppression works?

5.36 Given the function $\cos 16\pi x + 0.8 \cos 15\pi x - 0.6 \cos 18\pi x$, perform a filtered Fourier transform on the interval $[-16, 16]$ using 1024 points. Use Parzen, hanning, Hamming, and Welch filters.

5.37 Repeat Exercise 5.36, but use $\sin 16\pi x + 0.8 \cos 8\pi x - 0.3 \sin 12\pi x$.

5.38 Repeat Exercise 5.36, but use the constant function $f(x) = 1$. Compare the shapes of the peaks of the filtered transforms in Exercise 5.36 with the shape of each peak in this exercise. Explain the similarity.

5.39 The disk accompanying this book contains the function data file, *signal1*, which is a received signal that has been corrupted by noise. The originally transmitted signal was a pulse of amplitude 1 containing a single fundamental frequency. Determine this frequency and reconstruct, at least approximately, the original signal. (Use the interval $[-16, 16]$ and 1024 points.)

5.40 The disk accompanying this book contains the function data file, *signal2*, which is a received signal that has been corrupted by noise. The originally transmitted signal was a pulse of amplitude 1 containing two fundamental frequencies. Determine these frequencies. (Use the interval $[-16, 16]$ and 1024 points.)

5.41 The disk accompanying this book contains the function data file, *signal3*, which is a received signal that has been corrupted by noise. The originally transmitted signal was a sequence of pulses all having amplitude 1 and a single fundamental frequency. These pulses represent a sequence of bits. Determine this sequence. Repeat this exercise for the signals *signal4* and *signal5*. (All these signals are defined over the interval $[-16, 16]$ using 1024 points.)

5.42 The disk accompanying this book contains the function data file, *signal6*, which is a received signal that has been corrupted by noise. The originally transmitted signal was a pulse of amplitude 1 containing a single fundamental frequency equal to 10. Suppose that the received signal was a radar signal received after bouncing off an object. Determine the fundamental frequency in the received signal. Is the object moving away from or towards the transmitter? (Use the interval $[-16, 16]$ and 1024 points.)

REMARK 5.10: Exercise 5.42 illustrates the basic idea behind Doppler radar. See also Exercise 5.48.

5.43 The disk accompanying this book contains the function data file, *signal7*, which is a received signal that has been corrupted by noise. Use the method of stochastic resonance to find the fundamental frequency of the signal. (Use the interval $[-1, 1]$ and 1024 points.)

5.44 Repeat Exercise 5.43 for the signal *signal8*.

5.45 The disk accompanying this book contains the function data file, *signal9*, which is a noisy signal of the form

$$f(x) = \cos(2\pi v x - \phi)e^{-10\pi x^2} + n(x)$$

defined over the interval $[-0.5, 0.5]$ using 1024 points. Using the Fourier transform filtering method, determine the fundamental frequency v in the received signal. Observe that the power spectrum $|\hat{f}(u)|^2$ does not indicate the value of the phase shift ϕ. If the original signal was $s(x) = 4\cos(50\pi x)e^{-10\pi x^2}$, then try to estimate the phase shift.

5.46 Repeat Exercise 5.45, but use the signal contained in the file *signal10*.

5.47 The disk accompanying this book contains the signals *signal11*, *signal12*, ..., *signal16*, all of which are defined over the interval $[-1, 1]$ using 2048 points. Using either of the two methods described in the text, determine a single characteristic frequency for each signal.

5.48 A signal pulse has been transmitted by a radar device with a frequency of 75 for a time interval of length 1. The disk accompanying this text contains the function data files *signal17*, *signal18*, *signal19*, and *signal20*. These signals are the readings of a receiver for successive time intervals of length 1. Which of these signals contains the returning echo of the transmitted signal? Is the object that reflected the pulse moving away from or toward the receiver?

Section 9

5.49 Show that (5.122) and (5.123) are true. Also, find the exact sums of

$$\sum_{n=0}^{\infty} \frac{1}{1+n^2} \quad \text{and} \quad \sum_{n=1}^{\infty} \frac{1}{1+n^2}.$$

5.50 Using Poisson summation, show that for $t > 0$

$$\sum_{n=-\infty}^{\infty} e^{-2n^2\pi^2 t} e^{i2\pi nx} = \frac{1}{(2\pi t)^{\frac{1}{2}}} \sum_{n=-\infty}^{\infty} e^{-(x-n)^2/2t} \qquad (5.207)$$

The identity (5.207) is known as *Jacobi's identity*.

5.51 Approximate $\sum_{n=-\infty}^{\infty} e^{-2n^2\pi^2 t} e^{i2\pi nx}$ for $t = 0.001$, using a filtered Fourier series partial sum. Using *FAS*, graph this approximation over $[-1/2, 1/2]$ with 1024 points.

5.52 Show that $\sum_{n=-\infty}^{\infty} e^{-2n^2\pi^2 t} \cos 2\pi nx > 0$ for all $x \in \mathbf{R}$ and all $t > 0$.

5.53 Find the exact value of $\sum_{n=1}^{\infty} 1/(a^2 + n^2)$ for all $a > 0$.

Section 10

5.54 Use Theorem 5.15 to give another proof that the Cesàro kernel is a summation kernel over $[-\pi, \pi]$.

5.55 Show that the dlVP kernel V_M is a summation kernel over $[-\pi, \pi]$. [The kernel V_M is defined in (4.54) and (4.55) in Chapter 4.]

5.56 Suppose that F is continuous and even, $F(0) = 1$, $F = 0$ outside of $[-1, 1]$ *and* the derivatives F' and F'' are both continuous. Show that (5.127) and (5.128) hold for this function F, hence $\mathcal{P}_M(x) = \sum_{n=-M}^{M} F(n/M)e^{inx}$ is a summation kernel. [*Hint:* use the inequality in Exercise 5.6 to bound $|\hat{F}(u)|$.]

5.57 Given a positive constant $\rho > 0$, the *Riesz kernel* is defined by

$$\mathcal{P}_M(x) = \sum_{n=-M}^{M} \left[1 - \left(\frac{n}{M}\right)^2 \right]^{\rho} e^{inx}$$

Use the result of Exercise 5.56 to show that the Riesz kernel is a summation kernel when $\rho \geq 2$.

REMARK 5.11: It is known that the Riesz kernel is a summation kernel when $\rho > 0$. See [St-W, Chapter 7]. ∎

Section 11

5.58 Show that the function $f(x) = \mathrm{sinc}^2(2x)\cos(4\pi x)$ has a Nyquist rate of 8. Using *FAS*, graph the approximations to $f(x)$ given by the sampling series partial sums

$$S_M(x) = \sum_{k=-M}^{M} f\left(\frac{k}{8}\right) \mathrm{sinc}\,(8x - k)$$

for $M = 32, 48$, and 64. Use 1024 points over the interval $[-4, 4]$. Using Sup-Norm and 2-Norm differences, estimate the accuracy of these partial sums as approximations of f.

5.59 Determine the Nyquist sampling rates of the following functions. Also compute the Sup-Norm differences between these functions and the indicated partial sums of their Nyquist sampling series over the given intervals.

(a) $f(x) = 2\,\mathrm{sinc}^2(4x) + \mathrm{sinc}^2(4x - 12) + \mathrm{sinc}^2(4x + 12)$; compare with the Nyquist partial sum $S_{32}(x)$ over $[-4, 4]$.

(b) $f(x) = \mathrm{sinc}^2(8x) - \mathrm{sinc}^2(8x - 2) - \mathrm{sinc}^2(8x + 2)$; compare with the Nyquist partial sum $S_{16}(x)$ over $[-1, 1]$.

(c) $f(x) = \mathrm{sinc}^4(2x)$, compare with the Nyquist partial sum $S_{32}(x)$ over $[-4, 4]$.

5.60 Suppose that $\hat{f}(u) = 0$ for $|u| > L$. Explain why, if $g(x) = 2L\,\mathrm{sinc}\,(2Lx)$, then $f * g = f$. Using *FAS*, check this result for $f(x) = \mathrm{sinc}^2(x)$ over the interval $[-8, 8]$. Use 1024 points.

5.61 Show that $\{\,\mathrm{sinc}\,(2Lx - n)\}_{n=-\infty}^{\infty}$ satisfies the following orthogonality relation:

$$2L \int_{-\infty}^{\infty} \mathrm{sinc}\,(2Lx - m)\,\mathrm{sinc}\,(2Lx - n)\,dx = \begin{cases} 0 & \text{if } m \neq n \\ 1 & \text{if } m = n. \end{cases}$$

(*Hint:* use Parseval's equalities.) Conclude that if f is band limited, with $\hat{f} = 0$ outside of $[-L, L]$, then

$$f\left(\frac{n}{2L}\right) = 2L \int_{-\infty}^{\infty} f(x)\,\mathrm{sinc}\,(2Lx - n)\,dx = \int_{-\infty}^{\infty} f(x)\,\frac{\sin 2L\pi(x - n/2L)}{\pi(x - n/2L)}\,dx.$$

5.62 The disk that accompanies this book contains a function data file called *samples1*. Retrieve these data using an interval of $[-4, 4]$ and 1024 points. The non-zero function values are samples $\{f(k/8)\}_{k=-32}^{k=32}$ of a signal f, using a sampling rate of 8. Assuming that the Nyquist rate of f is 8, reconstruct f using a sinc-function sampling series.

5.63 The disk that accompanies this book contains a function data file called *samples2*. Retrieve these data using an interval of $[-4, 4]$ and 1024 points. The non-zero function values are samples $\{f(k/8)\}_{k=-32}^{k=32}$ of a signal f, using a sampling rate of 8. Assuming that the Nyquist rate of f is 4, reconstruct f in the following two ways: (a) using the Nyquist kernel $S_1(x) = \mathrm{sinc}\,(4x)$, (b) using the kernel $S_2(x) = \mathrm{sinc}\,(8x)$ at twice the Nyquist rate. Find the Sup-Norm difference between these two reconstructions.

Section 12

5.64 For a function f satisfying (5.162), and kernel $\mathbf{S}$ whose window W satisfies (5.176a) to (5.176c), show that the following inequality holds:

$$\left| \mathcal{A}_f(x) - \sum_{n=-M}^{M} f\left(\frac{n}{2L}\right) \mathbf{S}\left(x - \frac{n}{2L}\right) \right| \leq \left\| \hat{f}_\mathbf{P} - S_M^\mathbf{P} \right\|_2 \|W\|_2$$

where $\|W\|_2 = \left[\int_{-L}^{L} |W(u)|^2 \, du \right]^{\frac{1}{2}}$ and $S_M^\mathbf{P}$ is the M-harmonic partial sum of the Fourier series for $\hat{f}_\mathbf{P}$. Using this result, show that

$$\left| f(x) - \sum_{n=-M}^{M} f\left(\frac{n}{2L}\right) \mathbf{S}\left(x - \frac{n}{2L}\right) \right| \leq \frac{4A}{\alpha} \frac{1}{(L/2 + 1)^\alpha} + \left\| \hat{f}_\mathbf{P} - S_M^\mathbf{P} \right\|_2 \sqrt{2L}.$$

5.65 Using *FAS*, graph the approximations to $f(x) = 0.5 \exp(-0.25\pi x \wedge 2)$ given by

$$S_M(x) = \sum_{k=-M}^{M} f(k) \operatorname{sinc}(x - k)$$

for $M = 8, 16$, and 32. Use the interval $[-8, 8]$ and 1024 points. Find the Sup-Norm differences between f and S_M for each M. [*Note:* if $-8 \leq x \leq 8$ defines the allowed x-values, then only $M = 8$ is a valid choice for M.]

5.66 Repeat Exercise 5.65, but use $f(x) = (1/\pi)[1/(1 + x \wedge 2)]$.

5.67 Suppose $f(x) = 0.5 \exp(-0.25\pi x^2) \cos^2(\pi x/4)$ for $-2 \leq x \leq 2$ and $f(x) = 0$ for $|x| > 2$. Using *FAS* graph

$$\mathcal{A}_f(x) = \sum_{k=-4L}^{4L} f\left(\frac{k}{2L}\right) \operatorname{sinc}(2Lx - k)$$

for $L = 4, 8$, and 16, over the interval $[-2, 2]$. Estimate the Sup-Norm differences between f and $\mathcal{A}_f$ for each L. Repeat these computations, but use the kernel $\mathbf{S}$ defined in Formula (5.185).

5.68 Suppose that $\hat{f}(u) = 0$ for $|u| \leq L/2$ and for $|u| > L$. Show that f can be reconstructed using the samples $\{f(k/L)\}_{k=-\infty}^{\infty}$ with reconstruction kernel $\mathbf{S}(x) = \operatorname{sinc}(Lx)[2\cos(L\pi x) - 1]$. (*Note:* The sampling rate here is *half* the Nyquist rate.)

5.69 Repeat Example 5.23, but use the function

$$f(x) = \begin{cases} e^{-0.1\pi x^2} \cos(5\pi x/8) & \text{if } |x| \leq 4 \\ 0 & \text{if } |x| > 4. \end{cases}$$

5.70 Given that the window W is defined by

$$W(u) = \begin{cases} 1 & \text{if } |u| \leq L/2 \\ \sin^2(\pi u/L) & \text{if } L/2 < |u| \leq L \\ 0 & \text{if } |u| > L. \end{cases}$$

Show that W satisfies (5.176a) to (5.176c) and that its kernel $\mathbf{S}$ satisfies

$$S(x) = \frac{1}{8}[2 \operatorname{sinc}(Lx) + 4 \operatorname{sinc}(2Lx) + \operatorname{sinc}(Lx - 1) + \operatorname{sinc}(Lx + 1)$$

$$-2 \operatorname{sinc}(2Lx - 2) - 2 \operatorname{sinc}(2Lx + 2)].$$

Hint: you may want to use the trigonometric identity $2 \sin^2 \theta = 1 - \cos(2\theta)$.

5.71 Using the functions f_1 and f_2 in Example 5.23, compute additional columns for Tables 5.1 and 5.2 using the kernel $\mathbf{S}$ in Exercise 5.70.

Section 13

5.72 Compile a table like Table 5.3, but use the function

$$f(x) = \begin{cases} e^{-0.1\pi|x|} \sin \pi x & \text{if } |x| \leq 4 \\ 0 & \text{if } |x| > 4. \end{cases}$$

over the interval $[-4, 4]$. Use a sampling rate of 8 for $\mathbf{S}_1$ and a sampling rate of 16 for both $\mathbf{S}_2$ and $\mathbf{S}_3$.

5.73 Repeat Exercise 5.72, but use the function

$$f(x) = \begin{cases} e^{-0.1\pi|x|} \sin^2 \pi x & \text{if } |x| \leq 4 \\ 0 & \text{if } |x| > 4. \end{cases}$$

over the interval $[-4, 4]$.

5.74 Repeat Exercise 5.72, but use the function

$$f(x) = \begin{cases} \sum_{i=-3}^{3}(-1)^i \operatorname{sinc}^2(2x - 2i) & \text{if } |x| \leq 4 \\ 0 & \text{if } |x| > 4. \end{cases}$$

over the interval $[-4, 4]$.

5.75 Add a fourth column to Table 5.3 in the text, but use the kernel $\mathbf{S}$ given in Exercise 5.70 with $2L = 8$. Does this kernel $\mathbf{S}$ appear to have a shorter response time than the kernel $\mathbf{S}_3$ ($\mathbf{S}_2$, $\mathbf{S}_1$)?

5.76 *Fast evaluation of sampling series.* (a) Show that if $x_k = k/(2L)$, then the sampling series partial sum S_M in Formula (5.161) satisfies

$$S_M(x_j) = \sum_{k=-M}^{M} f(x_k)S(x_{j-k}). \tag{5.208}$$

(b) Show that the sum in (5.208) can be approximated using FFTs, for $N = 2^R$ points $\{x_j\}$ from a given interval $[-\Omega, \Omega]$.

5.77 The method involving FFTs introduced in the previous exercise is programmed into *FAS* in the procedure entitled *Interpolation (sampling)* in the *Series* menu (see the User's Manual in Appendix A for more details). Use this *FAS* procedure to graph sampling series partial sums given the following data (in each case use 1024 points), and find the relative 2-Norm error between the function and its *FAS* computed sampling series partial sum (the relative 2-Norm error equals $\|f - S^F_M\|_2 / \|f\|_2$ where S^F_M is the *FAS*-computed sampling series).

(a) Interval, $[-4, 4]$. Kernel, sinc $(4x)$. Function, $\exp(-(x/2)^{20})$. Sampling increment, $1/(2L) = 0.25$. Order of partial sum, $M = 16$.

(b) Interval, $[-4, 4]$. Kernel, sinc $(8x)$. Function, $\exp(-(x/2)^{20})$. Sampling increment, $1/(2L) = 0.125$. Order of partial sum, $M = 32$.

(c) Interval, $[-4, 4]$. Kernel, 0.75 sinc $(6x)$ sinc $(2x)$. Function, $\exp(-(x/2)^{20})$. Sampling increment, $1/(2L) = 0.125$. Order of partial sum, $M = 32$.

(d) Interval, $[-4, 4]$. Kernel, sinc (x). Function, $\exp(-(x/3)^{20})$. Sampling increment, $1/(2L) = 1.0$. Order of partial sum, $M = 4$.

(e) Interval, $[-4, 4]$. Kernel, 0.875 sinc $(7x)$ sinc (x). Function, $\exp(-(x/3)^{20})$. Sampling increment, $1/(2L) = 0.125$. Order of partial sum, $M = 32$.

Calculate the explicit sampling series partial sums for each of the four cases above, and find the relative 2-Norm error between these explicit sums and the *FAS*-computed sums (the relative 2-Norm error equals $\|S_M - S^F_M\|_2 / \|S_M\|_2$ where S^F_M is the *FAS*-computed sum and S_M is the explicitly computed sum). For which of these kernels does the fast method provide acceptable results (that is, a small relative 2-Norm error between the explicit partial sum and the *FAS*-computed partial sum)?

5.78 *Periodic reconstruction kernels.* (a) Over the interval $[-4, 4]$, using 1024 points, plot the following partial sum of a sampling series:

$$f(x) = \text{sumk}[\exp(-(a/2) \wedge 20) \text{ sinc } (4t)] \setminus k = -16,16 \setminus a = k/4 \setminus z = x - k/4$$

$$\setminus t = u - w/2 \setminus u = v - w \text{gri}(v/w) \setminus v = z + w/2 \setminus w = 8 \quad (5.209)$$

(*Note:* this formula can be loaded into *FAS* using the formula file *per_sser* included with the disk accompanying this book.) The kernel in the sampling series in (5.209) is the periodic extension of sinc $(4x)$, period 4 (where sinc $(4x)$ is graphed over $[-4, 4]$). (b) Find the Sup-Norm difference between the function graphed in (a) and the graph of the *FAS*-computed sampling series partial sum S^F_{16} obtained in part (a) of Exercise 5.77. (Your answer should be approximately 7.42×10^{-14}, which shows that the two sums are virtually identical.) (c) Compute similar partial sums using periodic kernels and compare them with the *FAS* computed sampling series partial sums obtained in parts (b) to (e) of Exercise 5.77. In every case, the two sums should be virtually identical.

5.79 Explain the results of Exercise 5.78.

Section 14

5.80 Using *FAS*, graph the Fourier sine and cosine transforms of the following functions (use the interval [0, 16] and 512 points).

    (a) rect $(x - 2)$

    (b) $1/[4 + (x - 6)^2]$

    (c) $e^{-0.4x} \cos 12\pi x$

    (d) $e^{-0.4x} \sin 12\pi x$

5.81 Explain why a Fourier sine transform and a Fourier cosine transform can be approximated by a fast sine transform and a fast cosine transform, respectively.

6

Fourier Optics

In this chapter, we will discuss some applications of Fourier analysis to optics. We will show how *FAS* can be used to analyze some important problems in optics, such as the behavior of diffraction gratings and imaging with lenses.

6.1 Introduction—Diffraction and Coherency of Light

In this section we will derive the fundamental formula of diffraction theory. Although we do not have space for a completely rigorous discussion (see [Go] or [Bo-W]), our treatment should capture the main ideas. If the presentation below is found too obscure, some readers might wish *to take Formula (6.22) for granted* (it is a classical formula in diffraction theory); however, in that case, do read Definition 6.1 where the notation used throughout this chapter is given.

We shall assume that unpolarized light of wavelength λ traveling from a point P to a point Q along a curve $\mathbf{C}$ can be described as follows. The light amplitude $\psi(Q, t)$ at the point Q is given by

$$\psi(Q, t) = \frac{\psi(P, t)}{\lambda |\mathbf{C}|} e^{i\frac{2\pi}{\lambda} \int_{\mathbf{C}} \eta(s)\, ds} \qquad (6.1)$$

where $\psi(P, t)$ is the light amplitude at P, $|\mathbf{C}|$ is the length of the curve $\mathbf{C}$, and $\int_{\mathbf{C}} \eta(s)\, ds$ is the line integral (using the arc-length differential ds) of the index of refraction $\eta(s)$ along the curve $\mathbf{C}$. Since s represents arc length and $\eta(s)$ is the index of refraction, $\int_{\mathbf{C}} \eta(s)\, ds$ is often called the *optical length* of $\mathbf{C}$ (as opposed to the length of $\mathbf{C}$ which equals $\int_{\mathbf{C}} 1\, ds$). We should note that we are omitting a *time delay* $t_{\mathbf{C}}$ in (6.1), where $t_{\mathbf{C}}$ is the time it takes light to travel from P to Q along the curve $\mathbf{C}$. In other words, to be absolutely precise, instead of $\psi(P, t)$ in Formula (6.1) we should have $\psi(P, t - t_{\mathbf{C}})$. We omit the time delay $t_{\mathbf{C}}$ for the sake of simplicity. At the distances we shall consider, the speed of light is so great that $t_{\mathbf{C}} \approx 0$.

Although Formula (6.1) may appear strange, it will allow us to quickly derive the fundamental formulas of optical diffraction and imaging. Formula (6.1) is related to Fermat's principle and to the Feynman path integral method in optics (which is a more rigorous development of the approach described here).

A couple of examples should help to clarify the meaning of Formula (6.1). First, though, we need to adopt some convenient notation. To enable us to write our formulas compactly, we shall adopt the following *vector* notation.

DEFINITION 6.1:   *A vector* **x** *will have two components* **x** $= (x, y)$ *and the point* (x, y, D) *will be denoted by* $(\mathbf{x}, D)$. *When we want to express integrals like*

$$\int_{-\infty}^{\infty} \int_{-\infty}^{\infty} f(x, y) \, dx \, dy \qquad \text{or} \qquad \int_{-\infty}^{\infty} \int_{-\infty}^{\infty} g(u, v) \, du \, dv$$

then we will write

$$\int_{\mathbf{R}^2} f(\mathbf{x}) \, d\mathbf{x} \qquad \text{or} \qquad \int_{\mathbf{R}^2} g(\mathbf{u}) \, d\mathbf{u}.$$

And, we will write $|\mathbf{u} - \mathbf{x}|$ *for the distance* $\left[(u - x)^2 + (v - y)^2\right]^{1/2}$ *between* **u** *and* **x**.

Example 6.1:
 Suppose that light of wavelength λ is shone onto an opaque screen with a single point $P = (\mathbf{x}, 0)$ cut out of it. See Figure 6.1. Assume that the index of refraction

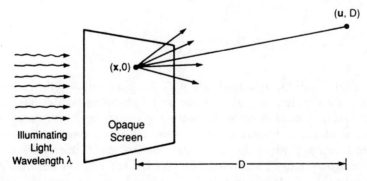

FIGURE 6.1
Diffraction from a single point.

η is a *constant*, say $\eta = 1$ for simplicity. Then the light at a point $Q = (\mathbf{u}, D)$, on a parallel plane D units in front of the screen, will be caused by the light traveling along the straight line segment $\mathbf{C}$ from $(\mathbf{x}, 0)$ to $(\mathbf{u}, D)$. (See, however, Remark 6.1 below.) Since $\int_{\mathbf{C}} 1 \, ds = |\mathbf{C}|$, we obtain from (6.1) that

$$\psi(\mathbf{u}, D, t) = \frac{\psi(\mathbf{x}, 0, t)}{\lambda |\mathbf{C}|} e^{i \frac{2\pi}{\lambda} |\mathbf{C}|}. \tag{6.2}$$

Now, we define the *intensity* $I(\mathbf{u}, D)$ of the light at $(\mathbf{u}, D)$ by

$$I(\mathbf{u}, D) = \lim_{T \to \infty} \frac{1}{T} \int_0^T |\psi(\mathbf{u}, D, t)|^2 dt. \tag{6.3}$$

Equation (6.3) is a mathematical idealization (since we cannot actually observe $T \to \infty$). Consequently, we shall use instead

$$I(\mathbf{u}, D) = \frac{1}{T} \int_0^T |\psi(\mathbf{u}, D, t)|^2 dt. \tag{6.4}$$

where T is taken so large that the limit in (6.3) can be assumed to be attained. For example, T might be the time involved in taking a photograph, which is huge relative to the fluctuations of visible light.

Using (6.2) in (6.4), we obtain (since $|e^{i\phi}| = 1$ for all real ϕ)

$$I(\mathbf{u}, D) = \frac{I(\mathbf{x}, 0)}{\lambda^2 |\mathbf{C}|^2}. \tag{6.5}$$

Formula (6.5) says that *the light intensity decreases in inverse proportion to the square of the distance, $|\mathbf{C}|$, from $(\mathbf{x}, 0)$ to $(\mathbf{u}, D)$.* This result is consistent with light being an electromagnetic phenomenon, and it shows the purpose of dividing by $|\mathbf{C}|$ in Formula (6.1). [The reason for also dividing by λ will only be clear after our work in the next section.]

REMARK 6.1: In the example above we made the assumption that only light traveling along the *straight line segment* $\mathbf{C}$ from $(\mathbf{x}, 0)$ to $(\mathbf{u}, D)$ contributes to the light $\psi(\mathbf{u}, D, t)$ at $(\mathbf{u}, D)$. This assumption is consistent with classical geometrical optics, since the light rays are straight in a medium with constant index of refraction. However, a more rigorous approach like Feynman's path integral method would assume that the photons of light can travel *all* possible paths $\mathbf{C}$ from $(\mathbf{x}, 0)$ to $(\mathbf{u}, D)$. Then, an integral of quantities similar to those in (6.2) would be formed over *all* paths $\mathbf{C}$ from $(\mathbf{x}, 0)$ to $(\mathbf{u}, D)$. We will not pursue this approach here; because *when a long time average is taken* (as we did above) the results of this

more rigorous method end up reducing to formulas like the ones described in this
section. For a fascinating nontechnical discussion of Feynman's method, see [Fe].
∎

This last example showed why the factor $1/|\mathbf{C}|$ is in Formula (6.1). Our next
example will show why the *phase factor,*

$$e^{i\frac{2\pi}{\lambda}\int_{\mathbf{C}}\eta(s)\,ds}$$

is in Formula (6.1).

Example 6.2:
Suppose that light of wavelength λ is shone onto an opaque screen with two points
cut out of it. Let's say these points are at $(\mathbf{x}, 0)$ and $(\mathbf{z}, 0)$. Assuming again that
we have a constant index of refraction $\eta = 1$, we now must sum the contributions
of the light from $(\mathbf{x}, 0)$ and $(\mathbf{z}, 0)$. If we let $\mathbf{C_X}$ denote the line segment from $(\mathbf{x}, 0)$
to $(\mathbf{u}, D)$, then (6.2) generalizes to

$$\psi(\mathbf{u}, D, t) = \frac{\psi(\mathbf{x}, 0, t)}{\lambda|\mathbf{C_X}|}e^{i\frac{2\pi}{\lambda}|\mathbf{C_X}|} + \frac{\psi(\mathbf{z}, 0, t)}{\lambda|\mathbf{C_Z}|}e^{i\frac{2\pi}{\lambda}|\mathbf{C_Z}|}. \qquad (6.6)$$

We will now calculate the intensity $I(\mathbf{u}, D)$ using Formula (6.4). First, we observe
that

$$|\psi(\mathbf{u}, D, t)|^2 = \frac{|\psi(\mathbf{x}, 0, t)|^2}{\lambda^2|\mathbf{C_X}|^2} + \frac{|\psi(\mathbf{z}, 0, t)|^2}{\lambda^2|\mathbf{C_Z}|^2}$$

$$+2\text{Re}\left[\frac{\psi(\mathbf{x}, 0, t)\psi^*(\mathbf{z}, 0, t)}{\lambda^2|\mathbf{C_X}|\cdot|\mathbf{C_Z}|}e^{i\frac{2\pi}{\lambda}(|\mathbf{C_X}|-|\mathbf{C_Z}|)}\right]. \qquad (6.7)$$

Since our next step will be to compute $1/T$ times the integral $\int_0^T dt$ of each side
of (6.7), the following definition will be useful. The *coherency function* $\Gamma(\mathbf{x}, \mathbf{z})$ is
defined by the identity

$$\Gamma(\mathbf{x}, \mathbf{z})\sqrt{I(\mathbf{x}, 0)I(\mathbf{z}, 0)} = \frac{1}{T}\int_0^T \psi(\mathbf{x}, 0, t)\psi^*(\mathbf{z}, 0, t)\,dt. \qquad (6.8)$$

(*Note:* in the special case that either $I(\mathbf{x}, 0) = 0$ or $I(\mathbf{z}, 0) = 0$, we define
$\Gamma(\mathbf{x}, \mathbf{z}) = 0$.) With this definition in mind, we now use (6.4) to calculate the

intensity $I(\mathbf{u}, D)$ from (6.7). We obtain

$$I(\mathbf{u}, D) = \frac{I(\mathbf{x}, 0)}{\lambda^2 |C_{\mathbf{x}}|^2} + \frac{I(\mathbf{z}, 0)}{\lambda^2 |C_{\mathbf{z}}|^2}$$

$$+ 2\mathrm{Re} \left[\frac{\Gamma(\mathbf{x}, \mathbf{z})\sqrt{I(\mathbf{x}, 0)I(\mathbf{z}, 0)}}{\lambda^2 |C_{\mathbf{x}}| \cdot |C_{\mathbf{z}}|} e^{i\frac{2\pi}{\lambda}(|C_{\mathbf{x}}| - |C_{\mathbf{z}}|)} \right]. \qquad (6.9)$$

Formula (6.9) is very complicated, reflecting the fact that we placed no constraints upon the nature of the light radiating onto the opaque screen (other than it is of one wavelength λ). There are two fundamental constraints that can be used to simplify (6.9). The first is called *spatial incoherency* (*incoherency,* for short). The light will be called *incoherent* if the coherency function Γ satisfies

$$\Gamma(\mathbf{x}, \mathbf{z}) = 0 \quad \text{for } \mathbf{x} \neq \mathbf{z} \qquad [\textit{Incoherency}]. \qquad (6.10)$$

In this case, (6.9) simplifies to

$$I(\mathbf{u}, D) = \frac{I(\mathbf{x}, 0)}{\lambda^2 |C_{\mathbf{x}}|^2} + \frac{I(\mathbf{z}, 0)}{\lambda^2 |C_{\mathbf{z}}|^2} \qquad (6.11)$$

Thus, *in the case of incoherent illumination, the intensity at* $(\mathbf{u}, D)$ *is the sum of the intensities that would result from the points* $(\mathbf{x}, 0)$ *and* $(\mathbf{z}, 0)$ *individually.*

The second form of constraint is called *spatial coherency* (*coherency,* for short). It is coherency that we will deal with most frequently in this chapter. In order to define coherency, we first note that $\Gamma(\mathbf{x}, \mathbf{z})$ satisfies the following inequality:

$$|\Gamma(\mathbf{x}, \mathbf{z})| \leq 1. \qquad (6.12)$$

This inequality is proved using the Schwarz inequality (see Exercise 6.7). Inequality (6.12) says that $\Gamma(\mathbf{x}, \mathbf{z})$ cannot have values of magnitude greater than 1. Coherency consists in $\Gamma(\mathbf{x}, \mathbf{z})$ attaining this maximum value of 1. That is, the light will be said to be *coherent* if $\Gamma(\mathbf{x}, \mathbf{z})$ satisfies

$$\Gamma(\mathbf{x}, \mathbf{z}) = 1 \qquad [\textit{Coherency}]. \qquad (6.13)$$

If the light is coherent, then using (6.13) in (6.9) yields

$$I(\mathbf{u}, D) = \frac{I(\mathbf{x}, 0)}{\lambda^2 |\mathbf{C_x}|^2} + \frac{I(\mathbf{z}, 0)}{\lambda^2 |\mathbf{C_z}|^2}$$

$$+ \frac{2\sqrt{I(\mathbf{x}, 0)I(\mathbf{z}, 0)}}{\lambda^2 |\mathbf{C_x}| \cdot |\mathbf{C_z}|} \cos\left[\frac{2\pi}{\lambda}(|\mathbf{C_x}| - |\mathbf{C_z}|)\right]. \quad (6.14)$$

Thus, in the case of coherent illumination, we see from this formula that the intensity $I(\mathbf{u}, D)$ contains an *oscillatory term:*

$$\frac{2\sqrt{I(\mathbf{x}, 0)I(\mathbf{z}, 0)}}{\lambda^2 |\mathbf{C_x}| \cdot |\mathbf{C_z}|} \cos\left[\frac{2\pi}{\lambda}(|\mathbf{C_x}| - |\mathbf{C_z}|)\right].$$

As the point $(\mathbf{u}, D)$ varies, this term will vary between

$$\frac{2\sqrt{I(\mathbf{x}, 0)I(\mathbf{z}, 0)}}{\lambda^2 |\mathbf{C_x}| \cdot |\mathbf{C_z}|}$$

and

$$\frac{-2\sqrt{I(\mathbf{x}, 0)I(\mathbf{z}, 0)}}{\lambda^2 |\mathbf{C_x}| \cdot |\mathbf{C_z}|}$$

thus creating *interference fringes* superimposed upon the background intensity described by

$$\frac{I(\mathbf{x}, 0)}{\lambda^2 |\mathbf{C_x}|^2} + \frac{I(\mathbf{z}, 0)}{\lambda^2 |\mathbf{C_z}|^2}.$$

A reader who desires to graph examples of Formulas (6.11) and (6.14) might begin by looking in the *Textbook formulas* entry in the on-line help for the function creation procedure of *FAS*.

Much experimental work has verified that *all* types of illumination will satisfy the coherency condition (at least to a good approximation) *provided* the points $(\mathbf{x}, 0)$ and $(\mathbf{z}, 0)$ are sufficiently close (how close is called the *coherency interval* of the light). Some light has such a small coherency interval that it can be assumed incoherent, while other kinds of light satisfy the coherency condition (6.13) over a wide range. For example, in Figure 6.2 we show a diagram illustrating how coherent light can be created. If light from a strong source is channeled through a point drilled in an opaque screen (a pinhole filter), and then collimated by a lens sitting at a focal distance from the point in the screen, then the light immediately behind the collimating lens is coherent. (This will be shown in Section 6.8; see Exercise 6.49). Often a laser is used as the initial source, since the light from the laser is nearly coherent to begin with.

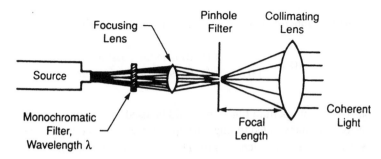

FIGURE 6.2
Creation of coherent, monochromatic light.

We shall now determine the intensity $I(\mathbf{u}, D)$ generated from an *aperture* in an opaque screen that is illuminated by coherent light. See Figure 6.3. In this case, we

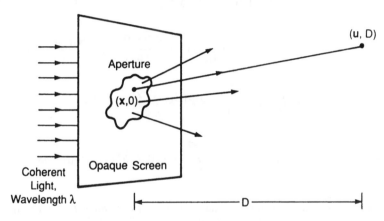

FIGURE 6.3
Diffraction from an aperture.

generalize (6.2) and (6.6), by forming an *integral* that consists of *superpositioning all contributions* along light rays $\mathbf{C_x}$ from $(\mathbf{x}, 0)$ to $(\mathbf{u}, D)$. Thus, we have

$$\psi(\mathbf{u}, D, t) = \int_{\mathbf{R}^2} A(\mathbf{x}) \frac{\psi(\mathbf{x}, 0, t)}{\lambda |\mathbf{C_x}|} e^{i\frac{2\pi}{\lambda}|\mathbf{C_x}|} \, d\mathbf{x}. \tag{6.15}$$

In (6.15) we have introduced the *aperture function* $A(\mathbf{x})$ to account for the finite extent of the aperture. In particular, we assume that

$$A(\mathbf{x}) = 0, \qquad \text{for } |\mathbf{x}| > R \tag{6.16}$$

where R is a finite positive number which marks off the extent of the aperture. The type of aperture function that we will typically look at has the form

$$A(\mathbf{x}) = \begin{cases} 1 & \text{if } (\mathbf{x}, 0) \text{ is in the aperture} \\ 0 & \text{if } (\mathbf{x}, 0) \text{ is not in the aperture.} \end{cases} \qquad (6.17)$$

But, other types of aperture functions could be used to take into account varying amounts of light transmitted through the aperture as well as other effects.

Now, we calculate the intensity $I(\mathbf{u}, D)$ using (6.4) and (6.15):

$$
\begin{aligned}
I(\mathbf{u}, D) &= \frac{1}{T} \int_0^T \psi(\mathbf{u}, D, t) \psi^*(\mathbf{u}, D, t) \, dt \\
&= \frac{1}{T} \int_0^T \left[\int_{\mathbf{R}^2} A(\mathbf{x}) \frac{\psi(\mathbf{x}, 0, t)}{\lambda |\mathbf{C}_{\mathbf{x}}|} e^{i \frac{2\pi}{\lambda} |\mathbf{C}_{\mathbf{x}}|} \, d\mathbf{x} \right] \\
&\qquad\qquad \left[\int_{\mathbf{R}^2} A^*(\mathbf{z}) \frac{\psi^*(\mathbf{z}, 0, t)}{\lambda |\mathbf{C}_{\mathbf{z}}|} e^{-i \frac{2\pi}{\lambda} |\mathbf{C}_{\mathbf{z}}|} \, d\mathbf{z} \right] dt \\
&= \int_{\mathbf{R}^2} \int_{\mathbf{R}^2} \frac{A(\mathbf{x})}{\lambda |\mathbf{C}_{\mathbf{x}}|} e^{i \frac{2\pi}{\lambda} |\mathbf{C}_{\mathbf{x}}|} \frac{A^*(\mathbf{z})}{\lambda |\mathbf{C}_{\mathbf{z}}|} e^{-i \frac{2\pi}{\lambda} |\mathbf{C}_{\mathbf{z}}|} \\
&\qquad\qquad \left[\frac{1}{T} \int_0^T \psi(\mathbf{x}, 0, t) \psi^*(\mathbf{z}, 0, t) \, dt \right] d\mathbf{x} \, d\mathbf{z}. \quad (6.18)
\end{aligned}
$$

Using the coherency assumption, that $\Gamma(\mathbf{x}, \mathbf{z}) = 1$ *over the extent of the aperture*, the last expression in brackets in (6.18) equals $\sqrt{I(\mathbf{x}, 0) I(\mathbf{z}, 0)}$. Hence,

$$I(\mathbf{u}, D) = \int_{\mathbf{R}^2} \int_{\mathbf{R}^2} \frac{A(\mathbf{x})}{\lambda |\mathbf{C}_{\mathbf{x}}|} e^{i \frac{2\pi}{\lambda} |\mathbf{C}_{\mathbf{x}}|} \frac{A^*(\mathbf{z})}{\lambda |\mathbf{C}_{\mathbf{z}}|} e^{-i \frac{2\pi}{\lambda} |\mathbf{C}_{\mathbf{z}}|} \sqrt{I(\mathbf{x}, 0) I(\mathbf{z}, 0)} \, d\mathbf{x} \, d\mathbf{z}.$$

The integrals then *separate out* and we obtain

$$I(\mathbf{u}, D) = \left[\int_{\mathbf{R}^2} \frac{A(\mathbf{x}) \sqrt{I(\mathbf{x}, 0)}}{\lambda |\mathbf{C}_{\mathbf{x}}|} e^{i \frac{2\pi}{\lambda} |\mathbf{C}_{\mathbf{x}}|} \, d\mathbf{x} \right] \left[\int_{\mathbf{R}^2} \frac{A(\mathbf{z}) \sqrt{I(\mathbf{z}, 0)}}{\lambda |\mathbf{C}_{\mathbf{z}}|} e^{i \frac{2\pi}{\lambda} |\mathbf{C}_{\mathbf{z}}|} \, d\mathbf{z} \right]^*.$$

Substituting $\mathbf{x}$ in place of $\mathbf{z}$ in the second integral above, we get

$$I(\mathbf{u}, D) = \left| \int_{\mathbf{R}^2} \frac{A(\mathbf{x}) \sqrt{I(\mathbf{x}, 0)}}{\lambda |\mathbf{C}_{\mathbf{x}}|} e^{i \frac{2\pi}{\lambda} |\mathbf{C}_{\mathbf{x}}|} \, d\mathbf{x} \right|^2. \qquad (6.19)$$

If we define the *transmittance function* $T(\mathbf{x})$ through the aperture by

$$T(\mathbf{x}) = A(\mathbf{x})\sqrt{I(\mathbf{x}, 0)}$$

then (6.19) can be rewritten as

$$I(\mathbf{u}, D) = \left| \int_{\mathbf{R}^2} \frac{T(\mathbf{x})}{\lambda |\mathbf{Cx}|} e^{i \frac{2\pi}{\lambda} |\mathbf{Cx}|} d\mathbf{x} \right|^2. \tag{6.20}$$

For simplicity, we shall usually assume that the light intensity $I(\mathbf{x}, 0)$ *is a constant over the aperture.* In fact, to simplify as much as possible, we shall assume that

$$I(\mathbf{x}, 0) = 1 \qquad \text{for } \mathbf{x} \text{ in the aperture.} \tag{6.21}$$

Using (6.21), we find that $T(\mathbf{x})$ equals $A(\mathbf{x})$, hence Formula (6.20) becomes

$$I(\mathbf{u}, D) = \left| \int_{\mathbf{R}^2} \frac{A(\mathbf{x})}{\lambda |\mathbf{Cx}|} e^{i \frac{2\pi}{\lambda} |\mathbf{Cx}|} d\mathbf{x} \right|^2. \tag{6.22}$$

Formula (6.22) is a classic formula in diffraction theory. For most of the rest of this chapter we will examine (6.22) for various types of apertures.

REMARK 6.2: A rigorous treatment of our subject (see [Go, Chapter 3] or [Bo-W]) would include an *obliquity factor* in the integrand of (6.22). However, for the type of illumination illustrated in Figure 6.2, this obliquity factor is approximately 1. This approximation by 1 is a standard approximation in diffraction theory. ∎

We close this section by introducing a convenient notation for the integral on the right side of Formula (6.8), since it will appear frequently in subsequent formulas.

DEFINITION 6.2: *The **cross-correlation function** $\gamma(\mathbf{x}, \mathbf{z})$ is defined by*

$$\gamma(\mathbf{x}, \mathbf{z}) = \lim_{T \to \infty} \frac{1}{T} \int_0^T \psi(\mathbf{x}, 0, t) \psi^*(\mathbf{z}, 0, t) \, dt. \tag{6.23a}$$

For calculational purposes, however, we shall use

$$\gamma(\mathbf{x}, \mathbf{z}) = \frac{1}{T} \int_0^T \psi(\mathbf{x}, 0, t) \psi^*(\mathbf{z}, 0, t) \, dt \tag{6.23b}$$

where T is the same large time value used to calculate the intensity in Formula (6.4).

Based on Definition 6.2 we can simplify the definition of the coherency function Γ to the following:

$$\Gamma(\mathbf{x}, \mathbf{z}) = \frac{\gamma(\mathbf{x}, \mathbf{z})}{\sqrt{\gamma(\mathbf{x}, \mathbf{x})\gamma(\mathbf{z}, \mathbf{z})}}. \tag{6.24}$$

6.2 Fresnel Diffraction

In this section we will discuss *Fresnel diffraction,* which is one of the fundamental types of diffraction in optics. Fresnel diffraction is based on an integral approximating the one in Formula (6.22) for moderate distances D from the aperture.

Formula (6.22) says that the intensity $I(\mathbf{u}, D)$ is given by

$$I(\mathbf{u}, D) = \left| \int_{\mathbf{R}^2} \frac{A(\mathbf{x})}{\lambda |\mathbf{C_x}|} e^{i\frac{2\pi}{\lambda} |\mathbf{C_x}|} d\mathbf{x} \right|^2. \tag{6.25}$$

Since $\mathbf{C_x}$ is the line segment from $(\mathbf{x}, 0)$ to $(\mathbf{u}, D)$ we have

$$|\mathbf{C_x}| = \left[|\mathbf{u} - \mathbf{x}|^2 + D^2 \right]^{\frac{1}{2}}. \tag{6.26}$$

We will now show that $\lambda |\mathbf{C_x}|$ in the denominator inside the integral in (6.25) can be replaced by λD. To show this we look at the ratio

$$\frac{\lambda |\mathbf{C_x}|}{\lambda D} = \left[1 + |\mathbf{u} - \mathbf{x}|^2 / D^2 \right]^{\frac{1}{2}} \tag{6.27}$$

and *we assume that $|\mathbf{u} - \mathbf{x}|/D$ is small enough that*

$$\left[1 + |\mathbf{u} - \mathbf{x}|^2 / D^2 \right]^{\frac{1}{2}} \approx 1. \tag{6.28}$$

In fact, the approximation in (6.28) will hold to within 95% accuracy provided

$$|\mathbf{u} - \mathbf{x}|/D \le 0.32. \tag{6.29}$$

For a small aperture (relative to D), this inequality is not too restrictive on what $\mathbf{u}$-values we allow in (6.25).

Because of (6.28) and (6.27) we can say that $\lambda|\mathbf{C_x}| \approx \lambda D$ and then (6.25) simplifies to

$$I(\mathbf{u}, D) \approx \left| \frac{1}{\lambda D} \int_{\mathbf{R}^2} A(\mathbf{x}) e^{i\frac{2\pi}{\lambda}|\mathbf{C_x}|} d\mathbf{x} \right|^2. \tag{6.30}$$

We will now show how to approximate the phase factor $e^{i2\pi|\mathbf{C_x}|/\lambda}$ in the integrand in (6.30). Here we have to approach the approximation differently than above, since for most light (e.g., visible light) the reciprocal of the wavelength, $1/\lambda$, is *enormous*. Consequently, $2\pi D/\lambda$ is a terrible approximation of $2\pi|\mathbf{C_x}|/\lambda$. To obtain a useful approximation we multiply (6.27) by $2\pi D/\lambda$, obtaining

$$\frac{2\pi}{\lambda}|\mathbf{C_x}| = \frac{2\pi D}{\lambda} \left[1 + |\mathbf{u} - \mathbf{x}|^2/D^2 \right]^{\frac{1}{2}}. \tag{6.31}$$

Using the series expansion from calculus

$$(1+a)^{\frac{1}{2}} = 1 + \frac{1}{2}a - \frac{1}{8}a^2 + \dots \tag{6.32}$$

to rewrite the right side of (6.31), we have

$$\frac{2\pi}{\lambda}|\mathbf{C_x}| = \frac{2\pi D}{\lambda} + \frac{\pi}{\lambda D}|\mathbf{u} - \mathbf{x}|^2 - \frac{\pi}{4\lambda D^3}|\mathbf{u} - \mathbf{x}|^4 + \dots. \tag{6.33}$$

Now, let's assume that $|\mathbf{u} - \mathbf{x}|/D$ is small enough that $\pi|\mathbf{u} - \mathbf{x}|^4/4\lambda D^3$, and higher power terms, can be neglected in (6.33). Then (6.33) simplifies to

$$\frac{2\pi}{\lambda}|\mathbf{C_x}| \approx \frac{2\pi D}{\lambda} + \frac{\pi}{\lambda D}|\mathbf{u} - \mathbf{x}|^2. \tag{6.34}$$

From (6.34) we obtain

$$e^{i\frac{2\pi}{\lambda}|\mathbf{C_x}|} \approx e^{i\frac{2\pi}{\lambda}D} e^{\frac{i\pi}{\lambda D}|\mathbf{u} - \mathbf{x}|^2}. \tag{6.35}$$

Substituting the right side of (6.35) into (6.30), and factoring the constant outside the integral, we have

$$I(\mathbf{u}, D) \approx \left| e^{i\frac{2\pi}{\lambda}D} \frac{1}{\lambda D} \int_{\mathbf{R}^2} A(\mathbf{x}) e^{\frac{i\pi}{\lambda D}|\mathbf{u} - \mathbf{x}|^2} d\mathbf{x} \right|^2. \tag{6.36}$$

We have now established two things. First, since $|e^{i\frac{2\pi}{\lambda}D}|^2 = 1$ we have

$$I(\mathbf{u}, D) \approx \left| \frac{1}{\lambda D} \int_{\mathbf{R}^2} A(\mathbf{x}) e^{\frac{i\pi}{\lambda D}|\mathbf{u}-\mathbf{x}|^2} \, d\mathbf{x} \right|^2. \tag{6.37}$$

Formula (6.37) is the *Fresnel diffraction formula.*

Second, all of our approximations remain valid for the integral describing $\psi(\mathbf{u}, D, t)$ in (6.15). Thus, from (6.15) we obtain

$$\psi(\mathbf{u}, D, t) \approx \frac{e^{i\frac{2\pi}{\lambda}D}}{\lambda D} \int_{\mathbf{R}^2} A(\mathbf{x}) \psi(\mathbf{x}, 0, t) e^{\frac{i\pi}{\lambda D}|\mathbf{u}-\mathbf{x}|^2} \, d\mathbf{x}. \tag{6.38}$$

Formula (6.38) will prove useful later, when we discuss imaging by a lens.

Here is an example of how *FAS* can be used to analyze the Fresnel diffraction formula (6.37).

Example 6.3:
Suppose that coherent light of wavelength $\lambda = 5 \times 10^{-4}$ mm (greenish-blue) light is shined onto a square aperture with aperture function $A(\mathbf{x}) = A(x, y)$ given by

$$A(x, y) = \begin{cases} 1 & \text{if } |x| < 1 \text{ mm and } |y| < 1 \text{ mm} \\ 0 & \text{if } |x| > 1 \text{ mm or } |y| > 1 \text{ mm}. \end{cases}$$

Graph $I(\mathbf{u}, D)$ for $D = 100$ mm, 200 mm, and 300 mm.

SOLUTION   It is not too difficult to see that

$$A(x, y) = \text{rect}\left(\frac{x}{2}\right) \text{rect}\left(\frac{y}{2}\right). \tag{6.39}$$

For $\mathbf{u} = (u, v)$ we have

$$e^{\frac{i\pi}{\lambda D}|\mathbf{u}-\mathbf{x}|^2} = e^{\frac{i\pi}{\lambda D}\left[(u-x)^2 + (v-y)^2\right]}$$

$$= e^{\frac{i\pi}{\lambda D}(u-x)^2} e^{\frac{i\pi}{\lambda D}(v-y)^2}. \tag{6.40}$$

Thus, using (6.39) and (6.40) in the integral in (6.37), we see that the integral over $\mathbf{R}^2$ separates into iterated integrals from $-\infty$ to ∞ in each variable x and y.

Therefore, we obtain

$$I(\mathbf{u}, D) = \left| \frac{1}{(\lambda D)^{\frac{1}{2}}} \int_{-\infty}^{\infty} \text{rect}\left(\frac{x}{2}\right) e^{\frac{i\pi}{\lambda D}(u-x)^2} dx \right|^2 .$$

$$\times \left| \frac{1}{(\lambda D)^{\frac{1}{2}}} \int_{-\infty}^{\infty} \text{rect}\left(\frac{y}{2}\right) e^{\frac{i\pi}{\lambda D}(v-y)^2} dy \right|^2 . \qquad (6.41)$$

Because of (6.41) we see that it suffices to graph $I_1(u, D)$, defined by

$$I_1(u, D) = \left| \frac{1}{(\lambda D)^{\frac{1}{2}}} \int_{-\infty}^{\infty} \text{rect}\left(\frac{x}{2}\right) e^{\frac{i\pi}{\lambda D}(u-x)^2} dx \right|^2 \qquad (6.42)$$

since the second factor in (6.41) is just $I_1(v, D)$, which is obtained from $I_1(u, D)$ by changing variables from x to y and u to v. In particular, we have

$$I(\mathbf{u}, D) = I_1(u, D)I_1(v, D). \qquad (6.43)$$

Now, the integral in (6.42)

$$\frac{1}{(\lambda D)^{\frac{1}{2}}} \int_{-\infty}^{\infty} \text{rect}\left(\frac{x}{2}\right) e^{\frac{i\pi}{\lambda D}(u-x)^2} dx \qquad (6.44)$$

can be expressed in real and imaginary parts as

$$\int_{-\infty}^{\infty} \text{rect}\left(\frac{x}{2}\right) \left[\frac{1}{(\lambda D)^{\frac{1}{2}}} \cos \frac{\pi}{\lambda D}(u-x)^2 \right] dx$$

$$+i \int_{-\infty}^{\infty} \text{rect}\left(\frac{x}{2}\right) \left[\frac{1}{(\lambda D)^{\frac{1}{2}}} \sin \frac{\pi}{\lambda D}(u-x)^2 \right] dx. \qquad (6.45)$$

Therefore, we can graph the real and imaginary parts of the integral in (6.44) by doing two convolutions. First, we convolve $\text{rect}(x/2)$ with $(\lambda D)^{-1/2} \cos[(\pi/\lambda D)x^2]$ to produce the real part. Second, we convolve $\text{rect}(x/2)$ with $(\lambda D)^{-1/2}$ $\sin[(\pi/\lambda D)x^2]$ to produce the imaginary part. If $g1$ stands for the real part and $g2$ stands for the imaginary part, then $g1(x) \wedge 2 + g2(x) \wedge 2$ will create the intensity $I_1(u, D)$ in Formula (6.42). We show graphs of $I_1(u, D)$ for $\lambda = 5 \times 10^{-4}$ and

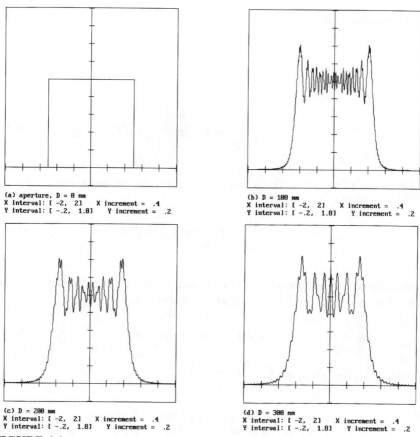

(a) aperture, D = 0 mm
X interval: [-2, 2]   X increment = .4
Y interval: [-.2, 1.8]   Y increment = .2

(b) D = 100 mm
X interval: [-2, 2]   X increment = .4
Y interval: [-.2, 1.8]   Y increment = .2

(c) D = 200 mm
X interval: [-2, 2]   X increment = .4
Y interval: [-.2, 1.8]   Y increment = .2

(d) D = 300 mm
X interval: [-2, 2]   X increment = .4
Y interval: [-.2, 1.8]   Y increment = .2

FIGURE 6.4
Graphs of the intensity function $I_1(u, D)$ from Example 6.3.

$D = 100, 200$, and 300 in Figure 6.4. These graphs were produced over an initial interval of $[-4, 4]$ using 4096 points.

By changing variables in (6.42) from x to y and from u to v, we obtain $I_1(v, D)$, the other factor in Formula (6.43) for $I(\mathbf{u}, D)$. The intensity function $I(\mathbf{u}, D)$ then generates a diffraction pattern like the one shown in Figure 6.5. We see in Figure 6.5 the classic checkerboard (or plaid) pattern of a Fresnel diffraction pattern from a square aperture. ∎

REMARK 6.3:   (a) It is clear from Figure 6.5 that the Fresnel diffraction pattern represents a *distorted image* of the original aperture. (b) It is also worth noting the similarity between Fresnel diffraction and the electron diffraction that

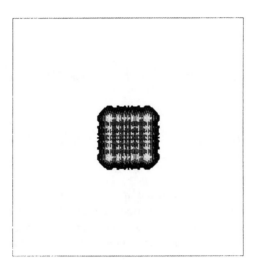

FIGURE 6.5
Fresnel diffraction from a square aperture; computer-generated graph.

we described in Section 5.7 of Chapter 5. In fact, if we put

$$\lambda D = 4\pi t (\hbar/2m) \tag{6.46}$$

then the integral in (6.44) differs from $(f * {}_t F)(u)$ in Formula (5.87) of Chapter 5, for $f(x) = \text{rect}(x/2)$, by only a factor of $i^{1/2}$. Since $|i^{1/2}|^2 = 1$, we can ignore this difference, *which does not occur anyway when we graph the expression in* (6.42). Equation (6.46) represents more than just a mathematical relationship. Because electrons (like all particles in quantum mechanics) have wavelengths, these particles diffract through apertures in much the same way as light does. In fact, everything we discuss in this chapter on coherency, diffraction, and imaging applies to *electron optics,* as in *electron microscopy,* and *electron diffraction.* Furthermore, *neutron diffraction spectroscopy* uses similar ideas. ∎

As we can see from the example above, *FAS* can be used to graph the individual factors of the intensity function $I(\mathbf{u}, D)$ whenever the aperture function factors in the following way:

$$A(\mathbf{x}) = A_1(x)A_2(y). \tag{6.47}$$

When this happens, we have

$$I(u, v, D) = I_1(u, D)I_2(v, D) \tag{6.48}$$

where

$$I_1(u, D) = \left| \frac{1}{(\lambda D)^{\frac{1}{2}}} \int_{-\infty}^{\infty} A_1(x) e^{\frac{i\pi}{\lambda D}(u-x)^2} \, dx \right|^2 \qquad (6.49)$$

$$I_2(v, D) = \left| \frac{1}{(\lambda D)^{\frac{1}{2}}} \int_{-\infty}^{\infty} A_2(y) e^{\frac{i\pi}{\lambda D}(v-y)^2} \, dy \right|^2. \qquad (6.50)$$

In such a situation, we shall call the aperture function A *separable*. There are a number of separable aperture functions that are important cases for study, for example, aperture functions for rectangular apertures or aperture functions that are Gaussian exponentials. These examples, and the case of *edge diffraction,* are treated in the exercises.

6.3 Fraunhofer Diffraction

In this section we begin our discussion of *Fraunhofer diffraction,* which plays a very important role in applications of diffraction theory. This type of diffraction occurs as a limiting case of Fresnel diffraction, either when the distance from the aperture is very large or when the dimensions of the aperture are very small.

Fraunhofer diffraction theory is essential for understanding the diffraction of X-rays from crystals. X-ray diffraction photographs from crystals are used to determine the underlying crystal structure. For example, the structure of DNA as well as many protein structures have been determined from diffraction patterns. Another important application is Fraunhofer diffraction from diffraction gratings. This diffraction plays a role in physical chemistry, where it is used to identify chemical compounds. It is also used in astronomy, where it provides a way of identifying the chemical constituents of stars.

We will now show how Fraunhofer diffraction arises as a limiting case of Fresnel diffraction. From Formula (6.37) we have for the diffracted light intensity

$$I(\mathbf{u}, D) \approx \left| \frac{1}{\lambda D} \int_{\mathbf{R}^2} A(\mathbf{x}) e^{\frac{i\pi}{\lambda D}|\mathbf{u}-\mathbf{x}|^2} \, d\mathbf{x} \right|^2. \qquad (6.51)$$

Expanding the exponential in (6.51), we get

$$e^{\frac{i\pi}{\lambda D}|\mathbf{u}-\mathbf{x}|^2} = e^{\frac{i\pi}{\lambda D}|\mathbf{u}|^2} e^{\frac{-i2\pi}{\lambda D}\mathbf{u}\cdot\mathbf{x}} e^{\frac{i\pi}{\lambda D}|\mathbf{x}|^2}. \qquad (6.52)$$

Recall the assumption in (6.16) that

$$A(\mathbf{x}) = 0, \qquad \text{for } |\mathbf{x}| > R. \tag{6.53}$$

We will show that

$$e^{\frac{i\pi}{\lambda D}|\mathbf{x}|^2} \approx 1, \qquad \text{for } |\mathbf{x}| \le R \tag{6.54}$$

in either of the following two cases:

$$D \to \infty \qquad \text{(far-field case)} \tag{6.54a}$$

$$R \to 0 \qquad \text{(small aperture case).} \tag{6.54b}$$

For the case of (6.54a), we observe that, for $|x| \le R$, as $D \to \infty$

$$0 \le \frac{\pi}{\lambda D}|\mathbf{x}|^2 \le \frac{\pi R^2}{\lambda D} \approx 0. \tag{6.55}$$

Hence,

$$e^{\frac{i\pi}{\lambda D}|\mathbf{x}|^2} \approx e^{i0} = 1$$

so (6.54) holds. On the other hand, for the case of (6.54b) we still have (6.55) *provided R is sufficiently small (and D stays fixed).* Thus, (6.54) holds again.

Now, assuming that (6.54) holds, we obtain from (6.52) that

$$e^{\frac{i\pi}{\lambda D}|\mathbf{u}-\mathbf{x}|^2} \approx e^{\frac{i\pi}{\lambda D}|\mathbf{u}|^2} e^{-i\frac{2\pi}{\lambda D}\mathbf{u}\cdot\mathbf{x}}, \qquad \text{for } |\mathbf{x}| \le R. \tag{6.56}$$

Since we have (6.53), the integral in (6.51) is actually taken over only those $\mathbf{x}$ for which $|\mathbf{x}| \le R$. So we can use (6.56) to simplify the integral in (6.51). Substituting the right side of (6.56) into the integrand in (6.51), and factoring the exponential involving $|\mathbf{u}|^2$ outside the integral, we obtain

$$I(\mathbf{u}, D) \approx \left| \frac{e^{\frac{i\pi}{\lambda D}|\mathbf{u}|^2}}{\lambda D} \int_{\mathbf{R}^2} A(\mathbf{x}) e^{-i\frac{2\pi}{\lambda D}\mathbf{u}\cdot\mathbf{x}} \, d\mathbf{x} \right|^2. \tag{6.57}$$

Since the complex exponential involving $|\mathbf{u}|^2$ has magnitude 1, Formula (6.57) becomes

$$I(\mathbf{u}, D) \approx \left| \frac{1}{\lambda D} \int_{\mathbf{R}^2} A(\mathbf{x}) e^{-i\frac{2\pi}{\lambda D}\mathbf{u}\cdot\mathbf{x}} \, d\mathbf{x} \right|^2. \tag{6.58}$$

Formula (6.58) is the *Fraunhofer diffraction formula.* The integral in (6.58) can be viewed as a two-dimensional Fourier transform. The two-dimensional Fourier transform of $A(\mathbf{x})$ is denoted by $\hat{A}(\mathbf{u})$, where $\hat{A}(\mathbf{u})$ is defined by

$$\hat{A}(\mathbf{u}) = \int_{\mathbf{R}^2} A(\mathbf{x}) e^{-i2\pi \mathbf{u} \cdot \mathbf{x}} \, d\mathbf{x}. \tag{6.59}$$

Using Formula (6.59), we can rewrite Formula (6.58) as

$$I(\mathbf{u}, D) \approx \left| \frac{1}{\lambda D} \hat{A} \left(\frac{1}{\lambda D} \mathbf{u} \right) \right|^2 \tag{6.60}$$

which shows that *the intensity function from Fraunhofer diffraction is the modulus-squared of the (scaled) Fourier transform of the aperture function.*

Example 6.4:
Discuss Fraunhofer diffraction from a rectangular aperture, having aperture function for $\mathbf{x} = (x, y)$

$$A(\mathbf{x}) = \begin{cases} 1 & \text{if } |x| < 0.1 \text{ mm and } |y| < .05 \text{ mm} \\ 0 & \text{if } |x| > 0.1 \text{ mm or } |y| > .05 \text{ mm.} \end{cases}$$

SOLUTION   It is not too hard to see that $A(x, y) = \text{rect}(5x) \, \text{rect}(10y)$. Substituting this expression for $A(x, y)$ into (6.59) and separating into two integrals of x and y, we obtain [for $\mathbf{u} = (u, v)$ and $\mathbf{u} \cdot \mathbf{x} = ux + vy$]

$$\hat{A}(u, v) = \int_{\mathbf{R}^2} \text{rect}(5x) \, \text{rect}(10y) e^{-i2\pi(ux+vy)} \, dx \, dy$$

$$= \int_{-\infty}^{\infty} \text{rect}(5x) e^{-i2\pi ux} \, dx \int_{-\infty}^{\infty} \text{rect}(10y) e^{-i2\pi vy} \, dy$$

$$= \frac{1}{5} \text{sinc}\left(\frac{u}{5}\right) \frac{1}{10} \text{sinc}\left(\frac{v}{10}\right).$$

Thus, from (6.60) we obtain

$$I(u, v, D) \approx \frac{1}{(50\lambda D)^2} \text{sinc}^2\left(\frac{u}{5\lambda D}\right) \text{sinc}^2\left(\frac{v}{10\lambda D}\right). \tag{6.61}$$

In Figure 6.6(a) we show a computer generated graph of $I(u, v, D)$; notice how well it corresponds to the photograph of such a diffraction pattern in Figure 6.6(b).

From (6.61) it is easy to calculate where there is *zero intensity* in the diffraction pattern. There will be zero intensity whenever $\text{sinc}^2[u/(5\lambda D)] = 0$ or $\text{sinc}^2[v/(10\lambda D)] = 0$. Hence the equations for *zero intensity* are

$$u = \pm 5\lambda D, \; \pm 10\lambda D, \ldots \qquad \text{and} \qquad v = \pm 10\lambda D, \; \pm 20\lambda D, \ldots.$$

Notice that these equations define *lines* of zero intensity in the u-v plane, and they are located at positions that are proportional to λD. ∎

Example 6.5:

Describe the Fraunhofer diffraction pattern from a *vertical slit*.

SOLUTION A vertical slit is a very thin rectangular aperture. Hence, its aperture function is

$$A(x, y) = \text{rect}\left(\frac{x}{a}\right) \text{rect}\left(\frac{y}{b}\right) \tag{6.62}$$

where the height b is many times larger than the width a. The transform of A is $\hat{A}(u, v) = ab \, \text{sinc}(au) \, \text{sinc}(bv)$. Therefore, the intensity function for the diffraction pattern is

$$I(u, v, D) = \left(\frac{ab}{\lambda D}\right)^2 \text{sinc}^2\left(\frac{au}{\lambda D}\right) \text{sinc}^2\left(\frac{bv}{\lambda D}\right). \tag{6.63}$$

A computer-generated graph of such a function is shown in Figure 6.7. The diffraction pattern has zero intensity at the lines

$$u = \pm\frac{\lambda D}{a}, \; \pm\frac{2\lambda D}{a}, \; \pm\frac{3\lambda D}{a}, \ldots \qquad \text{and} \qquad v = \pm\frac{\lambda D}{b}, \; \pm\frac{2\lambda D}{b}, \; \pm\frac{3\lambda D}{b}, \ldots. \tag{6.64}$$

Since b is much larger than a, the pattern is visible mainly along the u-axis. There is a strip of high intensity along the u-axis (of width $2\lambda D/b$), marked off by spots of low intensity around the points $u = \pm\lambda D/a, \; \pm 2\lambda D/a, \; \pm 3\lambda D/a, \ldots$ ∎

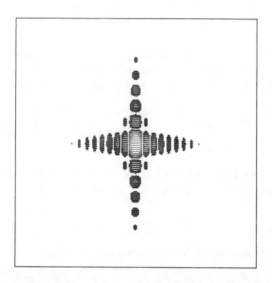

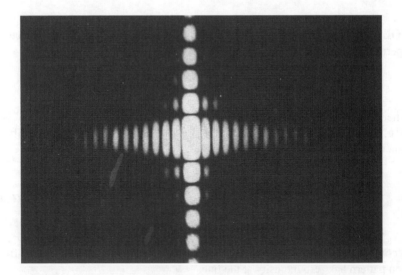

FIGURE 6.6
Fraunhofer diffraction by a rectangular aperture. (a) Computer-generated
graph of the intensity $I(u, v, D)$ (using a logarithmic intensity scale and setting $\lambda D = 1$). (b) Photograph of a Fraunhofer diffraction pattern from a
rectangular aperture. (From Walker, J.S., *Fourier Analysis*, Oxford University Press, Oxford, 1988. With permission.)

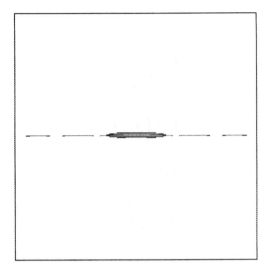

FIGURE 6.7
Diffraction pattern from a vertical slit (negative image).

6.4 Circular Apertures

In this section we will describe Fraunhofer diffraction from circular apertures. The diffraction integral for a circular aperture is usually described in terms of Bessel functions (see [Wa, Chapter 7.6]). The advantage of the method described below is that no special knowledge of Bessel functions is needed.

For a circular aperture, the aperture function is

$$A(\mathbf{x}) = \begin{cases} 1 & \text{if } |\mathbf{x}| < R \\ 0 & \text{if } |\mathbf{x}| > R \end{cases} \tag{6.65}$$

where R is the radius of the aperture. If we use (6.65) in Formula (6.59) we have

$$\hat{A}(\mathbf{u}) = \int_{\{\mathbf{x}:|\mathbf{x}|<R\}} e^{-i2\pi \mathbf{u} \cdot \mathbf{x}} \, d\mathbf{x}. \tag{6.66}$$

Now, if we change to *polar coordinates*

$$\mathbf{x} = (r\cos\theta, r\sin\theta), \qquad \mathbf{u} = (\rho\cos\phi, \rho\sin\phi) \tag{6.67}$$

then (6.66) becomes

$$\hat{A}(\rho \cos \phi, \rho \sin \phi) = \int_0^{2\pi} \int_0^R e^{-i2\pi r\rho(\cos\theta \cos\phi + \sin\theta \sin\phi)} \, r \, dr \, d\theta$$

$$= \int_0^R r \left[\int_0^{2\pi} e^{-i2\pi r\rho \cos(\theta - \phi)} \, d\theta \right] dr. \tag{6.68}$$

Substituting θ for $\theta - \phi$, we obtain

$$\hat{A}(\rho \cos \phi, \rho \sin \phi) = \int_0^R r \left[\int_{-\phi}^{2\pi - \phi} e^{-i2\pi r\rho \cos\theta} \, d\theta \right] dr. \tag{6.69}$$

Because of the periodicity of $\cos\theta$, the inner integrand in (6.69) has period 2π. Therefore, it can be integrated over the interval $[-\pi, \pi]$ instead of over the interval $[-\phi, 2\pi - \phi]$. Thus,

$$\hat{A}(\rho \cos \phi, \rho \sin \phi) = \int_0^R r \left[\int_{-\pi}^{\pi} e^{-i2\pi r\rho \cos\theta} \, d\theta \right] dr. \tag{6.70}$$

It is important to note that the right side of (6.70) has no dependence on ϕ. It is just a function of ρ (the variables r and θ being integrated out). For simplicity, then, we shall just write $\hat{A}(\rho)$ instead of $\hat{A}(\rho \cos \phi, \rho \sin \phi)$. And so, Formula (6.70) can be rewritten as

$$\hat{A}(\rho) = \int_0^R r \left[\int_{-\pi}^{\pi} e^{-i2\pi r\rho \cos\theta} \, d\theta \right] dr. \tag{6.71}$$

Substituting $s = 2\pi \rho r$ into the outer integral in (6.71) we get

$$\hat{A}(\rho) = \frac{1}{(2\pi\rho)^2} \int_0^{2\pi\rho R} s \left[\int_{-\pi}^{\pi} e^{-is \cos\theta} \, d\theta \right] ds. \tag{6.72}$$

In order to simplify (6.72) we define the function $H(x)$ by

$$H(x) = \int_0^x s \left[\int_{-\pi}^{\pi} e^{-is \cos\theta} \, d\theta \right] ds. \tag{6.73}$$

We will show at the end of this section why $H(x)$ satisfies the following identity:

$$H(x) = 2x^2 \int_{-1}^{1} \left[1 - w^2 \right]^{\frac{1}{2}} e^{-ixw} \, dw. \tag{6.74}$$

Making use of (6.74) for now, we obtain from (6.72),

$$\hat{A}(\rho) = \frac{1}{(2\pi\rho)^2} H(2\pi\rho R)$$

$$= 2R^2 \int_{-1}^{1} \left[1 - w^2\right]^{\frac{1}{2}} e^{-i2\pi\rho R w} \, dw. \tag{6.75}$$

Making a final change of variables, $x = Rw$, we can rewrite (6.75) as

$$\hat{A}(\rho) = \int_{-R}^{R} 2R \left[1 - \left(\frac{x}{R}\right)^2\right]^{\frac{1}{2}} e^{-i2\pi\rho x} \, dx. \tag{6.76}$$

The integral on the right side of (6.76) *is a one-dimensional Fourier transform.* If we define the function $a(x)$ by

$$a(x) = \begin{cases} 2R\left[1 - (x/R)^2\right]^{\frac{1}{2}} & \text{for } |x| \le R \\ 0 & \text{for } |x| > R \end{cases} \tag{6.77}$$

then we can express (6.76) as

$$\hat{A}(\rho) = \hat{a}(\rho). \tag{6.78}$$

Using (6.78) we can express the intensity function $I(\mathbf{u}, D)$ as follows:

$$I(\mathbf{u}, D) = \left| \frac{1}{\lambda D} \hat{a} \left(\frac{\rho}{\lambda D} \right) \right|^2, \qquad (\rho = |\mathbf{u}|). \tag{6.79}$$

REMARK 6.4: Formula (6.78) shows that the radial dependence of the transform $\hat{A}$ can be determined by Fourier transforming the function $a(x)$ in Formula (6.77). *When using FAS to do this, it is best to use the following formula for* $a(x)$:

$$a(x) = 2R \, \text{rect}[x/(2R)] \, \text{sqr}[\text{abs}(1 - (x/R) \wedge 2)]. \tag{6.80}$$

Using the abs function in this formula for $a(x)$ avoids possible run-time errors that can occur when using the sqr function. ∎

Example 6.6:
Suppose that light of wavelength λ illuminates a circular aperture of radius 1 mm. Describe the Fraunhofer diffraction pattern that results.

SOLUTION   The advantage of Formula (6.79) is that *FAS* can be used to approximate it. Assuming units of millimeters, we use *FAS* to compute the *power spectrum* $|\hat{a}(\rho)|^2$ where [see Formula (6.80)]

$$a(x) = 2 \operatorname{rect}(x/2) \operatorname{sqr}[\operatorname{abs}(1 - x \wedge 2)]. \tag{6.81}$$

By labeling coordinates by multiples of λD we obtain the graph of $I(\mathbf{u}, D)$. This graph is shown in Figure 6.8(a) as a function of the radial coordinate ρ. To obtain Figure 6.8(a) we used 1024 points and an initial interval of $[-16, 16]$. Notice that it corresponds nicely to the diffraction pattern shown in Figure 6.8(b). This pattern is called an *Airy pattern*. The alternating bright and dark rings are called *Airy rings*. ∎

Example 6.7:
 Suppose that an aperture is shaped like an annulus. That is, it has an aperture function

$$A(\mathbf{x}) = \begin{cases} 1 & \text{if } 2 \text{ mm } < |\mathbf{x}| < 5 \text{ mm} \\ 0 & \text{if } |\mathbf{x}| < 2 \text{ mm or } |\mathbf{x}| > 5 \text{ mm}. \end{cases} \tag{6.82}$$

Graph the intensity function $I(\mathbf{u}, D)$.

SOLUTION   The aperture function is the difference of two circular aperture functions (one using 5 mm and the other using 2 mm). Thus, using Formula (6.80), we write $a(x)$ as

$$a(x) = 10 \operatorname{rect}(x/10) \operatorname{sqr}[\operatorname{abs}(1 - (x/5) \wedge 2)]$$

$$- 4 \operatorname{rect}(x/4) \operatorname{sqr}[\operatorname{abs}(1 - (x/2) \wedge 2)].$$

Using this formula, we compute the power spectrum $|\hat{a}(\rho)|^2 = I$, and label co-ordinates by multiples of λD. See Figure 6.9. To obtain this figure, we used an initial interval of $[-64, 64]$ and 1024 points. ∎

We conclude this section by establishing (6.74). We begin by showing that the *derivatives* of both sides of (6.74) are equal. That is, that

$$H'(x) = \frac{d}{dx}\left[2x^2 \int_{-1}^{1} \left[1 - w^2\right]^{\frac{1}{2}} e^{-ixw}\, dw \right]. \tag{6.83}$$

Computing $H'(x)$ is done by applying the Fundamental Theorem of Calculus

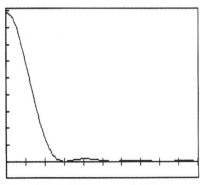

(a) intensity as a function of radius

X interval: $[0, 2\lambda D]$ X increment $= .2\lambda D$

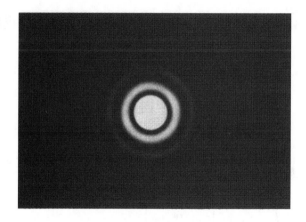

(b) Photograph of diffraction pattern

FIGURE 6.8
Fraunhofer diffraction by a circular aperture of radius 1. (Photograph reproduced from Walker, J. S., *Fourier Analysis,* **Oxford University Press, Oxford, 1988. With permission.)**

to (6.73). In this way we obtain

$$H'(x) = x \int_{-\pi}^{\pi} e^{-ix\cos\theta} \, d\theta.$$

Because $\cos\theta$ is even, we then have

$$H'(x) = 2x \int_{0}^{\pi} e^{-ix\cos\theta} \, d\theta.$$

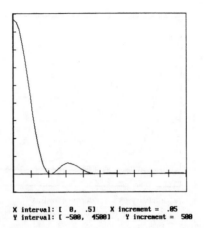

X interval: [0, .5]   X increment = .05
Y interval: [-500, 4500]   Y increment = 500

FIGURE 6.9
Diffraction by an annulus. Intensity as a function of radius.

Substituting $w = \cos\theta$, we get

$$H'(x) = 2x \int_{-1}^{1} \left[1 - w^2\right]^{-\frac{1}{2}} e^{-ixw} \, dw. \tag{6.84}$$

We now turn to the right side of (6.83). Using the product rule and also differentiating inside the integral, we have

$$\frac{d}{dx}\left[2x^2 \int_{-1}^{1}\left[1 - w^2\right]^{\frac{1}{2}} e^{-ixw}\, dw\right]$$

$$= 4x \int_{-1}^{1}\left[1 - w^2\right]^{\frac{1}{2}} e^{-ixw}\, dw - 2ix^2 \int_{-1}^{1} w\left[1 - w^2\right]^{\frac{1}{2}} e^{-ixw}\, dw.$$

Bringing $-ix$ inside the last integral above, as a factor on e^{-ixw}, we obtain

$$\frac{d}{dx}\left[2x^2 \int_{-1}^{1}\left[1 - w^2\right]^{\frac{1}{2}} e^{-ixw}\, dw\right]$$

$$= 4x \int_{-1}^{1}\left[1 - w^2\right]^{\frac{1}{2}} e^{-ixw}\, dw + 2x \int_{-1}^{1} w\left[1 - w^2\right]^{\frac{1}{2}} \frac{d}{dw}\left(e^{-ixw}\right)\, dw.$$

Integrating by parts in the last integral above, and then combining integrals, we have

$$\frac{d}{dx}\left[2x^2\int_{-1}^{1}\left[1-w^2\right]^{\frac{1}{2}}e^{-ixw}\,dw\right]$$

$$= 4x\int_{-1}^{1}\left\{\left[1-w^2\right]^{\frac{1}{2}}-\frac{1}{2}\left[1-w^2\right]^{\frac{1}{2}}+\frac{1}{2}w^2\left[1-w^2\right]^{-\frac{1}{2}}\right\}e^{-ixw}\,dw.$$

Performing simple algebra then yields

$$\frac{d}{dx}\left[2x^2\int_{-1}^{1}\left[1-w^2\right]^{\frac{1}{2}}e^{-ixw}\,dw\right] = 2x\int_{-1}^{1}\left[1-w^2\right]^{-\frac{1}{2}}e^{-ixw}\,dw. \quad (6.85)$$

Comparing (6.85) with (6.84), we see that (6.83) holds. Consequently,

$$H(x) = 2x^2\int_{-1}^{1}\left[1-w^2\right]^{\frac{1}{2}}e^{-ixw}\,dw + C \quad (6.86)$$

where C is an unknown constant. Putting $x = 0$ in both sides of (6.86) we get $0 = 0 + C$. Hence $C = 0$, which proves (6.74).

REMARK 6.5: There is an alternate derivation of Formula (6.76) which is more straightforward (although it makes use of one of Green's identities from vector calculus). See Exercise 6.26. ∎

6.5 Interference

In this section we will examine Fraunhofer diffraction from two identical apertures, separated by some finite distance. This situation results in an *interference* phenomenon which is easily described using Fourier transforms.

Suppose two identical apertures, each having aperture function $A_0(\mathbf{x})$, are separated from each other by a finite distance (see Figure 6.10). Assuming that the apertures are separated along the horizontal x-axis, we can write the aperture function $A(\mathbf{x})$ as

$$A(x, y) = A_0\left(x - \frac{1}{2}\delta, y\right) + A_0\left(x + \frac{1}{2}\delta, y\right) \quad (6.87)$$

where δ is a positive constant.

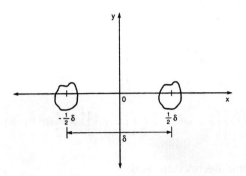

FIGURE 6.10
Two identical apertures.

Fourier transforming both sides of (6.87), and applying shifting and linearity properties, we obtain

$$\hat{A}(u, v) = \hat{A}_0(u, v)e^{-i\pi\delta u} + \hat{A}_0(u, v)e^{+i\pi\delta u}$$

$$= \hat{A}_0(u, v)\, 2\cos(\pi\delta u).$$

Thus, the intensity function $I(u, v, D)$ is given by

$$I(u, v, D) = \left| \frac{1}{\lambda D} \hat{A}_0 \left(\frac{u}{\lambda D}, \frac{v}{\lambda D} \right) \right|^2 \left[4\cos^2 \frac{\pi\delta u}{\lambda D} \right]. \qquad (6.88)$$

Or, using $I_0(u, v, D)$ to denote the intensity from a single aperture

$$I(u, v, D) = I_0(u, v, D) \left[4\cos^2 \frac{\pi\delta u}{\lambda D} \right]. \qquad (6.89)$$

The factor $4\cos^2(\pi\delta u/\lambda D)$ is the *interference factor*. It will be *zero* (destructive interference) when

$$u = \pm \frac{\lambda D}{2\delta}, \ \pm\frac{3\lambda D}{2\delta}, \dots, \ \pm\frac{(2k+1)\lambda D}{2\delta}, \dots, \qquad (6.90)$$

and it will achieve a maximum value of 4 (constructive interference) when

$$u = 0, \ \pm\frac{\lambda D}{\delta}, \ \pm\frac{2\lambda D}{\delta}, \dots, \ \pm\frac{n\lambda D}{\delta}, \dots. \qquad (6.91)$$

Since the equations in (6.90) represent *vertical lines* in the u-v plane, it follows from (6.89) that the diffraction pattern from the two separated apertures will look like the pattern from a single aperture *overlayed with alternating dark vertical strips of low intensity* (interference fringes). See, for example, Figure 6.11, where

FIGURE 6.11
Diffraction pattern from two horizontally separated, identical circular apertures. (Reproduced from Walker, J. S., *Fourier Analysis,* Oxford University Press, Oxford, 1988. With permission.)

we have shown the diffraction pattern from two horizontally separated circular apertures.

An important fact to notice from (6.90) is that the distance between the interference fringes is *inversely* proportional to the distance between the apertures.

In the next section we will examine a more complicated interference phenomenon, the interference from diffraction gratings. The basic principle, however, is contained in the discussion we have given above of interference from two apertures.

6.6 Diffraction Gratings

In this section we begin a discussion of diffraction gratings. Diffraction gratings are an important tool in the *spectral analysis* of light, the separation of light into its component frequencies.

Diffraction gratings are an example of a *linear array* of apertures. Suppose that M identical apertures are evenly spaced by a distance δ along the x-axis, as shown in Figure 6.12. If we choose the origin of coordinates $(0, 0)$ to be within

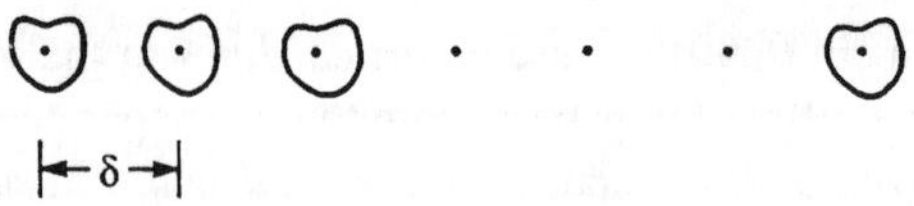

FIGURE 6.12
Linear array of identical apertures.

the left-most aperture, and denote the aperture function for this aperture by $A_0(\mathbf{x})$, then the aperture function $A(\mathbf{x})$ for the entire array is given by

$$A(x, y) = \sum_{m=0}^{M-1} A_0(x - m\delta, y). \qquad (6.92)$$

Consequently, by linearity and shifting properties, the Fourier transform $\hat{A}$ satisfies

$$\hat{A}(u, v) = \sum_{m=0}^{M-1} \hat{A}_0(u, v) e^{-i2\pi m\delta u}. \qquad (6.93)$$

Factoring out $\hat{A}_0(u, v)$ we get

$$\hat{A}(u, v) = \hat{A}_0(u, v) \sum_{m=0}^{M-1} e^{-i2\pi m\delta u}. \qquad (6.94)$$

The sum on the right side of (6.94) can be treated as a finite geometric series, and its exact sum found by using Formula (2.7) from Chapter 2 with $r = e^{-i2\pi\delta u}$. Doing this, we have

$$\sum_{m=0}^{M-1} e^{-i2\pi m\delta u} = \sum_{m=0}^{M-1} \left[e^{-i2\pi\delta u} \right]^m$$

$$= \frac{1 - e^{-i2\pi M\delta u}}{1 - e^{-i2\pi\delta u}}.$$

By factoring out $e^{-i\pi M\delta u}$ from the numerator of the fraction above and $e^{-i\pi\delta u}$ from its denominator, we obtain

$$\sum_{m=0}^{M-1} e^{-i2\pi m\delta u} = \frac{e^{-i\pi M\delta u}}{e^{-i\pi\delta u}} \frac{e^{i\pi M\delta u} - e^{-i\pi M\delta u}}{e^{i\pi\delta u} - e^{-i\pi\delta u}}$$

$$= e^{-i\pi(M-1)\delta u} \frac{\sin(\pi M\delta u)}{\sin(\pi\delta u)}.$$

Using this last result in (6.94), we have

$$\hat{A}(u, v) = \hat{A}_0(u, v) \frac{\sin(\pi M\delta u)}{\sin(\pi\delta u)} e^{-i\pi(M-1)\delta u}. \qquad (6.95)$$

The phase factor $e^{-i\pi(M-1)\delta u}$ in (6.95) is a result of our choice of origin of coordinates. In fact, by the shifting property, if we shift this origin by $(M-1)\delta/2$ to the right, then *the phase factor will be canceled* and we will have

$$\hat{A}(u, v) = \hat{A}_0(u, v) \frac{\sin(\pi M\delta u)}{\sin(\pi\delta u)}. \qquad (6.96)$$

In any case, since $|e^{-i\pi(M-1)\delta u}| = 1$, we will have for the intensity function $I(u, v, D)$ of the array

$$I(u, v, D) = I_0(u, v, D) \frac{\sin^2(\pi M\delta u/\lambda D)}{\sin^2(\pi\delta u/\lambda D)} \qquad (6.97)$$

where

$$I_0(u, v, D) = \left| \frac{1}{\lambda D} \hat{A}_0 \left(\frac{u}{\lambda D}, \frac{v}{\lambda D} \right) \right|^2 \qquad (6.98)$$

is the intensity function from one aperture of the array.

Formula (6.97) says that the diffraction pattern intensity from a linear array of M apertures is equal to the diffraction pattern intensity from a single aperture (called the *form factor,* or *shape factor*) multiplied by a *structure factor* $S_M(u)$ defined by

$$S_M(u) = \frac{\sin^2(\pi M\delta u/\lambda D)}{\sin^2(\pi\delta u/\lambda D)} = M^2 \frac{\text{sinc}^2(M\delta u/\lambda D)}{\text{sinc}^2(\delta u/\lambda D)}. \qquad (6.99)$$

[The second formula for $S_M(u)$ is the one to use for graphing by *FAS.*]

In Figure 6.13 we have graphed the structure factor $S_M(u)$. The pattern created *in the u-v plane* would be a series of vertical strips of high intensity, M^2, with

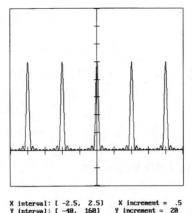

X interval: [-2.5, 2.5]   X increment = .5
Y interval: [-40, 160]   Y increment = 20

FIGURE 6.13
Graph of the structure factor from a linear array of 10 identical apertures.

some very faint side fringes from the lower intensity peaks. The width of these strips is inversely proportional to $1/M$.

A common type of linear array is a *slit diffraction grating*. The slit diffraction grating is a linear array of M vertical slits. In this case, the shape factor $I_0(u, v, D)$ has a pattern like the one shown in Figure 6.7. When M is very large, the very thin, high-intensity, vertical strips from the structure factor lay across the horizontal band from the shape factor, creating a sequence of thin vertical line segments (see Figure 6.14). In this case, *the graph of $I(u, 0, D) = I_0(u, 0, D)\mathcal{S}_M(u)$*

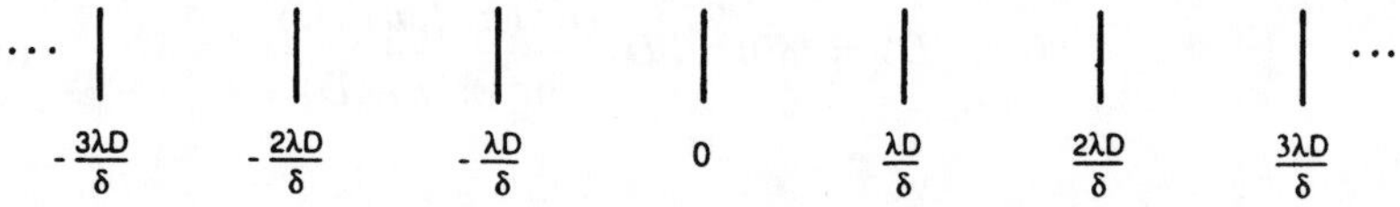

$$\cdots \quad \Big| \qquad \Big| \qquad \Big| \qquad \Big| \qquad \Big| \qquad \Big| \qquad \Big| \quad \cdots$$

$$-\frac{3\lambda D}{\delta} \quad -\frac{2\lambda D}{\delta} \quad -\frac{\lambda D}{\delta} \quad\quad 0 \quad\quad \frac{\lambda D}{\delta} \quad\quad \frac{2\lambda D}{\delta} \quad\quad \frac{3\lambda D}{\delta}$$

FIGURE 6.14
Diffraction pattern from a linear array of vertical slits.

along the u-axis is a good indicator of the form of $I(u, v, D)$. We will refer to $I(u, 0, D)\mathcal{S}_M(u)$ as the *instrument function* for the diffraction grating. And we label the instrument function by $\mathcal{I}(u)$, thus

$$\mathcal{I}(u) = I(u, 0, D)\mathcal{S}_M(u), \qquad (instrument\ function). \qquad (6.100)$$

Notice, in particular, that the thin lines corresponding to the main peaks in the instrument function are located at the points

$$u = 0, \ \pm\frac{\lambda D}{\delta}, \ \pm\frac{2\lambda D}{\delta}, \ldots, \ \pm\frac{m\lambda D}{\delta}, \ldots.$$

We now show how to use *FAS* to model the slit diffraction grating, and generalizations of it.

Example 6.8: SLIT DIFFRACTION GRATING

Use *FAS* to graph the instrument function for a slit diffraction grating.

SOLUTION First, we generate a graph like the one in Figure 6.15, which is a good model for a 31-slit diffraction grating. To do this we choose 4096 points and the interval $[-32, 32]$, and then enter the following formula:

$$f(x) = \text{sumk}(\text{rect}[(x - 0.9k)/0.2]) \setminus k=-15,15. \qquad (6.101)$$

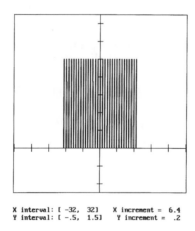

<pre>
X interval: [-32, 32]   X increment = 6.4
Y interval: [-.5, 1.5]   Y increment = .2
</pre>

FIGURE 6.15
One-dimensional, 31-slit diffraction grating.

As you can see from Figure 6.15, this produces a series of thin rectangular slits centered on the points $x = 0, \pm 0.9, \pm 1.8, \ldots, \pm 13.5$. Each slit has width 0.2 and the separation constant δ equals 0.9. Formula (6.101) is a special case of the following formula,

$$f(x) = \text{sumk}[g(x - k\delta)] \qquad (6.102)$$

which produces replicas of $g(x)$ centered on the points $x = k\delta$ (where k ranges over some set of consecutive integers). The function $g(x)$ is called the *unit function* for the diffraction grating. In Formula (6.101) the unit function is $\text{rect}(x/0.2)$. The constant δ is called the *separation constant*.

After producing the graph of the diffraction grating function $f(x)$, we then have *FAS* compute its power spectrum $|\hat{f}(u)|^2$. For the function in (6.101) this produces

the graph shown in Figure 6.16(a). This is the graph of the instrument function
$\mathcal{I}(u)$.

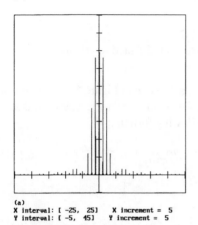

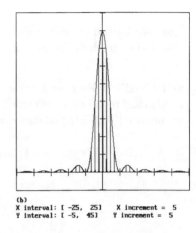

(a)
X interval: [-25, 25]   X increment = 5
Y interval: [-5, 45]   Y increment = 5

(b)
X interval: [-25, 25]   X increment = 5
Y interval: [-5, 45]   Y increment = 5

FIGURE 6.16
**(a) Instrument function for a 31-slit grating. (b) Instrument function and a
multiple of the form factor for a 31-slit grating.**

To see the relation of the form factor to the instrument function, you can do the
following. Compute the power spectrum of $31 \, \mathrm{rect}(x/0.2)$, where 31 is the number
of slits. This produces a multiple of the form factor which acts as an envelope for
the structure factor; see Figure 6.16(b).   ∎

REMARK 6.6:   In the example above, we could have produced a plot of the
instrument function by making use of Formulas (6.98) to (6.100). Assuming that
the units of u are expressed in multiples of λD and that $M = 31$, then $\mathcal{I}(u)$ for the
31-slit grating defined by (6.101) is

$$\mathcal{I}(u) = [31(0.2) \, \mathrm{sinc} \, (0.2u)]^2 \, \frac{\mathrm{sinc}^2(31(0.9)u)}{\mathrm{sinc}^2(0.9u)} . \qquad (6.103)$$

We can graph $\mathcal{I}$ by entering the formula

$$f(x) = [31(0.2) \, \mathrm{sinc} \, (0.2x) \, \mathrm{sinc} \, [31(0.9)x]/ \, \mathrm{sinc} \, (0.9x)] \wedge 2. \qquad (6.104)$$

This produces a graph that appears almost identical to the one shown in Figure 6.16(a). The method of graphing the instrument function described in Example 6.8 enjoys a number of advantages over this second method. Its principal
advantages are the following: (1) it does not require an explicit formula for $\mathcal{I}(u)$
and (2) it can be used to study diffraction gratings which are not *ideal* (see Exercises 6.34 to 6.37).   ∎

6.7   Spectral Analysis with Diffraction Gratings

The importance of diffraction gratings consists in their ability to *spatially separate* light of distinct wavelengths. This process is called *spectral analysis.*

Suppose that light composed of two distinct wavelengths, say $\lambda_1 < \lambda_2$, illuminates in a coherent way a diffraction grating consisting of M vertical slits. In addition to the spatial coherency condition (6.13), we shall assume that the *intensities from the separate wavelengths simply add together* with no interference. (The technical term for this is *time incoherency.* A mathematical formulation of this is similar to the one we used to define spatial incoherency in Section 6.1, but the details would take us too far afield; see [Bo-W].) As can be seen from Figure 6.14, the location of the lines (intensity peaks) in the diffraction pattern are *wavelength dependent.* Thus, as shown in Figure 6.17, these wavelengths correspond to distinct lines, *spectral lines*, in the diffraction pattern.

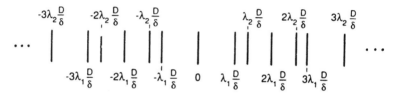

FIGURE 6.17
Spectral lines from two wavelengths.

The ability to separate distinct wavelengths depends upon how close the wavelengths are together. Since each spectral line has a certain width, it is not possible to separate wavelengths that are much closer together than this width. One criterion that is used to determine when separation of wavelengths is possible is *Rayleigh's criterion.* Rayleigh's criterion says that to ensure separation one should have the intensity peak for the larger wavelength λ_2 lying to the right of the first zero value of the intensity function for the smaller wavelength λ_1 (see Figure 6.18).

For a vertical slit grating there are several wavelength-dependent spectral lines. Notice that in Figure 6.14 there are spectral lines at $\lambda D/\delta$, $2\lambda D/\delta$, ... Each of these lines will give rise to wavelength separation (the one at 0 will not, because it does not depend on wavelength). In Figure 6.17 we show the spectral lines resulting from the two wavelengths λ_1 and λ_2. Suppose we concentrate on the lines at $2\lambda_1 D/\delta$ and $2\lambda_2 D/\delta$ which belong to the *second-order spectrum.* The first zero of the intensity function for λ_1 to the right of $2\lambda_1 D/\delta$ is $2\lambda_1 D/\delta + \lambda_1 D/M\delta$. Hence, Rayleigh's criterion says that the spectral lines for λ_1 and λ_2 can be distinguished from each other if

$$\frac{2\lambda_2 D}{\delta} \geq \frac{2\lambda_1 D}{\delta} + \frac{\lambda_1 D}{M\delta}$$

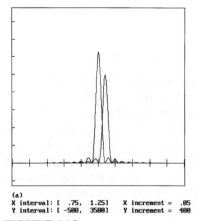 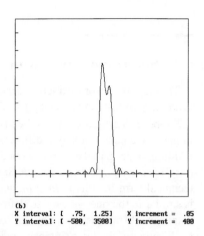

(a)
X interval: [.75, 1.25]   X increment = .05
Y interval: [-500, 3500]   Y increment = 400

(b)
X interval: [.75, 1.25]   X increment = .05
Y interval: [-500, 3500]   Y increment = 400

FIGURE 6.18
(a) Intensity functions for two spectral lines, sufficiently separated according to Rayleigh's criterion. (b) Sum of these two intensities; two peaks are visible.

Thus, we will have separation if

$$\frac{\lambda_2 - \lambda_1}{\lambda_1} \geq \frac{1}{2M}. \tag{6.105}$$

Formula (6.105) describes the minimum *relative separation* needed in the second-order spectrum between λ_1 and λ_2 according to Rayleigh's criterion. Notice that the magnitude of this minumum relative separation is decreased if M is increased. When the slit distance δ is kept constant, increasing M corresponds to increasing the length ($\approx M\delta$) of the grating. This gives the general rule that *the resolution* (ability to separate wavelengths) *of a vertical slit grating is proportional to the length of the grating.* If, for practical reasons, the length of the grating cannot be increased, then the only way to increase resolution is to *increase the number M of slits* (in other words, increase the number per unit length of the slits).

There are several other criterion for measuring resolution of diffraction gratings. For example, *Sparrow's criterion* (see [Mi-T] or [RDPT]). We will not go into these other criterion here.

Vertical slit gratings are not the only type of diffraction grating. Another popular type is the *sinusoidal* (or *holographic*) grating. This grating has an aperture function of the form (here we describe just the x-direction component, as discussed in Section 6.6)

$$A(x) = \text{rect}\left(\frac{x}{\beta}\right)[(1 - \alpha) + \alpha \cos(2\pi N x)]. \tag{6.106}$$

Here, β is a positive constant, α is a constant satisfying $0 < \alpha < 0.5$, and N is a positive integer called the *frequency* of the grating.

For example, in Figure 6.19(a) and (b) we graph the instrument function resulting from choosing $\beta = 20$, $\alpha = 0.4$, and $N = 5$. The main feature of the instrument function for the sinusoidal grating is that essentially all the light away from the central peak at the origin is chanelled into *one* intense peak on each side (instead of several peaks on each side for the vertical slit grating). This feature, which removes the difficulty of distinguishing several spectral orders and also can yield more intense spectral lines, has made sinusoidal gratings very popular.

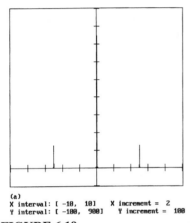

(a)
X interval: [-10, 10]   X increment = 2
Y interval: [-100, 900]   Y increment = 100

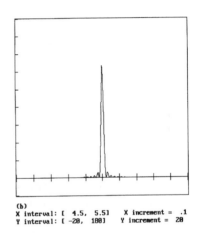

(b)
X interval: [4.5, 5.5]   X increment = .1
Y interval: [-20, 180]   Y increment = 20

FIGURE 6.19
Diffraction from a sinusoidal grating. (a) Graph of the instrument function. (b) Close up of a spectral line, centered on $N\lambda D$.

We will now work out the exact transform of the function $A(x)$ in (6.106). Rewriting $\cos(2\pi Nx)$ as a sum of exponentials, we have

$$A(x) = (1 - \alpha)\text{rect}\left(\frac{x}{\beta}\right) + \frac{\alpha}{2}\text{rect}\left(\frac{x}{\beta}\right)e^{i2\pi Nx} + \frac{\alpha}{2}\text{rect}\left(\frac{x}{\beta}\right)e^{-i2\pi Nx}.$$

Applying the scaling property to transform rect (x/β) and using the modulation property, we have

$$\hat{A}(u) = (1 - \alpha)\beta\text{sinc}\,(\beta u) + \frac{\alpha\beta}{2}\text{sinc}\,[\beta(u - N)]$$

$$+ \frac{\alpha\beta}{2}\text{sinc}\,[\beta(u + N)]. \qquad (6.107)$$

From (6.107) we can approximate $|\hat{A}(u)|^2$ by *ignoring* all the non-square terms.

The approximation

$$|\hat{A}(u)|^2 \approx (1-\alpha)^2\beta^2\text{sinc}^2\,(\beta u) + \frac{\alpha^2\beta^2}{4}\text{sinc}^2\,[\beta(u-N)]$$

$$+ \frac{\alpha^2\beta^2}{4}\text{sinc}^2\,[\beta(u+N)]. \quad (6.108)$$

will be valid when N is large enough that the cross terms are approximately zero. From (6.108) we obtain for the instrument function $\mathcal{I}(u)$

$$\mathcal{I}(u) \approx (1-\alpha)^2\beta^2\text{sinc}^2\left(\frac{\beta u}{\lambda D}\right) + \frac{\alpha^2\beta^2}{4}\text{sinc}^2\left[\beta\left(\frac{u}{\lambda D} - N\right)\right]$$

$$+ \frac{\alpha^2\beta^2}{4}\text{sinc}^2\left[\beta\left(\frac{u}{\lambda D} + N\right)\right]. \quad (6.109)$$

In Figure 6.19(b) we show this instrument function near $u = N\lambda D$. For u near $N\lambda D$ there is also the approximation

$$\mathcal{I}(u) \approx \frac{\alpha^2\beta^2}{4}\text{sinc}^2\left[\beta\left(\frac{u}{\lambda D} - N\right)\right], \quad (u \approx N\lambda D). \quad (6.110)$$

Using the spectral lines to the right of 0 created by this instrument function, we have by Rayleigh's criterion that

$$\frac{\lambda_2 - \lambda_1}{\lambda_1} \geq \frac{1}{N\beta} \quad (6.111)$$

is needed to ensure separation of distinct wavelengths. Thus, *the resolution of a sinusoidal grating is increased by either increasing the length β of the grating or by increasing the frequency N of the grating.*

REMARK 6.7:   Besides having advantages in spectral analysis, sinusoidal gratings are easily manufactured. By exposing a photoreactive substance to two interfering laser beams, a sinusoidal interference pattern (as described in Section 6.5) of light intensity is converted into a sinusoidal pattern of chemically transformed material. If the transformed material absorbs more light than the original material, then a transmittance intensity similar to the sinusoid $(1 - \alpha) + \alpha\cos(2\pi N x)$ is produced. Finally, by placing the wafer in an opaque holder, an aperture function like (6.106) is produced. ∎

6.8 The Phase Transformation Induced by a Thin Lens

In this section we begin our discussion of the effect of a thin lens on monochromatic light. Understanding the effect of lenses is obviously of fundamental importance in optics.

To understand the effect of a lens, we must modify our discussion in Sections 6.1 and 6.2. Suppose that light travels along a path $\mathbf{C}$ from the point $(\mathbf{x}, 0)$ to the point $(\mathbf{u}, D + \epsilon)$, where $(\mathbf{u}, D + \epsilon)$ is located at the *exit plane* of a thin lens (see Figure 6.20). As shown in Figure 6.20, we will assume that the lens is thin enough

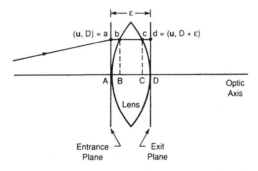

FIGURE 6.20
Light ray through a thin lens.

that the light path from the entrance plane to the exit plane can be approximated by a horizontal path (parallel to the optic axis). In particular, the path $\mathbf{C}$ intersects the entrance plane at the point $(\mathbf{u}, D)$.

We now need to reexamine the following quantity from (6.1):

$$\frac{1}{\lambda |\mathbf{C}|} e^{i \frac{2\pi}{\lambda} \int_{\mathbf{C}} \eta(s)\, ds}. \qquad (6.112)$$

If we let $\mathbf{C}_1$ denote the line segment from $(\mathbf{x}, 0)$ to $(\mathbf{u}, D)$, and let $\mathbf{C}_2$ denote the line segment from $(\mathbf{u}, D)$ to $(\mathbf{u}, D + \epsilon)$, then

$$\frac{1}{\lambda |\mathbf{C}|} e^{i \frac{2\pi}{\lambda} \int_{\mathbf{C}} \eta(s)\, ds} = \frac{1}{\lambda |\mathbf{C}|} e^{i \frac{2\pi}{\lambda} \left[\int_{\mathbf{C}_1} \eta(s)\, ds + \int_{\mathbf{C}_2} \eta(s)\, ds \right]}$$

$$= \frac{1}{\lambda |\mathbf{C}|} e^{i \frac{2\pi}{\lambda} \int_{\mathbf{C}_1} \eta(s)\, ds}\, e^{i \frac{2\pi}{\lambda} \int_{\mathbf{C}_2} \eta(s)\, ds}. \qquad (6.113)$$

We now assume that the index of refraction $\eta(s)$ is the constant 1 outside the lens, and that within the lens $\eta(s)$ is a different constant, which we denote by η. For example, η might be the index of refraction of glass (relative to air). In any case, we require that $\eta > 1$.

Since $\eta(s) = 1$ over $\mathbf{C}_1$ we have

$$\frac{1}{\lambda|\mathbf{C}|}e^{i\frac{2\pi}{\lambda}\int_{\mathbf{C}_1}\eta(s)\,ds} = \frac{1}{\lambda|\mathbf{C}|}e^{i\frac{2\pi}{\lambda}|\mathbf{C}_1|}. \tag{6.114}$$

Making the same approximation as we described in discussing Fresnel diffraction in Section 6.2, we get

$$\frac{1}{\lambda|\mathbf{C}|}e^{i\frac{2\pi}{\lambda}|\mathbf{C}_1|} \approx \frac{e^{i\frac{2\pi}{\lambda}D}}{\lambda|\mathbf{C}|}e^{i\frac{\pi}{\lambda D}|\mathbf{u}-\mathbf{x}|^2}.$$

Thus, (6.113) becomes

$$\frac{1}{\lambda|\mathbf{C}|}e^{i\frac{2\pi}{\lambda}\int_{\mathbf{C}}\eta(s)\,ds} \approx \frac{e^{i\frac{2\pi}{\lambda}D}}{\lambda|\mathbf{C}|}e^{i\frac{\pi}{\lambda D}|\mathbf{u}-\mathbf{x}|^2}e^{i\frac{2\pi}{\lambda}\int_{\mathbf{C}_2}\eta(s)\,ds}. \tag{6.115}$$

Since $|\mathbf{C}| = |\mathbf{C}_1| + |\mathbf{C}_2| = |\mathbf{C}_1| + \epsilon$, we obtain (by similar reasoning as we used in Section 6.2, since ϵ is very small)

$$\frac{1}{\lambda|\mathbf{C}|} = \frac{1}{\lambda|\mathbf{C}_1| + \lambda\epsilon} \approx \frac{1}{\lambda D + \lambda\epsilon} \approx \frac{1}{\lambda D}.$$

So (6.115) becomes

$$\frac{1}{\lambda|\mathbf{C}|}e^{i\frac{2\pi}{\lambda}\int_{\mathbf{C}}\eta(s)\,ds} \approx \frac{e^{i\frac{2\pi}{\lambda}D}}{\lambda D}e^{i\frac{\pi}{\lambda D}|\mathbf{u}-\mathbf{x}|^2}e^{i\frac{2\pi}{\lambda}\int_{\mathbf{C}_2}\eta(s)\,ds}. \tag{6.116}$$

Now, we examine the last factor in (6.116):

$$e^{i\frac{2\pi}{\lambda}\int_{\mathbf{C}_2}\eta(s)\,ds}. \tag{6.117}$$

We break up the line segment $\mathbf{C}_2$ into three parts (see Figure 6.20). The segment from a to b (denoted by ab), the segment bc through the lens itself, and the segment cd. Therefore,

$$e^{i\frac{2\pi}{\lambda}\int_{\mathbf{C}_2}\eta(s)\,ds} = e^{i\frac{2\pi}{\lambda}\left(\int_{ab}1\,ds+\int_{bc}\eta\,ds+\int_{cd}1\,ds\right)}$$

$$= e^{i\frac{2\pi}{\lambda}(|ab|+\eta|bc|+|cd|)}. \tag{6.118}$$

By projecting down vertically to the optic axis (see Figure 6.20), we have $|ab| = |AB|$, $|bc| = |BC|$, and $|cd| = |CD|$. Thus, from (6.118) we get

$$e^{i\frac{2\pi}{\lambda}\int_{C_2}\eta(s)\,ds} = e^{i\frac{2\pi}{\lambda}(|AB|+\eta|BC|+|CD|)}. \qquad (6.119)$$

We now determine each of the lengths on the right side of (6.119). We assume, as shown in Figure 6.20, that the lens has the shape of a *double convex lens*. This is a very common type of lens, although our analysis can be adapted to cover any of the basic lens shapes. Let's suppose that the front and back surfaces of the lens are both spherical with radii of curvature R_1 and R_2, respectively (see Figure 6.21).

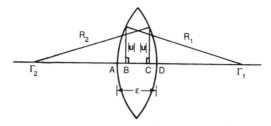

FIGURE 6.21
Geometry of a double convex lens with spherical surfaces.

Denoting the center of curvature for the front surface by Γ_1, and for the back surface by Γ_2, we see from Figure 6.21 that

$$|B\Gamma_1| = \left[R_1^2 - |\mathbf{u}|^2 \right]^{\frac{1}{2}}, \qquad |\Gamma_2 C| = \left[R_2^2 - |\mathbf{u}|^2 \right]^{\frac{1}{2}},$$

$$|A\Gamma_1| = R_1, \qquad |\Gamma_2 D| = R_2, \qquad |AD| = \epsilon.$$

From all these equalities it is not hard to deduce that

$$|AB| = R_1 - \left[R_1^2 - |\mathbf{u}|^2 \right]^{\frac{1}{2}}$$

$$|CD| = R_2 - \left[R_2^2 - |\mathbf{u}|^2 \right]^{\frac{1}{2}}$$

$$|BC| = \epsilon - |AB| - |CD|. \qquad (6.120)$$

We now make the approximations

$$\left[R_1^2 - |\mathbf{u}|^2 \right]^{\frac{1}{2}} \approx R_1 - \frac{|\mathbf{u}|^2}{2R_1}, \qquad \left[R_2^2 - |\mathbf{u}|^2 \right]^{\frac{1}{2}} \approx R_2 - \frac{|\mathbf{u}|^2}{2R_2} \qquad (6.121)$$

which are similar to the approximations we made in discussing Fresnel diffraction in Section 6.2. These approximations will certainly be valid when R_1 and R_2 are relatively large in comparison to the radius of the lens aperture.

Using (6.121) in (6.120), we obtain

$$|AB| \approx \frac{|\mathbf{u}|^2}{2R_1}, \qquad |CD| \approx \frac{|\mathbf{u}|^2}{2R_2}, \qquad |BC| = \epsilon - |AB| - |CD|. \quad (6.122)$$

Employing (6.122), we have

$$|AB| + \eta|BC| + |CD| = \epsilon\eta + (1 - \eta)\,(|AB| + |CD|)$$

$$\approx \epsilon\eta + (1 - \eta)\left[\frac{1}{R_1} + \frac{1}{R_2}\right]\frac{|\mathbf{u}|^2}{2}. \quad (6.123)$$

Before we make use of this result, we make the following definition. For a *double convex lens* with spherical surfaces having radii of curvature R_1 and R_2, we define the *focal length* f via the formula

$$\frac{1}{f} = (\eta - 1)\left[\frac{1}{R_1} + \frac{1}{R_2}\right]. \quad (6.124)$$

From (6.124), we can express (6.123) as

$$|AB| + \eta|BC| + |CD| \approx \epsilon\eta - \frac{|\mathbf{u}|^2}{2f}. \quad (6.125)$$

Combining (6.125) with (6.119), we have

$$e^{i\frac{2\pi}{\lambda}\int_{\mathbf{C}_2}\eta(s)\,ds} \approx e^{i\frac{2\pi}{\lambda}\epsilon\eta}e^{\frac{-i\pi}{\lambda f}|\mathbf{u}|^2}. \quad (6.126)$$

Substituting the right side of (6.126) in place of the last factor in (6.116) yields

$$\frac{1}{\lambda|\mathbf{C}|}e^{i\frac{2\pi}{\lambda}\int_{\mathbf{C}}\eta(s)\,ds} \approx \frac{e^{i\frac{2\pi}{\lambda}(D+\epsilon\eta)}}{\lambda D}e^{\frac{i\pi}{\lambda D}|\mathbf{u}-\mathbf{x}|^2}e^{\frac{-i\pi}{\lambda f}|\mathbf{u}|^2}. \quad (6.127)$$

As a final simplification, we define δ to be the constant $\frac{2\pi}{\lambda}(D+\epsilon\eta)$, and then (6.127) becomes

$$\frac{1}{\lambda|\mathbf{C}|}e^{i\frac{2\pi}{\lambda}\int_{\mathbf{C}}\eta(s)\,ds} \approx \frac{e^{i\delta}}{\lambda D}e^{\frac{i\pi}{\lambda D}|\mathbf{u}-\mathbf{x}|^2}e^{\frac{-i\pi}{\lambda f}|\mathbf{u}|^2}. \quad (6.128)$$

Using (6.128) we can express the light $\psi(\mathbf{u}, D + \epsilon, t)$ at the exit plane of the lens as

$$\psi(\mathbf{u}, D + \epsilon, t) \approx \frac{e^{i\delta}}{\lambda D} \int_{\mathbf{R}^2} A(\mathbf{x})\psi(\mathbf{x}, 0, t)e^{\frac{i\pi}{\lambda D}|\mathbf{u}-\mathbf{x}|^2} \, d\mathbf{x} \, e^{\frac{-i\pi}{\lambda f}|\mathbf{u}|^2} P(\mathbf{u}). \quad (6.129)$$

The aperture function $A(\mathbf{x})$ was introduced in Section 6.1. The factor $P(\mathbf{u})$ is the *pupil function of the lens,* whose main purpose is to express the fact that the lens has a *finite extent.* Most often, we will assume that $P(\mathbf{u})$ is the simplest type of aperture function, i.e.,

$$P(\mathbf{u}) = \begin{cases} 1 & \text{if } \mathbf{u} \text{ lies inside the lens aperture} \\ 0 & \text{if } \mathbf{u} \text{ lies outside the lens aperture.} \end{cases} \quad (6.130)$$

Formula (6.129) consists of two parts. The first part is the integral

$$\frac{e^{i\delta}}{\lambda D} \int_{\mathbf{R}^2} A(\mathbf{x})\psi(\mathbf{x}, 0, t)e^{\frac{i\pi}{\lambda D}|\mathbf{u}-\mathbf{x}|^2} \, d\mathbf{x} \quad (6.131)$$

which describes *Fresnel diffraction from the aperture to the entrance plane of the lens.* The second part,

$$e^{\frac{-i\pi}{\lambda f}|\mathbf{u}|^2} P(\mathbf{u}) \quad (6.132)$$

describes *the phase transformation due to the lens.* After the light described by (6.131) has passed through the lens, it has been multiplied by the *phase factor* $e^{-i\pi|\mathbf{u}|^2/\lambda f}$ and by the lens pupil function $P(\mathbf{u})$.

The pupil function $P(\mathbf{u})$ in Formula (6.130) describes a lens having *no aberrations* (this is also called the *diffraction-limited* case). The pupil function plays the role of an aperture function when light diffracts through the lens aperture. When there are no aberrations, then the only limitation to perfect imaging is the finite extent of the lens, through which light from the original aperture must diffract. (Since we are dealing with monochromatic light, we are ignoring *chromatic aberration,* which affects all simple lenses.)

In subsequent sections, we will examine the profound implications of Formula (6.129).

6.9 Imaging with a Single Lens

In this section, we will derive the general formula for imaging with a single lens. This derivation shows that there is a *wave optical* reason for the classic lens equation. In subsequent sections we will discuss two specific cases of the imaging formula, the cases of *coherent* and *incoherent* illumination, which are the two fundamental types of illumination.

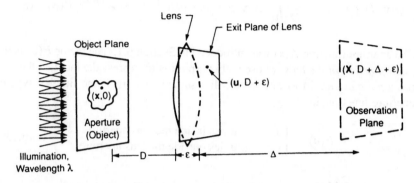

FIGURE 6.22
Geometry of single lens imaging.

Suppose that we have the setup shown in Figure 6.22, a lens situated D units in front of the aperture and an observation plane located Δ units behind the lens. From Formula (6.129) we have that the light emergent from the exit plane of the lens is described by

$$\psi(\mathbf{u}, D + \epsilon, t) \approx \frac{e^{i\delta}}{\lambda D} \int_{\mathbf{R}^2} A(\mathbf{x})\psi(\mathbf{x}, 0, t) e^{\frac{i\pi}{\lambda D}|\mathbf{u}-\mathbf{x}|^2}\, d\mathbf{x}\, e^{\frac{-i\pi}{\lambda f}|\mathbf{u}|^2} P(\mathbf{u}). \quad (6.133)$$

Our formulas in this section will be very unwieldy unless we do something to abbreviate them. Accordingly, we adopt the following conventions.

DEFINITION 6.3: *We will write $\int$ instead of $\int_{\mathbf{R}^2}$, all integrals will be over $\mathbf{R}^2$. Instead of $\psi(\mathbf{x}, 0, t)$ we will just write $\psi(\mathbf{x}, t)$, the vector $\mathbf{x}$ being used to indicate $(\mathbf{x}, 0)$. Similarly, we shall write $\psi(\mathbf{u}, t)$ in place of $\psi(\mathbf{u}, D + \epsilon, t)$, and $\psi(\mathbf{X}, t)$ in place of $\psi(\mathbf{X}, D + \Delta + \epsilon, t)$. The function $\psi(\mathbf{X}, t)$ represents the light at the observation plane.*

With these conventions, Formula (6.133) simplifies to

$$\psi(\mathbf{u}, t) \approx \frac{e^{i\delta}}{\lambda D} \int A(\mathbf{x})\psi(\mathbf{x}, t) e^{\frac{i\pi}{\lambda D}|\mathbf{u}-\mathbf{x}|^2}\, d\mathbf{x}\, e^{\frac{-i\pi}{\lambda f}|\mathbf{u}|^2} P(\mathbf{u}). \quad (6.134)$$

Applying the Fresnel diffraction formula (6.38) to $\psi(\mathbf{u}, t)$ instead of $A(\mathbf{x})\psi(\mathbf{x}, 0, t)$, we can express the light $\psi(\mathbf{X}, t)$ at the observation plane by

$$\psi(\mathbf{X}, t) \approx \frac{e^{i\frac{2\pi}{\lambda}\Delta}}{\lambda\Delta} \int \psi(\mathbf{u}, t) e^{\frac{i\pi}{\lambda\Delta}|\mathbf{X}-\mathbf{u}|^2} \, d\mathbf{u}. \tag{6.135}$$

Our next step will be to substitute the right side of (6.134) into the integrand in (6.135). To simplify notation we define $E(\mathbf{x}, \mathbf{u}, \mathbf{X})$ by

$$E(\mathbf{x}, \mathbf{u}, \mathbf{X}) = e^{\frac{i\pi}{\lambda D}|\mathbf{u}-\mathbf{x}|^2} e^{\frac{-i\pi}{\lambda f}|\mathbf{u}|^2} e^{\frac{i\pi}{\lambda\Delta}|\mathbf{X}-\mathbf{u}|^2}. \tag{6.136}$$

This quantity is what we get if we combine all the complex exponentials involved in (6.134) and (6.135). Now, substituting the right side of (6.134) into the integrand in (6.135), and combining all the complex exponentials so that we get $E(\mathbf{x}, \mathbf{u}, \mathbf{X})$, we obtain

$$\psi(\mathbf{X}, t) \approx \frac{e^{i\frac{2\pi}{\lambda}\Delta}e^{i\delta}}{\lambda^2 D\Delta} \int\int A(\mathbf{x})\psi(\mathbf{x}, t) E(\mathbf{x}, \mathbf{u}, \mathbf{X}) P(\mathbf{u}) \, d\mathbf{x} \, d\mathbf{u}. \tag{6.137}$$

To save further space, we define c by $c = \frac{2\pi}{\lambda}\Delta + \delta$, so that

$$\psi(\mathbf{X}, t) \approx \frac{e^{ic}}{\lambda^2 D\Delta} \int\int A(\mathbf{x})\psi(\mathbf{x}, t) E(\mathbf{x}, \mathbf{u}, \mathbf{X}) P(\mathbf{u}) \, d\mathbf{x} \, d\mathbf{u}. \tag{6.138}$$

Our next step is to simplify the expression for $E(\mathbf{x}, \mathbf{u}, \mathbf{X})$ in (6.136), which will help to make (6.138) more meaningful. If we expand the quantities $|\mathbf{u} - \mathbf{x}|^2$ and $|\mathbf{X} - \mathbf{u}|^2$ in (6.136) and recombine exponentials we obtain

$$E(\mathbf{x}, \mathbf{u}, \mathbf{X}) = e^{\frac{i\pi}{\lambda}\left[\frac{1}{D}+\frac{1}{\Delta}-\frac{1}{f}\right]|\mathbf{u}|^2} e^{-i\frac{2\pi}{\lambda}\mathbf{u}\cdot\left(\frac{\mathbf{x}}{D}+\frac{\mathbf{X}}{\Delta}\right)} e^{\frac{i\pi}{\lambda D}|\mathbf{x}|^2} e^{\frac{i\pi}{\lambda\Delta}|\mathbf{X}|^2}. \tag{6.139}$$

The first exponential factor in (6.139) will be completely eliminated if we assume that

$$\frac{1}{D} + \frac{1}{\Delta} - \frac{1}{f} = 0. \tag{6.140}$$

Formula (6.140) is the classic *imaging equation* for a thin lens.

Let's assume that (6.140) holds. Then the first factor in (6.139) is just $e^0 = 1$. Hence (6.139) becomes

$$E(\mathbf{x}, \mathbf{u}, \mathbf{X}) = e^{-i\frac{2\pi}{\lambda}\mathbf{u}\cdot\left(\frac{\mathbf{x}}{D}+\frac{\mathbf{X}}{\Delta}\right)} e^{\frac{i\pi}{\lambda D}|\mathbf{x}|^2} e^{\frac{i\pi}{\lambda\Delta}|\mathbf{X}|^2}. \tag{6.141}$$

Substituting the right side of (6.141) in place of $E(\mathbf{x}, \mathbf{u}, \mathbf{X})$ in (6.138), and factoring the complex exponential involving $|\mathbf{X}|^2$ outside the integrals, we get

$$\psi(\mathbf{X}, t) \approx \frac{e^{ic}}{\lambda^2 D \Delta} e^{\frac{i\pi}{\lambda\Delta}|\mathbf{X}|^2}$$

$$\int \int A(\mathbf{x})\psi(\mathbf{x}, t) P(\mathbf{u}) e^{-i\frac{2\pi}{\lambda}\mathbf{u}\cdot\left(\frac{\mathbf{x}}{D} + \frac{\mathbf{X}}{\Delta}\right)} e^{\frac{i\pi}{\lambda D}|\mathbf{x}|^2} d\mathbf{u}\, d\mathbf{x}. \quad (6.142)$$

We also interchanged the integrals to obtain (6.142).
Since

$$\int P(\mathbf{u}) e^{-i\frac{2\pi}{\lambda}\mathbf{u}\cdot\left(\frac{\mathbf{x}}{D} + \frac{\mathbf{X}}{\Delta}\right)} d\mathbf{u} = \hat{P}\left(\frac{\mathbf{x}}{\lambda D} + \frac{\mathbf{X}}{\lambda\Delta}\right)$$

we rewrite (6.142) as

$$\psi(\mathbf{X}, t) \approx \frac{e^{ic}}{\lambda^2 D \Delta} e^{\frac{i\pi}{\lambda\Delta}|\mathbf{X}|^2} \int A(\mathbf{x})\psi(\mathbf{x}, t) \hat{P}\left(\frac{\mathbf{x}}{\lambda D} + \frac{\mathbf{X}}{\lambda\Delta}\right) e^{\frac{i\pi}{\lambda D}|\mathbf{x}|^2} d\mathbf{x}. \quad (6.143)$$

Finally, suppose an observation of the intensity $I(\mathbf{X})$ defined by

$$I(\mathbf{X}) = \frac{1}{T}\int_0^T |\psi(\mathbf{X}, t)|^2\, dt$$

is made (for some large value of T). Then we obtain from (6.143)

$$I(\mathbf{X}) \approx \frac{1}{(\lambda^2 D \Delta)^2} \int \int \left\{ A(\mathbf{x}) e^{\frac{i\pi}{\lambda D}|\mathbf{x}|^2} \hat{P}\left(\frac{\mathbf{x}}{\lambda D} + \frac{\mathbf{X}}{\lambda\Delta}\right)\right\}$$

$$\times \left\{ A(\mathbf{z}) e^{\frac{i\pi}{\lambda D}|\mathbf{z}|^2} \hat{P}\left(\frac{\mathbf{z}}{\lambda D} + \frac{\mathbf{X}}{\lambda\Delta}\right)\right\}^* \gamma(\mathbf{x}, \mathbf{z})\, d\mathbf{x}\, d\mathbf{z}. \quad (6.144)$$

where $\gamma(\mathbf{x}, \mathbf{z})$ is the *cross-correlation function* defined in (6.23b).

Formula (6.144) is the *general equation for imaging with monochromatic light*. It is extremely complicated; and that is because we have made no assumptions about the initial illuminating light other than its being unpolarized and having a single wavelength λ.

In the next section we will discuss one method of simplifying (6.144). This method makes the assumption that we are dealing with coherent illumination. In Section 6.12 we will discuss the other basic method of simplifying (6.144), which makes the assumption that we are dealing with incoherent light.

6.10 Imaging with Coherent Light

In the previous section we derived the general formula for imaging. In this section we will discuss one of the two fundamental assumptions, *coherency of illumination*, that can be used to simplify that formula.

We assume that the light illuminating the aperture in Figure 6.22 is coherent. That is, the coherency function $\Gamma(\mathbf{x}, \mathbf{z})$ satisfies the following condition,

$$\Gamma(\mathbf{x}, \mathbf{z}) = 1 \qquad [\textit{Coherency}] \qquad (6.145)$$

for all points $\mathbf{x}$ and $\mathbf{z}$ *within the aperture*. Hence, based on the definition of the cross-correlation function $\gamma(\mathbf{x}, \mathbf{z})$ in (6.23b), along with Formula (6.8), we have

$$\gamma(\mathbf{x}, \mathbf{z}) = \sqrt{I(\mathbf{x}, 0)}\sqrt{I(\mathbf{z}, 0)} \qquad (6.146)$$

for all $\mathbf{x}$ and $\mathbf{z}$ within the aperture. For simplicity, we shall again assume, as we did in Section 6.1, that the light intensity is constant over the aperture. In fact, to simplify as much as possible let's assume that $I(\mathbf{x}, 0) = 1$ and $I(\mathbf{z}, 0) = 1$ for all $\mathbf{x}$ and $\mathbf{z}$ within the aperture. We then find that Formula (6.146) simplifies to

$$\gamma(\mathbf{x}, \mathbf{z}) = 1 \qquad (6.147)$$

for all $\mathbf{x}$ and $\mathbf{z}$ within the aperture. For $\mathbf{x}$ or $\mathbf{z}$ lying outside the aperture, we would have either $A(\mathbf{x}) = 0$ or $A(\mathbf{z}) = 0$, so we are free to use (6.147) for *all* $\mathbf{x}$ and $\mathbf{z}$ in the imaging Equation (6.144). Doing so, we find that the two integrals in (6.144) *separate into integrals over $\mathbf{x}$ and $\mathbf{z}$*, as follows:

$$I(\mathbf{X}) \approx \frac{1}{(\lambda^2 D\Delta)^2} \left\{ \int A(\mathbf{x}) e^{\frac{i\pi}{\lambda D}|\mathbf{x}|^2} \hat{P}\left(\frac{\mathbf{x}}{\lambda D} + \frac{\mathbf{X}}{\lambda \Delta}\right) d\mathbf{x} \right\}$$

$$\times \left\{ \int A(\mathbf{z}) e^{\frac{i\pi}{\lambda D}|\mathbf{z}|^2} \hat{P}\left(\frac{\mathbf{z}}{\lambda D} + \frac{\mathbf{X}}{\lambda \Delta}\right) d\mathbf{z} \right\}^* .$$

Substituting $\mathbf{x}$ in place of $\mathbf{z}$ in the second integral above, we get

$$I(\mathbf{X}) \approx \left| \frac{1}{\lambda^2 D\Delta} \int A(\mathbf{x}) e^{\frac{i\pi}{\lambda D}|\mathbf{x}|^2} \hat{P}\left(\frac{\mathbf{x}}{\lambda D} + \frac{\mathbf{X}}{\lambda \Delta}\right) d\mathbf{x} \right|^2 . \qquad (6.148)$$

To simplify (6.148) further, we make the *small object assumption* that, for all $\mathbf{x}$ lying in the object's aperture, $|\mathbf{x}|$ is small enough that

$$e^{\frac{i\pi}{\lambda D}|\mathbf{x}|^2} \approx 1, \qquad \text{for } \mathbf{x} \text{ within the object's aperture.} \qquad (6.149)$$

Formula (6.149) is a natural approximation to make in the field of microscopy, for example. Since $A(\mathbf{x}) = 0$ for $\mathbf{x}$ outside the aperture, we can use (6.149) for *all* $\mathbf{x}$ in (6.148). Then, (6.148) simplifies to

$$I(\mathbf{x}) \approx \left| \frac{1}{\lambda^2 D\Delta} \int A(\mathbf{x})\hat{P}\left(\frac{\mathbf{x}}{\lambda D} + \frac{\mathbf{X}}{\lambda \Delta}\right) d\mathbf{x} \right|^2. \qquad (6.150)$$

As a final simplification, we define the *degree of magnification M* by

$$M = \frac{\Delta}{D}. \qquad (6.151)$$

We then make the substitution $\mathbf{x} = -\mathbf{s}/M$ (and $d\mathbf{x} = d\mathbf{s}/M^2$), so that (6.150) becomes

$$I(\mathbf{X}) \approx \left| \int \frac{1}{M} A\left(\frac{-\mathbf{s}}{M}\right) \frac{1}{(\lambda\Delta)^2} \hat{P}\left(\frac{1}{\lambda\Delta}(\mathbf{X} - \mathbf{s})\right) d\mathbf{s} \right|^2. \qquad (6.152)$$

The integral in (6.152) is a two-dimensional convolution. The function

$$\frac{1}{M} A\left(\frac{-\mathbf{X}}{M}\right), \qquad \textit{(magnified, inverted object)} \qquad (6.153)$$

is convolved with

$$\frac{1}{(\lambda\Delta)^2} \hat{P}\left(\frac{1}{\lambda\Delta}\mathbf{X}\right), \qquad \textit{(point spread function).} \qquad (6.154)$$

The interpretation of the function in (6.153) as a magnified, inverted object is fairly clear (the factor $1/M$ on the outside is to preserve the 2-Norm squared, which measures intensity). In the following four examples, we will show how *FAS* can be used to understand the role of the function in (6.154), which is called the *point spread function* (or *PSF*).

Example 6.9:
Suppose that coherent light having wavelength $\lambda = 5 \times 10^{-4}$ mm illuminates an edge having aperture function

$$A(x, y) = \begin{cases} 1 & \text{if } x > 0 \\ 0 & \text{if } x < 0 \end{cases}$$

and this edge is imaged by a lens with a square aperture of side length 40 mm and focal length 1 m. Assume that $D = 2$ m, and $\Delta = 2$ m. Describe the image that results.

SOLUTION   We will use (6.152) to describe the image (even though, strictly speaking, the small object condition is not satisfied). By the scaling property,

$$P(\lambda \Delta \mathbf{x}) \xrightarrow{\mathcal{F}} \frac{1}{(\lambda \Delta)^2} \hat{P}\left(\frac{1}{\lambda \Delta} \mathbf{x}\right) \tag{6.155}$$

where, for this example, since $\lambda \Delta = 1$,

$$P(\lambda \Delta \mathbf{x}) = P(x, y) = \begin{cases} 1 & \text{if } |x| < 20 \text{ and } |y| < 20 \\ 0 & \text{if } |x| > 20 \text{ or } |y| > 20 \end{cases}$$

$$= \text{rect}\left(\frac{x}{40}\right) \text{rect}\left(\frac{y}{40}\right). \tag{6.156}$$

It follows that

$$\frac{1}{(\lambda \Delta)^2} \hat{P}\left(\frac{1}{\lambda \Delta} \mathbf{x}\right) = 40 \, \text{sinc}\, (40X) \, 40 \, \text{sinc}\, (40Y). \tag{6.157}$$

When this last function is convolved with $A(-X, -Y)$, the integral separates into integrals over X and Y separately. Since the integral along the Y-direction will always be

$$\int_{-\infty}^{\infty} 40 \, \text{sinc}\, (40Y) \, dY = \text{rect}\left(\frac{y}{40}\right)\Big|_{y=0}, \qquad \textit{(Fourier inversion)}$$

$$= 1$$

we only need to convolve $40 \, \text{sinc}\, (40X)$ with

$$f(X) = \begin{cases} 1 & \text{for } X > 0 \\ 0 & \text{for } X < 0 \end{cases}$$

to obtain a representation of the image (*Note:* by convolving with the original edge function, rather than its inversion through the origin, we follow the standard procedure in optics.) Using *FAS* to perform this convolution, we get the results shown in Figure 6.23. In Figure 6.23(a) we have the graph of 40 sinc (40*X*), and

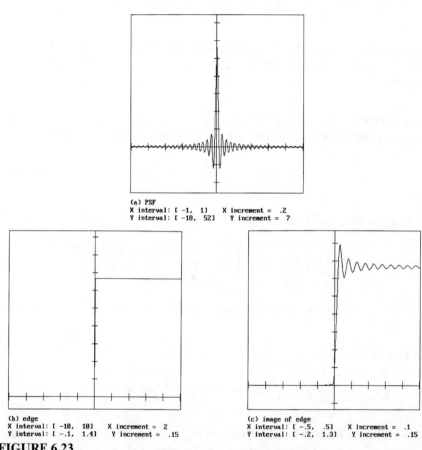

(a) PSF
X interval: [-1,   1]   X increment = .2
Y interval: [-18, 52]   Y increment = 7

(b) edge
X interval: [-10, 10]   X increment = 2
Y interval: [-.1, 1.4]   Y increment = .15

(c) image of edge
X interval: [-.5,   .5]   X increment = .1
Y interval: [-.2, 1.3]   Y increment = .15

FIGURE 6.23
Imaging of an edge using coherent light.

in Figure 6.23(b) we have the graph of the edge. The resulting image, obtained by convolving these two functions and then squaring, is shown in Figure 6.23(c). To obtain these figures we used an initial interval of [−10, 10] and 4096 points.

The graph shown in Figure 6.23(c) is a good prediction of the effects of coherent imaging. See [Go, p. 133] for a nice photograph illustrating this. In particular, the oscillation and Gibbs' phenomenon at the edge boundary which appear in Figure 6.23(c) are plainly visible in that photograph (the Gibbs' effect produces a bright band near the edge boundary). These are serious defects of coherent imaging (they can be mitigated, however, through *image processing;* see Exercise 6.63). A more subtle defect involves the displacement of the edge boundary in the image. If the *image edge boundary* is located at the point of half of maximum intensity, then (because of the Gibbs' effect) the boundary is located *slightly to the right of the true object edge boundary.* ∎

Example 6.10:
Suppose that coherent light having wavelength $\lambda = 5 \times 10^{-4}$ mm illuminates a small square with aperture function

$$A(x, y) = \begin{cases} 1 & \text{if } |x| < .001 \text{ mm and } |y| < .001 \text{ mm} \\ 0 & \text{if } |x| > .001 \text{ mm or } |y| > .001 \text{ mm} \end{cases}$$

and this square is imaged by a lens having focal length of $1/11$ m with a square aperture of side length 20 mm. Assume that $D = 0.1$ m, and $\Delta = 1$ m. Describe the image.

SOLUTION   In this example, the magnified inverted object is defined by

$$\frac{1}{10} A \left(\frac{-x}{10}, \frac{-y}{10} \right) = \begin{cases} 0.1 & \text{if } |x| < .01 \text{ mm and } |y| < .01 \text{ mm} \\ 0 & \text{if } |x| > .01 \text{ mm or } |y| > .01 \text{ mm} \end{cases}$$

$$= 0.1 \operatorname{rect}(50x) \operatorname{rect}(50y).$$

The *PSF* is the Fourier transform of (since $\lambda \Delta = 0.5$)

$$P(0.5x, 0.5y) = \begin{cases} 1 & \text{if } |x| < 20 \text{ and } |y| < 20 \\ 0 & \text{if } |x| > 20 \text{ or } |y| > 20 \end{cases}$$

$$= \operatorname{rect}\left(\frac{x}{40}\right) \operatorname{rect}\left(\frac{y}{40}\right).$$

Therefore, the convolution in (6.152) splits into two identical convolutions in x and y. The convolution with respect to x is

$$\int_{-\infty}^{\infty} \sqrt{0.1} \operatorname{rect}(50s) 40 \operatorname{sinc}\left[40(x - s)\right] ds.$$

In Figure 6.24 we show details of this convolution process. In Figure 6.24(a) we show the graph of $40 \operatorname{sinc}(40x)$, which corresponds to the *PSF* along the x-direction, while in Figure 6.24(b) we show the the graph of $\sqrt{0.1} \operatorname{rect}(50x)$, which corresponds to the magnified, inverted image. The convolution of these two

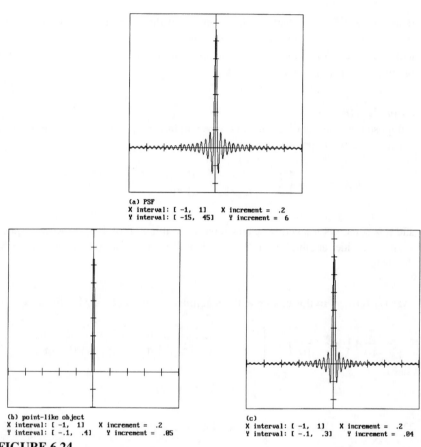

(a) PSF
X interval: [-1, 1]   X increment =   .2
Y interval: [-15, 45]   Y increment = 6

(b) point-like object
X interval: [-1, 1]   X increment =   .2
Y interval: [-.1, .4]   Y increment = .05

(c)
X interval: [-1, 1]   X increment =   .2
Y interval: [-.1, .3]   Y increment = .04

FIGURE 6.24
Coherent image of a point-like object.

functions, which is shown in Figure 6.24(c), is essentially just a reproduction of the *PSF* in Figure 6.24(a). For this reason, the function

$$\frac{1}{\lambda \Delta} \hat{P} \left(\frac{\mathbf{X}}{\lambda \Delta} \right)$$

is called the point spread function. The effect of the *PSF* is to spread out a point, in this case a very small square, or *pixel*. The convolution in (6.152) can be interpreted as a summation of the spreads of all the pixels composing the original image. (*Note:* the actual image will be the square of the convolution shown in Figure 6.24(c). We wanted to emphasize here, however, the reproduction of the *PSF* by a small point-like object.) ∎

REMARK 6.8:   These examples show some of the sensitivities of coherent imaging. Coherent imaging spreads out the images of very tiny features of an object; it also tends to brighten the edges of objects. For instance, dust particles will appear as *ringed objects* due to their tracing out of the *PSF* of the imaging lens (see Exercise 6.60). Other *false details* include enhancement and fringing of edges (as in our first example), and interference patterns due to the interference of nearby point spreads (see Exercise 6.61).   ∎

Our next two examples illustrate methods of image enhancement that are based on *modifying the transforms of objects*. These are just two examples taken from a wide variety of examples that belong to the category known as *spatial filtering* of images (i.e., modifying transforms of objects in order to enhance their images).

Example 6.11:
 In the field of microscopy there is a method known as *dark field imaging*. This method is used when intense background light obscures the details in an image. An opaque object (aperture stop) is placed over the center of the lens pupil. This aperture stop blocks out much of the background light and allows the previously obscured details to emerge.

In this example we describe a one-dimensional model for dark field imaging. In Figure 6.25(a) we show the graph of the following function (one-dimensional object):

$$A(x) = \exp(-(x/3) \wedge 10)[1 + .02\cos(6\pi x)] \tag{6.158}$$

Its transform is shown in Figure 6.25(b). The central spike is due mostly to the transform of $\exp(-(x/3) \wedge 10)$ which forms a kind of *bright background* on which is superimposed the small oscillations of $0.02\cos(6\pi x)$. The dark field method employs a pupil function of the form [see Figure 6.25(c)]:

$$P_{DF}(\lambda\Delta x) = \begin{cases} 0 & \text{if } |x| > 40 \\ 10 & \text{if } 2.5 < |x| < 40 \\ 0 & \text{if } |x| < 2.5. \end{cases} \tag{6.159}$$

The value of 10 is assigned for $3 < |x| < 40$ in order to model an *increase in intensity* of illumination (so as to make the image light intense enough to be recorded, say, by photographic film). After transforming this pupil function (using an interval of $[-128, 128]$ and 2048 points), convolving with the object function in (6.158), and then squaring, the dark field image shown in Figure 6.25(d) results. A comparison with the *bright field image* using the unblocked pupil

$$P_{BF}(\lambda\Delta x) = \begin{cases} 0 & \text{if } |x| > 40 \\ 1 & \text{if } |x| < 40 \end{cases} \tag{6.160}$$

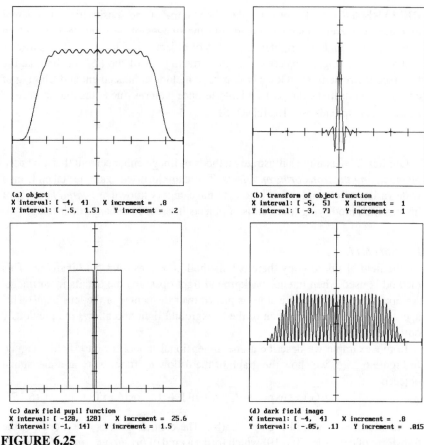

(a) object
X interval: [-4, 4]   X increment = .8
Y interval: [-.5, 1.5]   Y increment = .2

(b) transform of object function
X interval: [-5, 5]   X increment = 1
Y interval: [-3, 7]   Y increment = 1

(c) dark field pupil function
X interval: [-128, 128]   X increment = 25.6
Y interval: [-1, 14]   Y increment = 1.5

(d) dark field image
X interval: [-4, 4]   X increment = .8
Y interval: [-.05, .1]   Y increment = .015

FIGURE 6.25
Dark field imaging.

is shown in Figure 6.26. In order to measure the visibility of the oscillations in the two images, one calculates the *relative contrast* between the peaks and valleys by

$$\text{contrast} = \frac{I_{\max} - I_{\min}}{I_{\max} + I_{\min}}. \tag{6.161}$$

For the bright field image in Figure 6.26(a) we get a contrast of 0.04, or 4%. While for the dark field image the contrast is about 1, or 100%. Thus, the dark field method has improved the image contrast by a factor of about 25.

There are problems associated with dark field imaging. As the reader might have observed, in the previous example the contrast was dramatically improved with dark field imaging, but *the frequency of the oscillations was increased by a factor of 2*. This is a *false detail* in the dark field image. It is this problem of false

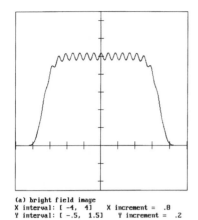

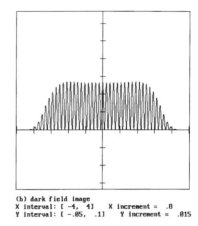

(a) bright field image
X interval: [-4, 4]   X increment = .8
Y interval: [-.5, 1.5]   Y increment = .2

(b) dark field image
X interval: [-4, 4]   X increment = .8
Y interval: [-.05, .1]   Y increment = .015

FIGURE 6.26
Comparison of bright field and dark field imaging of the object shown in Figure 6.25(a).

details which makes the interpretation of dark field images troublesome.

A method which alleviates some of the problems with false details is *gray field imaging*. In gray field imaging the aperture stop *partially transmits* a small fraction of the light at the center of the transform, thus providing a gray background in the image (against which more details of the object are visible).

Example 6.12:
Suppose that the object described by (6.158) is imaged, but now a *gray field* pupil function of the form

$$P_{GF}(\lambda \Delta x) = \begin{cases} 0 & \text{if } |x| > 40 \\ 10 & \text{if } 2.5 < |x| < 40 \\ 1 & \text{if } |x| \le 2.5 \end{cases} \qquad (6.162)$$

is used [see Figure 6.27(a)].

In this case, the gray field image shown in Figure 6.27(b) results. The false detail no longer appears. Also, the contrast is about 0.4, or 40%, which is a tenfold improvement over the contrast in the bright field image.

6.11 Fourier Transforming Property of a Lens

As we saw in Section 6.10, with the dark field and gray field imaging examples, sometimes images can be enhanced by manipulating the transform of the object.

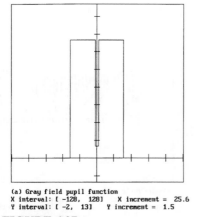 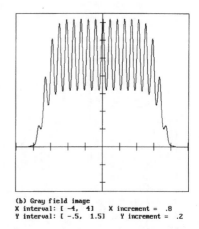

(a) Gray field pupil function
X interval: [-128, 128]   X increment = 25.6
Y interval: [-2, 13]   Y increment = 1.5

(b) Gray field image
X interval: [-4, 4]   X increment = .8
Y interval: [-.5, 1.5]   Y increment = .2

FIGURE 6.27
Gray field pupil function and gray field image of the object shown in Figure 6.25(a).

This requires, however, knowledge of the transform. While computer approximations can sometimes be used, it is an amazing fact that *when coherent illumination is used, nature produces the transform of the object automatically in the back focal plane of the imaging lens!* [See Formula (6.181).] In this section we will derive this important property mathematically.

If we put $\Delta = f$ in Formula (6.139), then we have

$$E(\mathbf{x}, \mathbf{u}, \mathbf{X}) = e^{\frac{i\pi}{\lambda D}|\mathbf{u}|^2} e^{-i\frac{2\pi}{\lambda}\mathbf{u}\cdot\left(\frac{\mathbf{X}}{D}+\frac{\mathbf{X}}{f}\right)} e^{\frac{i\pi}{\lambda D}|\mathbf{X}|^2} e^{\frac{i\pi}{\lambda f}|\mathbf{X}|^2}. \tag{6.163}$$

Substituting the right side of (6.163) into Formula (6.138) and rearranging the integrals, we obtain

$$\psi(\mathbf{X}, t) \approx \frac{e^{ic} e^{\frac{i\pi}{\lambda f}|\mathbf{X}|^2}}{\lambda^2 Df}$$

$$\int A(\mathbf{x})\psi(\mathbf{x}, t) e^{\frac{i\pi}{\lambda D}|\mathbf{X}|^2} \int P(\mathbf{u}) e^{\frac{i\pi}{\lambda D}|\mathbf{u}|^2} e^{-i\frac{2\pi}{\lambda D}\mathbf{u}\cdot\left(\mathbf{x}+\frac{D}{f}\mathbf{X}\right)} \, d\mathbf{u} \, d\mathbf{x}. \tag{6.164}$$

We now show how to simplify the integral

$$\frac{1}{\lambda D} \int P(\mathbf{u}) e^{\frac{i\pi}{\lambda D}|\mathbf{u}|^2} e^{-i\frac{2\pi}{\lambda D}\mathbf{u}\cdot\left(\mathbf{x}+\frac{D}{f}\mathbf{X}\right)} \, d\mathbf{u} \tag{6.165}$$

which is part of (6.164). First, we observe that

$$\left| \mathbf{u} - \left(\mathbf{x} + \frac{D}{f}\mathbf{X} \right) \right|^2 = |\mathbf{u}|^2 - 2\mathbf{u} \cdot \left(\mathbf{x} + \frac{D}{f}\mathbf{X} \right) + \left| \mathbf{x} + \frac{D}{f}\mathbf{X} \right|^2 . \tag{6.166}$$

It then follows from (6.166) that

$$e^{\frac{i\pi}{\lambda D}|\mathbf{u}|^2} e^{-i\frac{2\pi}{\lambda D}\mathbf{u} \cdot \left(\mathbf{x} + \frac{D}{f}\mathbf{X} \right)} = e^{\frac{i\pi}{\lambda D}\left| \mathbf{u} - \left(\mathbf{x} + \frac{D}{f}\mathbf{X} \right) \right|^2} e^{\frac{-i\pi}{\lambda D}\left| \mathbf{x} + \frac{D}{f}\mathbf{X} \right|^2} . \tag{6.167}$$

Hence (6.165) can be written as

$$\frac{1}{\lambda D} \int P(\mathbf{u}) e^{\frac{i\pi}{\lambda D}|\mathbf{u}|^2} e^{-i\frac{2\pi}{\lambda D}\mathbf{u} \cdot \left(\mathbf{x} + \frac{D}{f}\mathbf{X} \right)} d\mathbf{u}$$

$$= e^{\frac{-i\pi}{\lambda D}\left| \mathbf{x} + \frac{D}{f}\mathbf{X} \right|^2} \left[\frac{1}{\lambda D} \int P(\mathbf{u}) e^{\frac{i\pi}{\lambda D}\left| \mathbf{u} - \left(\mathbf{x} + \frac{D}{f}\mathbf{X} \right) \right|^2} d\mathbf{u} \right]. \tag{6.168}$$

The integral in brackets is a Fresnel integral, like the one we discussed in Section 6.2. In fact, as we pointed out in Remark 6.3, it represents a distorted image of the *lens* aperture. For example, suppose one uses a square lens aperture with pupil function

$$P(x, y) = \begin{cases} 1 & \text{if } |x| < 20 \text{ and } |y| < 20 \\ 0 & \text{if } |x| > 20 \text{ or } |y| > 20. \end{cases} \tag{6.169}$$

Then, we show in Figure 6.28(a) the graph of the real and imaginary parts of

$$\frac{1}{\lambda D} \int P(\mathbf{u}) e^{\frac{i\pi}{\lambda D}|\mathbf{u} - \mathbf{x}|^2} d\mathbf{u}$$

along one axis (using $\lambda D = 1$). Notice that there is some similarity to the pupil function defined in (6.169). This similarity is even greater near the origin; see Figure 6.28(b). In fact, for $\mathbf{x}$ near the origin, we have

$$\frac{1}{\lambda D} \int P(\mathbf{u}) e^{\frac{i\pi}{\lambda D}|\mathbf{u} - \mathbf{x}|^2} d\mathbf{u} \approx 0.7166 + 0.6976i$$

$$= e^{ib}, \qquad (b = 0.772).$$

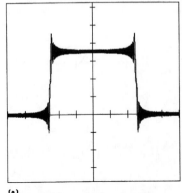

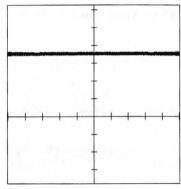

FIGURE 6.28
Fresnel diffraction from the pupil function defined in (6.169). (a) Graphs of the real and imaginary parts; (b) real and imaginary parts near the origin.

Hence, since $\mathbf{x}$ close to the origin implies $P(\mathbf{x}) = 1$, we can make the following approximation:

$$\frac{1}{\lambda D} \int P(\mathbf{u}) e^{\frac{i\pi}{\lambda D} |\mathbf{u} - \mathbf{x}|^2} \, d\mathbf{u} \approx e^{ib} P(\mathbf{x}). \tag{6.170}$$

Notice that (6.170) will also hold for $\mathbf{x}$ outside the lens aperture, since $P(\mathbf{x})$ is then 0 and [as you can see in Figure 6.28(a)] the left side of (6.170) is approximately 0. We shall assume that (6.170) holds for other pupil functions as well. Hence, replacing $\mathbf{x}$ by $\mathbf{x} + (D/f)\mathbf{X}$, we have the following approximation:

$$\frac{1}{\lambda D} \int P(\mathbf{u}) e^{\frac{i\pi}{\lambda D} \left| \mathbf{u} - \left(\mathbf{x} + \frac{D}{f} \mathbf{X} \right) \right|^2} \, d\mathbf{u} \approx e^{ib} P\left(\mathbf{x} + \frac{D}{f} \mathbf{X} \right). \tag{6.171}$$

Using the right side of (6.171) in (6.168) we have

$$\frac{1}{\lambda D} \int P(\mathbf{u}) \; e^{\frac{i\pi}{\lambda D} |\mathbf{u}|^2} e^{-i \frac{2\pi}{\lambda D} \mathbf{u} \cdot \left(\mathbf{x} + \frac{D}{f} \mathbf{X} \right)} \, d\mathbf{u}$$

$$\approx e^{ib} e^{\frac{-i\pi}{\lambda D} \left| \mathbf{x} + \frac{D}{f} \mathbf{X} \right|^2} P\left(\mathbf{x} + \frac{D}{f} \mathbf{X} \right). \tag{6.172}$$

Substituting the right side of (6.172) into (6.164) we get (writing e^{ia} in place of $e^{ic} e^{ib}$):

$$\psi(\mathbf{X}, t) \approx \frac{e^{ia}}{\lambda f} e^{\frac{i\pi}{\lambda f}|\mathbf{X}|^2}$$

$$\int A(\mathbf{x})\psi(\mathbf{x}, t) e^{\frac{i\pi}{\lambda D}|\mathbf{X}|^2} e^{\frac{-i\pi}{\lambda D}\left|\mathbf{x}+\frac{D}{f}\mathbf{X}\right|^2} P\left(\mathbf{x}+\frac{D}{f}\mathbf{X}\right) d\mathbf{x}. \quad (6.173)$$

Expanding $|\mathbf{x} + D\mathbf{X}/f|^2$, and factoring an exponential involving $|\mathbf{X}|^2$ outside the integral, we can rewrite (6.173) as

$$\psi(\mathbf{X}, t) \approx \frac{e^{ia}}{\lambda f} e^{\frac{i\pi}{\lambda f}\left(1-\frac{D}{f}\right)|\mathbf{X}|^2}$$

$$\int A(\mathbf{x})\psi(\mathbf{x}, t) P\left(\mathbf{x}+\frac{D}{f}\mathbf{X}\right) e^{-i\frac{2\pi}{\lambda f}\mathbf{x}\cdot\mathbf{X}} d\mathbf{x}. \quad (6.174)$$

From (6.174) we obtain, *using the coherency assumption (6.147) as we did in Section 6.10,*

$$I(\mathbf{X}) \approx \left| \frac{1}{\lambda f} \int A(\mathbf{x}) P\left(\mathbf{x}+\frac{D}{f}\mathbf{X}\right) e^{-i\frac{2\pi}{\lambda f}\mathbf{x}\cdot\mathbf{X}} d\mathbf{x} \right|^2. \quad (6.175)$$

The presence of the pupil factor $P(\mathbf{x} + D\mathbf{X}/f)$ in (6.175) is known as *vignetting*. The object $A(\mathbf{x})$ may not fully appear in the integrand if *the translated pupil* (translated by $-D\mathbf{X}/f$) does not completely contain the object.

One way of minimizing vignetting is to assume that the object described by the function A is very small relative to the lens pupil size. Here we are also assuming that the pupil function has the form described in in (6.130). Let R_A stand for the *radius of the object* A; in other words, R_A is the smallest positive number for which $A(\mathbf{x}) = 0$ whenever $|\mathbf{x}| < R_A$. Similarly, let R_P denote the radius of the lens aperture. Our *smallness assumption* is that $R_A \ll R_P$, in which case

$$A(\mathbf{x}) P\left(\mathbf{x}+\frac{D}{f}\mathbf{X}\right) = A(\mathbf{x}), \qquad \text{for } |\mathbf{X}| < \tfrac{f}{D}(R_P - R_A) \approx \tfrac{f}{D}R_P. \quad (6.176)$$

Using (6.176), Formula (6.175) simplifies to

$$I(\mathbf{X}) \approx \left| \frac{1}{\lambda f} \hat{A}\left(\frac{1}{\lambda f}\mathbf{X}\right) \right|^2, \qquad \text{for } |\mathbf{X}| < \tfrac{f}{D}(R_P - R_A) \approx \tfrac{f}{D}R_P. \quad (6.177)$$

Formula (6.177) shows that the intensity in the back focal plane of the lens is the modulus-squared of the (scaled) Fourier transform of the object function (provided one does not stray too far from the origin).

REMARK 6.9:   If we let $D \to 0$, then (6.175) becomes

$$I(\mathbf{X}) \approx \left| \frac{1}{\lambda f} \int A(\mathbf{x}) P(\mathbf{x}) e^{-i\frac{2\pi}{\lambda f}\mathbf{x}\cdot\mathbf{X}} d\mathbf{x} \right|^2. \qquad (6.178)$$

Formula (6.178) does not suffer from vignetting. In fact, we only need to assume that $R_A < R_P$ (not *much* less, just less) in order to obtain from (6.178)

$$I(\mathbf{X}) \approx \left| \frac{1}{\lambda f} \hat{A} \left(\frac{1}{\lambda f}\mathbf{X} \right) \right|^2 \qquad (6.179)$$

without the restrictions on $\mathbf{X}$ that are in (6.177). (*Note:* we also obtained Formula (6.179) in Exercise 6.53.) The practice of placing an object within the entrance aperture of a lens ($D = 0$ and $R_A < R_P$) is a very common practice, precisely because of the absence of vignetting effects. (See [HTW] for some beautiful examples.) ∎

Finally, we shall show that when an object is placed at a focal length in front of the lens, then an exact Fourier transform is obtained in the back focal plane. To see this, we note that when the object is placed at a distance of $D = f$ units in front of the imaging lens, Formula (6.174) simplifies to

$$\psi(\mathbf{X}, t) \approx \frac{e^{ia}}{\lambda f} \int A(\mathbf{x}) \psi(\mathbf{x}, t) P(\mathbf{x} + \mathbf{X}) e^{-i\frac{2\pi}{\lambda f}\mathbf{x}\cdot\mathbf{X}} d\mathbf{x}. \qquad (6.180)$$

If we make the smallness assumption $R_A \ll R_P$ again, then we get

$$\psi(\mathbf{X}, t) \approx \frac{e^{ia}}{\lambda f} \int A(\mathbf{x}) \psi(\mathbf{x}, t) e^{-i\frac{2\pi}{\lambda f}\mathbf{x}\cdot\mathbf{X}} d\mathbf{x}, \qquad \text{for } |\mathbf{X}| < R_P - R_A \approx R_P.$$

It is also true that when $|\mathbf{X}| > R_P + R_A$, then $A(\mathbf{x})P(\mathbf{x} + \mathbf{X}) = 0$. Hence

$$\psi(\mathbf{X}, t) = 0, \qquad \text{for } |\mathbf{X}| > R_P + R_A \approx R_P.$$

Since $R_P - R_A \approx R_P \approx R_P + R_A$, we combine these last two results by writing

$$\psi(\mathbf{X}, t) \approx \left[\frac{e^{ia}}{\lambda f} \int A(\mathbf{x}) \psi(\mathbf{x}, t) e^{-i\frac{2\pi}{\lambda f}\mathbf{x}\cdot\mathbf{X}} d\mathbf{x} \right] P(\mathbf{X}). \qquad (6.181)$$

Formula (6.181) shows that, except for the pupil factor $P(\mathbf{X})$, *the light $\psi(\mathbf{X}, t)$ at the back focal plane of the lens is precisely the (scaled) Fourier transform of $A(\mathbf{x})\psi(\mathbf{x}, t)$*. This result is the basis for *two-lens imaging*. A second lens is used to perform a second Fourier transform, thereby obtaining by Fourier inversion an inverted image of the object (see Exercise 6.67). Furthermore, because of (6.181), when modifications of the transform of the object function A are required (as in gray field imaging or Schlieren imaging, for instance), then *these modifications can be made in the back focal plane of the first lens*. For example, in gray field imaging, the partially transmitting material can be placed over the origin of the back focal plane of the first lens (instead of over the center of the lens aperture).

Since the basic theory and results are quite similar to what we described in Section 6.10, we will not discuss two-lens imaging. The interested reader can find good discussions in [Go] and [Ii]. Modifying the transform of an object in order to change the image is known as *spatial filtering*. For more details see [Go], [Ii], [Pi], [Mi], and [Me].

6.12 Imaging with Incoherent Light

The second main type of imaging is imaging under *incoherent illumination*. In Section 6.1 we gave one definition of incoherent illumination; see (6.10). Comparing this formula with Formula (6.24) we see that incoherency is also defined by

$$\gamma(\mathbf{x}, \mathbf{z}) = 0 \qquad \text{for } \mathbf{x} \neq \mathbf{z}. \tag{6.182}$$

From Formulas (6.23b) and (6.4) we also have

$$\gamma(\mathbf{x}, \mathbf{x}) = I(\mathbf{x}, 0). \tag{6.183}$$

Thus, if we write $I(\mathbf{x})$ in place of $I(\mathbf{x}, 0)$, the cross-correlation function $\gamma(\mathbf{x}, \mathbf{z})$ must satisfy the following two conditions:

$$\gamma(\mathbf{x}, \mathbf{x}) = I(\mathbf{x}) \tag{6.184a}$$

$$\gamma(\mathbf{x}, \mathbf{z}) = 0 \qquad \text{for } \mathbf{x} \neq \mathbf{z}. \tag{6.184b}$$

These conditions, however, lead to trouble if we use them in the imaging Equation (6.144). In fact, (6.184b) implies that the inner integral in (6.144) has an integrand that equals 0 except at the one point $\mathbf{x}$ that equals $\mathbf{z}$, hence (6.144) reduces to

$$I(\mathbf{X}) \approx \frac{1}{(\lambda^2 D \Delta)^2} \int 0 \times \left\{ A(\mathbf{z}) e^{\frac{i\pi}{\lambda D} |\mathbf{z}|^2} \hat{P} \left(\frac{\mathbf{z}}{\lambda D} + \frac{\mathbf{X}}{\lambda \Delta} \right) \right\}^* d\mathbf{z}.$$

Thus, $I(\mathbf{X}) \approx 0$, which is not a very useful result.

To extricate ourselves from this problem, we must more carefully formulate the notion of incoherency. We will examine what happens if we view the cross-correlation function γ in Formulas (6.184a) and (6.184b) as a limit. For example, if $\gamma(\mathbf{x}, \mathbf{z})$ is defined by the following limit:

$$\gamma(\mathbf{x}, \mathbf{z}) = \lim_{\epsilon \to 0+} I(\mathbf{x}) e^{-\pi |\mathbf{X}-\mathbf{z}|^2 / \epsilon^2} \tag{6.185}$$

then (6.184a) and (6.184b) are satisfied. The functions $I(\mathbf{x}) e^{-\pi |\mathbf{X}-\mathbf{z}|^2 / \epsilon^2}$ have an important property. If a function $g(\mathbf{z})$ is continuous at a point $\mathbf{x}$, and bounded for all values of $\mathbf{z}$, then

$$\int g(\mathbf{z}) I(\mathbf{x}) e^{-\pi |\mathbf{X}-\mathbf{z}|^2 / \epsilon^2} d\mathbf{z} \approx \epsilon^2 I(\mathbf{x}) g(\mathbf{x}) \tag{6.186}$$

provided ϵ is close enough to 0. For example, in Figure 6.29(a) we show the graph of $I(\mathbf{x}) e^{-\pi |\mathbf{X}-\mathbf{z}|^2 / \epsilon^2}$ for $\mathbf{x}$ restricted to the x-axis, $\mathbf{z} = (0, 0)$, $\epsilon = 0.05$, and $I(\mathbf{x})$ equal to the constant 400. When this function is integrated as a function of $\mathbf{z}$ against a function $g(\mathbf{z})$, like the one shown in Figure 6.29(b), a convolution is being performed. And, as you can see from Figure 6.29(c), the function g is closely approximated. This is what Formula (6.186) says should happen when $\epsilon = 0.05$ and $I(\mathbf{x}) = 400$.

We will demonstrate (6.186) at the end of this section. For now, let's just assume that (6.186) holds. Because of the limit in (6.185), let's also assume that ϵ is so close to 0 that $I(\mathbf{x}) e^{-\pi |\mathbf{X}-\mathbf{z}|^2 / \epsilon^2} \approx \gamma(\mathbf{x}, \mathbf{z})$, too. We can then rewrite (6.186) as

$$\int g(\mathbf{z}) \gamma(\mathbf{x}, \mathbf{z}) \, d\mathbf{z} \approx \epsilon^2 I(\mathbf{x}) g(\mathbf{x}). \tag{6.187}$$

Formula (6.187) is our new definition of incoherency. That is, illumination is *incoherent* if the approximation in (6.187) can be assumed to hold for each bounded function g that is continuous at $\mathbf{x}$.

If we apply (6.187) to the $\mathbf{z}$-integral in the imaging formula (6.144), replacing $g(\mathbf{z})$ by the function of $\mathbf{z}$ in the integrand of (6.144), we obtain

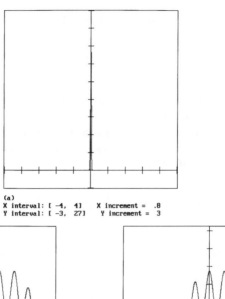

(a)
X interval: [-4, 4]   X increment = .8
Y interval: [-3, 27]   Y increment = 3

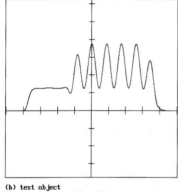

(b) test object
X interval: [-4, 4]   X increment = .8
Y interval: [-3, 5]   Y increment = .8

(c) convolution
X interval: [-4, 4]   X increment = .8
Y interval: [-3, 5]   Y increment = .8

FIGURE 6.29
Illustration of Formula (6.186).

$$\int \left[A(\mathbf{z}) e^{\frac{i\pi}{\lambda D}|\mathbf{z}|^2} \hat{P}\left(\frac{\mathbf{z}}{\lambda D} + \frac{\mathbf{X}}{\lambda \Delta} \right) \right]^* \gamma(\mathbf{x}, \mathbf{z}) \, d\mathbf{z}$$

$$\approx \epsilon^2 I(\mathbf{x}) \left[A(\mathbf{x}) e^{\frac{i\pi}{\lambda D}|\mathbf{x}|^2} \hat{P}\left(\frac{\mathbf{x}}{\lambda D} + \frac{\mathbf{X}}{\lambda \Delta} \right) \right]^*. \tag{6.188}$$

Substituting the right side of (6.188) in place of the $\mathbf{z}$-integral in the imaging equation (6.144), we obtain for the intensity $I(\mathbf{X})$ at the observation plane

$$I(\mathbf{X}) \approx \frac{\epsilon^2}{(\lambda^2 D \Delta)^2} \int \left| A(\mathbf{x}) \hat{P}\left(\frac{\mathbf{x}}{\lambda D} + \frac{\mathbf{X}}{\lambda \Delta} \right) \right|^2 I(\mathbf{x}) \, d\mathbf{x}. \tag{6.189}$$

Notice that the complex exponential factor does not appear in (6.189), since

$$\left| e^{\frac{i\pi}{\lambda D}|\mathbf{X}|^2} \right|^2 = 1.$$

Unlike coherent illumination, no small object approximation is needed when incoherent illumination is used.

For simplicity, we shall assume that the intensity $I(\mathbf{x})$ is uniform over the aperture, say, $I(\mathbf{x}) = B$ where B is a positive constant. Then, writing κ in place of $B\epsilon^2$, we can rewrite (6.189) as

$$I(\mathbf{X}) \approx \frac{\kappa}{(\lambda^2 D \Delta)^2} \int \left| A(\mathbf{x}) \hat{P} \left(\frac{\mathbf{x}}{\lambda D} + \frac{\mathbf{X}}{\lambda \Delta} \right) \right|^2 d\mathbf{x}. \tag{6.190}$$

We will generally ignore the constant κ, since we can treat the rest of Formula (6.190) very efficiently without worrying about the value of κ. (See Remark 6.10 below.)

As we did for the coherent case, we substitute $\mathbf{x} = -\mathbf{s}/M$ and $d\mathbf{x} = d\mathbf{s}/M^2$, where $M = \Delta/D$ is the degree of magnification, and we obtain from (6.190)

$$I(\mathbf{X}) \approx \int \left| A \left(\frac{-\mathbf{s}}{M} \right) \right|^2 \left| \frac{1}{(\lambda \Delta)^2} \hat{P} \left(\frac{1}{\lambda \Delta} (\mathbf{X} - \mathbf{s}) \right) \right|^2 d\mathbf{s}. \tag{6.191}$$

The integral in (6.191) is a two-dimensional convolution. As opposed to the coherent case, the function

$$\left| A \left(\frac{-\mathbf{X}}{M} \right) \right|^2 \tag{6.192}$$

is convolved with

$$\left| \frac{1}{(\lambda \Delta)^2} \hat{P} \left(\frac{1}{\lambda \Delta} \mathbf{X} \right) \right|^2. \tag{6.193}$$

The function in (6.192) is a magnified, inverted, form of $|A|^2$ which corresponds to the *intensity* of the aperture function (object). The function in (6.193) is the *PSF for incoherent illumination*. Notice that this *PSF* is the modulus-squared of the *PSF* for coherent illumination. The function in (6.193) is the *power spectrum* of the function $P(\lambda \Delta \mathbf{x})$. The operations involved are

$$P(\lambda \Delta \mathbf{x}) \xrightarrow{\mathcal{F}} \frac{1}{(\lambda \Delta)^2} \hat{P} \left(\frac{1}{\lambda \Delta} \mathbf{X} \right) \xrightarrow{|\cdot|^2} \left| \frac{1}{(\lambda \Delta)^2} \hat{P} \left(\frac{1}{\lambda \Delta} \mathbf{X} \right) \right|^2. \tag{6.194}$$

When using *FAS* there is a choice for computing a power spectrum automatically; it is part of the *Fourier transform* procedure.

The procedures for computing images under incoherent illumination are quite similar to those described in section 6.10, the main difference being that (6.194) is used in place of (6.155) to compute *PSFs*. We will limit ourselves to one example (some others are described in the exercises).

Example 6.13:
Suppose that incoherent light having wavelength $\lambda = 5 \times 10^{-4}$ mm illuminates an edge defined by

$$A(x, y) = \begin{cases} 1 & \text{if } x > 0 \\ 0 & \text{if } x < 0 \end{cases}$$

and this edge is imaged by a lens with a square aperture of side length 40 mm and focal length 1 m. Assume that $D = 2$ m and $\Delta = 2$ m. Describe the image that results.

SOLUTION Since $|A|^2 = A$, the only difference between this example and Example 6.9 is that the *PSF* is now (*note:* $\lambda\Delta = 1$)

$$\left| \frac{1}{(\lambda\Delta)^2} \hat{P}\left(\frac{1}{\lambda\Delta}\mathbf{X}\right) \right|^2 = [40 \operatorname{sinc}(40X)]^2 [40 \operatorname{sinc}(40Y)]^2.$$

In this case, using *Parseval's equality* [see (5.81a), Chapter 5]

$$\int_{-\infty}^{\infty} [40 \operatorname{sinc}(40Y)]^2 \, dY = \int_{-\infty}^{\infty} [\operatorname{rect}(y/40)]^2 \, dy$$

$$= \int_{-20}^{20} 1 \, dy = 40.$$

Therefore, we will convolve

$$f(X) = \begin{cases} 1 & \text{for } X > 0 \\ 0 & \text{for } X < 0 \end{cases}$$

with $g(X) = 40[40 \operatorname{sinc}(40X)]^2$. The result is shown in Figure 6.30(b); it was obtained using *FAS* with $[-4, 4]$ as the interval and 8192 points.

It is very interesting to compare this result for incoherent imaging of an edge with the result for coherent imaging from Example 6.9. For incoherent imaging there is no oscillation or Gibbs' effect at the boundary of the edge. *The absence of these defects is a major advantage that incoherent imaging has over coherent imaging.* There is also a more subtle advantage involving the location of the edge boundary. If the edge boundary is marked by half of maximum intensity, then

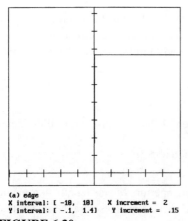

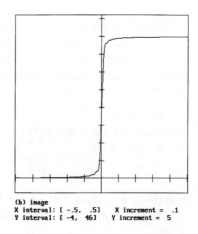

(a) edge
X interval: [-10, 10]   X increment = 2
Y interval: [-.1, 1.4]   Y increment = .15

(b) image
X interval: [-.5, .5]   X increment = .1
Y interval: [-4, 46]   Y increment = 5

FIGURE 6.30
Imaging of an edge using incoherent light.

the edge boundary in incoherent imaging nearly equals the edge boundary of the object. (This near equality can be proved theoretically; we leave this proof to the reader.) For coherent imaging, we saw in Example 6.9 that the edge boundary in the image is displaced by a small amount. ∎

Our remarks at the end of the example above show some of the nice features that an incoherent image of an edge has in comparison to the coherent image of an edge. This one example does not, however, tell the whole story. We will continue to compare these two types of imaging in the exercises. It is also important to note that *processing of images* is more difficult with incoherent light (there is no Fourier transform in the focal plane).

REMARK 6.10:   There is a small problem that occurs in the previous example. Notice that the *form* of the image appears to be a smooth version of an edge, but the vertical (intensity) scale is much larger for the image than for the original object. This discrepancy arises from the fact that we left the constant κ in Formula (6.190) unspecified (in effect, we set it equal to 1). In the Exercises, see Remark 6.16 preceding Exercises 6.74 to 6.77, we describe the standard method for handling this difficulty. ∎

We close this section by discussing the validity of (6.186). First, since $I(\mathbf{x})$ does not depend on $\mathbf{z}$, we have

$$\int g(\mathbf{z})I(\mathbf{x})e^{-\pi|\mathbf{x}-\mathbf{z}|^2/\epsilon^2}\,d\mathbf{z} = I(\mathbf{x})\int g(\mathbf{z})e^{-\pi|\mathbf{x}-\mathbf{z}|^2/\epsilon^2}\,d\mathbf{z}. \qquad (6.195)$$

Consequently, it suffices to show that whenever $g(\mathbf{z})$ is continuous at $\mathbf{x}$ and bounded for all values of $\mathbf{z}$, we have

$$\int g(\mathbf{z})e^{-\pi|\mathbf{X}-\mathbf{Z}|^2/\epsilon^2}\,d\mathbf{z} \approx \epsilon^2 g(\mathbf{x}) \qquad (6.196)$$

for ϵ close enough to 0. We now proceed to sketch the proof of (6.196). (A more complete discussion can be found in [Wa, sections 6.4 and 6.7].)

To establish (6.196) we make use of the following two facts:

$$e^{-\pi|\mathbf{S}|^2/\epsilon^2} \approx 0 \qquad \text{if } |\mathbf{s}| \geq \sqrt{\epsilon} \text{ and } \epsilon \approx 0 \qquad (6.197a)$$

$$\int e^{-\pi|\mathbf{S}|^2/\epsilon^2}\,d\mathbf{s} = \epsilon^2 \qquad \text{for each } \epsilon > 0. \qquad (6.197b)$$

The approximation in (6.197a) follows from $0 \leq e^{-\pi|\mathbf{S}|^2/\epsilon^2} \leq e^{-\pi/\epsilon}$ and $e^{-\pi/\epsilon} \to 0$ as $\epsilon \to 0+$. Formula (6.197b) holds because the integral on its left side splits into a product of two integrals, both of which are equal to $\int_{-\infty}^{\infty} e^{-\pi x^2/\epsilon^2}\,dx$. Making a change of variable, we have $\int_{-\infty}^{\infty} e^{-\pi x^2/\epsilon^2}\,dx = \epsilon \int_{-\infty}^{\infty} e^{-\pi x^2}\,dx$. Since $\int_{-\infty}^{\infty} e^{-\pi x^2}\,dx = 1$ (see Example 5.5 in Chapter 5), we have $\int_{-\infty}^{\infty} e^{-\pi x^2/\epsilon^2}\,dx = \epsilon$.

Now, by making the change of variable $\mathbf{z} = \mathbf{x} + \mathbf{s}$, $d\mathbf{z} = d\mathbf{s}$, we have

$$\int g(\mathbf{z})e^{-\pi|\mathbf{X}-\mathbf{Z}|^2/\epsilon^2}\,d\mathbf{z} = \int g(\mathbf{x}+\mathbf{s})e^{-\pi|\mathbf{S}|^2/\epsilon^2}\,d\mathbf{s}. \qquad (6.198)$$

Using (6.197a), we then have

$$\int g(\mathbf{x}+\mathbf{s})e^{-\pi|\mathbf{S}|^2/\epsilon^2}\,d\mathbf{s} \approx \int_{\{\mathbf{S}:|\mathbf{S}|\leq\sqrt{\epsilon}\}} g(\mathbf{x}+\mathbf{s})e^{-\pi|\mathbf{S}|^2/\epsilon^2}\,d\mathbf{s}. \qquad (6.199)$$

Since ϵ is close to 0, it follows that when $|\mathbf{s}| \leq \sqrt{\epsilon}$ we must have $\mathbf{x}+\mathbf{s} \approx \mathbf{x}$. Since g is continuous at $\mathbf{x}$, having $\mathbf{x}+\mathbf{s} \approx \mathbf{x}$ implies that $g(\mathbf{x}+\mathbf{s}) \approx g(\mathbf{x})$. Therefore,

$$\int_{\{\mathbf{S}:|\mathbf{S}|\leq\sqrt{\epsilon}\}} g(\mathbf{x}+\mathbf{s})e^{-\pi|\mathbf{S}|^2/\epsilon^2}\,d\mathbf{s} \approx g(\mathbf{x})\int_{\{\mathbf{S}:|\mathbf{S}|\leq\sqrt{\epsilon}\}} e^{-\pi|\mathbf{S}|^2/\epsilon^2}\,d\mathbf{s}. \qquad (6.200)$$

Combining (6.198) to (6.200), we obtain

$$\int g(\mathbf{z})e^{-\pi|\mathbf{X}-\mathbf{Z}|^2/\epsilon^2}\,d\mathbf{z} \approx g(\mathbf{x})\int_{\{\mathbf{S}:|\mathbf{S}|\leq\sqrt{\epsilon}\}} e^{-\pi|\mathbf{S}|^2/\epsilon^2}\,d\mathbf{s}. \qquad (6.201)$$

Using (6.197a) again, we have

$$\int_{\{\mathbf{S}:|\mathbf{S}|\le\sqrt{\epsilon}\}} e^{-\pi|\mathbf{S}|^2/\epsilon^2}\,d\mathbf{s} \approx \int e^{-\pi|\mathbf{S}|^2/\epsilon^2}\,d\mathbf{s} = \epsilon^2. \tag{6.202}$$

The equality holds because of (6.197b).

From (6.201) and (6.202) we conclude that

$$\int g(\mathbf{z})e^{-\pi|\mathbf{X}-\mathbf{Z}|^2/\epsilon^2}\,d\mathbf{z} \approx g(\mathbf{x})\epsilon^2$$

which proves (6.196). Combining (6.196) and (6.195) we conclude that (6.186) holds.

References

For more on Fourier optics, see [Go], [Ii], and [RDPT]. For more on image formation and image processing, see [Pi], [Me], [Mi], and [Mi-T]. Coherence is discussed in greater detail in [Bo-W] and [Go,2]. Many beautiful photographs can be found in [HTW].

Exercises

Section 1

6.1 Suppose that $\psi(\mathbf{x}, 0, t) = e^{i2\pi\nu t}$. [This is called *plane wave* illumination.] Show that plane wave illumination is coherent.

6.2 Suppose that $\psi(\mathbf{x}, 0, t) = e^{i\frac{2\pi}{\lambda}(\mathbf{a}\cdot\mathbf{x}-ct)}$ where $\mathbf{a}$ is a fixed vector in $\mathbf{R}^2$ and c is the speed of light. [This is also called *plane wave* illumination.] Show that

$$\Gamma(\mathbf{x}, \mathbf{z}) = \left[e^{i\frac{2\pi}{\lambda}\mathbf{a}\cdot\mathbf{x}} \right]\left[e^{i\frac{2\pi}{\lambda}\mathbf{a}\cdot\mathbf{z}} \right]^*.$$

6.3 Suppose that light diffracted from a single point source (as in Example 6.1) illuminates a parallel screen a distance D away. Show that, by choosing proper coordinates, the light $\psi(\mathbf{x}, 0, t)$ at the screen is described by

$$\psi(\mathbf{x}, 0, t) = \frac{S(t)e^{i\frac{2\pi}{\lambda}\left[D^2+|\mathbf{x}|^2\right]^{\frac{1}{2}}}}{\lambda\left[D^2 + |\mathbf{x}|^2\right]^{\frac{1}{2}}} \tag{6.203}$$

where $S(t)$ represents the light passing through the point source.

6.4 Show that for the function $\psi(\mathbf{x}, 0, t)$ from Exercise 6.3,

$$\Gamma(\mathbf{x}, \mathbf{z}) = \left[e^{i\frac{2\pi}{\lambda}\left[D^2 + |\mathbf{x}|^2\right]^{\frac{1}{2}}} \right] \left[e^{i\frac{2\pi}{\lambda}\left[D^2 + |\mathbf{z}|^2\right]^{\frac{1}{2}}} \right]^*.$$

6.5 *Generalized Coherence.* A more general definition of coherence than (6.13) is to assume that $\Gamma(\mathbf{x}, \mathbf{z})$ satisfies

$$\Gamma(\mathbf{x}, \mathbf{z}) = C(\mathbf{x})C^*(\mathbf{z}) \tag{6.204}$$

for some function C. Show that the following statements are true:

(a) If Γ satisfies (6.13), then it also satisfies (6.204).

(b) Each of the functions ψ described in Exercises 6.1, 6.2, and 6.3 are coherent in the sense of (6.204).

(c) Given the coherency defined in (6.204), Formula (6.20) becomes

$$I(\mathbf{u}, D) = \left| \int_{\mathbf{R}^2} \frac{T(\mathbf{x})}{\lambda |C_{\mathbf{x}}|} e^{i\frac{2\pi}{\lambda}|C_{\mathbf{x}}|} d\mathbf{x} \right|^2 \tag{6.205}$$

where $T(\mathbf{x}) = A(\mathbf{x})C(\mathbf{x})\sqrt{I(\mathbf{x}, 0)}$.

REMARK 6.11: The function $T(\mathbf{x})$ takes account of the aperture, via the aperture function $A(\mathbf{x})$, *and the nature of the light illuminating the aperture*, via the *coherency factor* $C(\mathbf{x})$ multiplied by the intensity factor $\sqrt{I(\mathbf{x}, 0)}$. In order to simplify our calculations in subsequent sections we will work with (6.22) where $A(\mathbf{x})$ is described by (6.17). This simplification can be realized approximately if a setup like the one shown in Figure 6.2 is used to generate coherent illumination of the aperture (see Exercise 6.49). ∎

6.6 Suppose that $\psi(\mathbf{u}, D, t)$ and $\psi(\mathbf{v}, D, t)$ are used in place of $\psi(\mathbf{x}, D, t)$ and $\psi(\mathbf{z}, D, t)$ in Formula (6.23b) [or (6.23a)]. Show that, if the light is coherent in the sense of formula (6.204) in Exercise 6.5, then $\gamma(\mathbf{u}, \mathbf{v}) = \kappa(\mathbf{u})\kappa^*(\mathbf{v})$ for some function κ. Consequently, $\Gamma(\mathbf{u}, \mathbf{v}) = K(\mathbf{u})K^*(\mathbf{v})$ for some function K. Thus, *light that is coherent over the aperture remains coherent after diffraction.*

6.7 Prove Inequality (6.12). *Hint:* use the Schwarz inequality (see Appendix C).

Section 2

6.8 Using *FAS*, graph approximations to $I_1(u, D)$ in (6.49) given

$$A_1(x) = \text{rect}\left(\frac{x-2}{2}\right) + \text{rect}\left(\frac{x+2}{2}\right)$$

and $\lambda = 7 \times 10^{-4}$ mm (red light), $D = 100$ mm, 200 mm, and 300 mm.

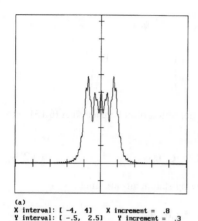

(a)
X interval: [-4, 4]   X increment = .8
Y interval: [-.5, 2.5]   Y increment = .3

(b)
X interval: [-4, 4]   X increment = .8
Y interval: [-.5, 2.5]   Y increment = .3

FIGURE 6.31

Fresnel diffraction pattern for a rectangle. (a) Along u-direction; (b) along v-direction.

6.9 Draw the diffraction patterns, along the u-direction and v-directions separately, for a rectangular aperture with aperture function

$$A(x, y) = \begin{cases} 1 & \text{if } |x| < 1 \text{ mm and } |y| < 2 \text{ mm} \\ 0 & \text{if } |x| > 1 \text{ mm or } |y| > 2 \text{ mm} \end{cases}$$

when $\lambda = 5 \times 10^{-4}$ mm, $D = 200$ mm, 300 mm, 400 mm. (Your graphs for 400 mm should look like those in Figure 6.31.)

6.10 Sketch the diffraction pattern when the aperture function is $A(x, y) = A_1(x) \text{rect}(y)$ where $A_1(x)$ is the function given in Exercise 6.8. Use the same values of λ and D as in Exercise 6.8.

6.11 Graph the intensity function $I_1(u, D)$ when the aperture function is $e^{-\pi(x^2+y^2)}$ and $\lambda = 5 \times 10^{-4}$ mm, $D = 100$ mm, 200 mm, 400 mm.

6.12 *Edge Diffraction.* Suppose that $A_1(x)$ in Formula (6.49) is given by

$$A_1(x) = \begin{cases} 1 & \text{if } 0 < x < 16 \\ 0 & \text{if } -16 < x < 0. \end{cases}$$

Using *FAS,* plot approximations to $I_1(u, D)$ in Formula (6.49). Use 8192 points, interval $[-16, 16]$, $\lambda = 5 \times 10^{-4}$ mm, and $D = 200$ mm. Confirm that $I_1(u, D)$ has the form shown in Figure 6.32.

6.13 Repeat Exercise 6.12 for $\lambda = 6 \times 10^{-4}$ mm and $\lambda = 7 \times 10^{-4}$ while keeping $D = 200$ mm. Compare your graphs to the result from Exercise 6.12, shown in Figure 6.32. What is the effect of increasing the wavelength?

6.14 *Fresnel integrals.* A *Fresnel cosine integral* F_C^β and *Fresnel sine integral* F_S^β are defined by

$$F_C^\beta(x) = \int_0^x \cos\left(\frac{\pi t^2}{\beta}\right) dt, \qquad F_S^\beta(x) = \int_0^x \sin\left(\frac{\pi t^2}{\beta}\right) dt \qquad (6.206)$$

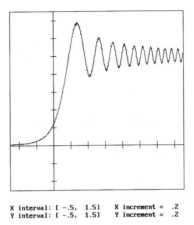

X interval: [-.5, 1.5]   X increment = .2
Y interval: [-.5, 1.5]   Y increment = .2

FIGURE 6.32
Fresnel diffraction by an edge.

where β is a positive constant. (a) Show that $y = F_C^\beta(x)$ solves the following initial-value problem: $y' = \cos(\pi x^2/\beta)$, $y(0) = 0$. And, show that $y = F_S^\beta(x)$ solves $y' = \sin(\pi x^2/\beta)$, $y(0) = 0$. (b) Use the results in (a) to graph $F_C^\beta(x)$ and $F_S^\beta(x)$, for $\beta = 2$, over the interval $[0, 10.24]$ using 1024 points [this is done similarly to the method described in Exercise 5.28]. (c) Consult a table (e.g., [A-S, page 321]) and verify that the *FAS* computed values for F_C^β and F_S^β are all accurate to 6 decimal places. (d) Show that if $f(x) = \text{rect}(x/2)$, then the intensity $I_1(u, 0)$ in (6.42) can be expressed as

$$I_1(u, D) = \frac{1}{\lambda D} \left\{ \left[F_C^{\lambda D}(u + 1) - F_C^{\lambda D}(u - 1) \right]^2 \right.$$

$$\left. + \left[F_S^{\lambda D}(u + 1) - F_S^{\lambda D}(u - 1) \right]^2 \right\}. \quad (6.207)$$

6.15 (a) Using (6.207) in Exercise 6.14, graph the intensity I_1 over the interval $[-4, 4]$ using 4096 points, letting $\lambda = 5 \times 10^{-4}$ mm and $D = 300$ mm. [*Note:* the functions that you use in formula (6.207) will have to be the *odd extensions* of the solutions to the initial-value problems defined in Exercise 6.14(a) for $0 \leq x \leq 5$ (*not* $0 \leq x \leq 4$).] (b) Compute the intensity I_1 by the convolution method described in Example 6.3, using the interval $[-4, 4]$ and 4096 points. (c) Find the Sup-Norm difference between the two versions of I_1 found in parts (a) and (b) *over the interval* $[-2, 2]$ (*not* over the interval $[-4, 4]$).

REMARK 6.12: The point of Exercises 6.14 and 6.15 is to show that (for rect functions, anyway) the Fresnel diffraction pattern can be computed accurately using Fresnel integrals, *and* that the convolution method that we have described gives similar results. The advantage of the convolution method is that it applies to a much wider variety of apertures than just rectangular ones (it applies to all the separable aperture functions, as we pointed out at the end of Section 6.2).

Section 3

6.16 Show that the two-dimensional Fourier transform defined in (6.59) satisfies the following properties [see Theorem 5.1 in Chapter 5].

(a) *Linearity.* For all complex numbers α and β

$$\alpha A(\mathbf{x}) + \beta B(\mathbf{x}) \xrightarrow{\mathcal{F}} \alpha \hat{A}(\mathbf{u}) + \beta \hat{B}(\mathbf{u}).$$

(b) *Scaling.* If ρ is a positive constant, then

$$A\left(\frac{1}{\rho}\mathbf{x}\right) \xrightarrow{\mathcal{F}} \rho^2 \hat{A}(\rho\mathbf{u}) \qquad \text{and} \qquad A(\rho\mathbf{x}) \xrightarrow{\mathcal{F}} \frac{1}{\rho^2} \hat{A}\left(\frac{1}{\rho}\mathbf{u}\right).$$

(c) *Shifting.* For each $\mathbf{c}$ in $\mathbf{R}^2$,

$$A(\mathbf{x} - \mathbf{c}) \xrightarrow{\mathcal{F}} \hat{A}(\mathbf{u})e^{-i2\pi\mathbf{c}\cdot\mathbf{u}}.$$

(d) *Modulation.* For each $\mathbf{c}$ in $\mathbf{R}^2$,

$$A(\mathbf{x})e^{i2\pi\mathbf{c}\cdot\mathbf{x}} \xrightarrow{\mathcal{F}} \hat{A}(\mathbf{u} - \mathbf{c}).$$

6.17 *Symmetry through the origin.* Most aperture functions are real valued (for an exception, see Exercise 6.27). Prove that if $A(\mathbf{x})$ is a real-valued aperture function, then its intensity function $I(\mathbf{u}, D)$ satisfies

$$I(-\mathbf{u}, D) = I(\mathbf{u}, D). \tag{6.208}$$

Formula (6.208) says that the Fraunhofer diffraction pattern from an aperture, with real-valued aperture function, is *symmetric about the origin* (unchanged after reflection through the origin).

6.18 *Translational symmetry.* Show that if an aperture is shifted by $\mathbf{c} = (c, d)$, then its diffraction pattern is *unchanged*.

6.19 *Rotational symmetry.* Suppose that an aperture is symmetrical in the sense that a rotation by an angle θ, about the origin in the plane of the aperture, leaves the aperture unchanged. Give a simple physical explanation for why the diffraction pattern from such an aperture possesses the same rotational symmetry.

6.20 The diffraction pattern from an aperture shaped like an equilateral triangle possesses six-fold rotational symmetry (i.e., a 60° rotation about the origin leaves the diffraction pattern unchanged). See Figure 6.33. The equilateral triangle, however, possesses only threefold rotational symmetry. Explain why the diffraction pattern has more symmetry than the aperture.

6.21 *Abbe's formula.* Suppose $\mathcal{R}$ is a bounded region in the plane where *Green's identity*

$$\int_{\mathcal{R}} \left(\frac{\partial^2 f}{\partial x^2} + \frac{\partial^2 f}{\partial y^2}\right) dx\, dy = \int_{\partial\mathcal{R}} \frac{\partial f}{\partial \eta}\, ds \tag{6.209}$$

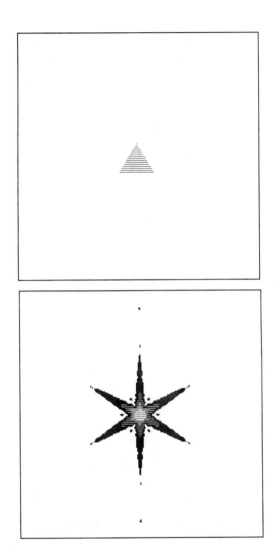

FIGURE 6.33
Equilateral triangle and its diffraction pattern.

holds. [Here, $\partial\mathcal{R}$ denotes the boundary of the region, η denotes the unit normal pointing outward from the boundary, and ds denotes the arc-length differential along the boundary (traveling in a direction so that the unit normal points to the right).] Show that for all $\mathbf{u}$, except $\mathbf{u} = (0, 0)$,

$$\int_{\mathcal{R}} e^{-i2\pi \mathbf{u}\cdot\mathbf{x}}\, d\mathbf{x} = \frac{-1}{(2\pi)^2|\mathbf{u}|^2} \int_{\partial\mathcal{R}} \frac{\partial}{\partial\eta}\left[e^{-i2\pi \mathbf{u}\cdot\mathbf{x}}\right] ds. \qquad (6.210)$$

REMARK 6.13: Formula (6.210) is called *Abbe's formula.* It shows that *if* $\mathcal{R}$ *is a region where Green's identity holds, then Fraunhofer diffraction is generated from the boundary of* $\mathcal{R}$. Green's identity holds for a wide variety of regions, such as circular discs, annuli, rectangles, triangles, and many other more complicated regions. ∎

6.22 The diffraction pattern from an equilateral triangle is shown in Figure 6.33. Using Formula (6.210) from Exercise 6.21, derive the intensity function $I(\mathbf{u}, D)$ for an equilateral triangular aperture. In particular, show why there are *arms* of higher intensity in the pattern at right angles to the three sides of the triangle.

Section 4

6.23 By choosing the x-interval appropriately, estimate the first three zeroes of $\hat{a}(\rho)$ where $a(x)$ is given in (6.81). And, estimate the values of the two local extreme values of $\hat{a}(\rho)$ lying to the right of the origin.

6.24 Graph the radial dependence of $I(\mathbf{u}, D)^{1/2}$ for a circular aperture of radius $R > 0$. You should be able to verify Figure 6.34.

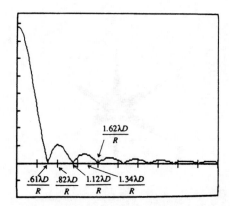

X interval: [0, 4]   X increment = .4
Y interval: [-.5, 3.5]   Y increment = .4

FIGURE 6.34
Radial dependence of the square root of intensity $I^{1/2}$ for diffraction from a circular aperture of radius R.

6.25 *Diffraction from annuli.* Suppose that coherent light of wavelength λ illuminates an annular aperture, with aperture function

$$A(\mathbf{x}) = \begin{cases} 1 & \text{if } R_1 < |\mathbf{x}| < R_2 \\ 0 & \text{if } R_1 > |\mathbf{x}| \text{ or } |\mathbf{x}| > R_2. \end{cases}$$

Graph $\hat{A}(\rho)$ and the radial dependencies of the intensity functions $I(\mathbf{u}, D)$ for the following radii:

(a) $R_1 = 1$, $R_2 = 2$

(b) $R_1 = 1$, $R_2 = 1.1$

(c) $R_1 = 1$, $R_2 = 1.05$.

6.26 *Alternate derivation of (6.76).* Let $\mathcal{R}$ stand for a disc of radius R centered at the origin. (a) Use Formula (6.210) in Exercise 6.21 to show that

$$\int_{\{\mathbf{X}:|\mathbf{X}|\leq R\}} e^{-i2\pi \mathbf{u}\cdot\mathbf{X}} \, d\mathbf{x} = \frac{-1}{(2\pi\rho)^2} \int_0^{2\pi} R^2 \frac{\partial}{\partial r} \left[e^{-i2\pi r\rho \cos(\theta-\phi)} \right]\Big|_{r=R} \, d\theta.$$

(6.211)

[*Hint:* use the identities $\partial/\partial\eta = R(\partial/\partial r)$ and $ds = R\,d\theta$. The first identity holds because $\eta = r/R$. The second identity holds because $s/R = \theta$ in radian measure.]
(b) Use formula (6.211) to derive Formula (6.76).

Section 5

6.27 Suppose that a sheet of mica is placed over the right aperture in Figure 6.10. Assuming that this multiplies the aperture function for that *single* aperture by a 90° *phase shift factor* $e^{i\pi/2} = i$, describe the diffraction pattern from the two apertures. Suppose the mica is placed over the left aperture instead; does the diffraction pattern differ from when it was placed over the right aperture?

6.28 Repeat Exercise 6.27, but assume now that the phase shift factor is $e^{i2\pi\tau}$ for some real constant τ.

6.29 Suppose that the aperture function for Figure 6.35(a) is

$$A(\mathbf{x}) = A_0(\mathbf{x} - \mathbf{c}) + A_0(\mathbf{x} + \mathbf{c})$$

where $\mathbf{c} = (a, b)$ is in $\mathbf{R}^2$, $|\mathbf{c}| = \frac{1}{2}\delta$, and A_0 is the aperture function for a single aperture. Show that

$$I(u, v, D) = I_0(u, v, D) \left[4\cos^2 \frac{2\pi}{\lambda D}(au + bv) \right]$$

and that the intensity I is 0 when

$$au + bv = \pm\frac{\lambda D}{4}, \ \pm\frac{3\lambda D}{4}, \ldots, \ \pm\frac{(2k+1)\lambda D}{4}, \ldots$$

Show that the distance between successive dark fringes is $\lambda D/\delta$. Why are the dark fringes perpendicular to the direction of $\mathbf{c}$?

6.30 Suppose that four circular apertures are arranged so that their centers are at the corners of a parallelogram (and they are spaced far enough apart so they do not overlap). Describe their Fraunhofer diffraction pattern.

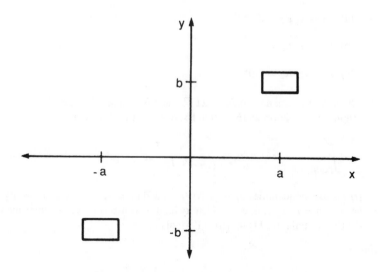

(a) apertures

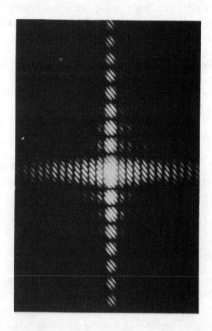

(b) diffraction pattern

FIGURE 6.35
**Diffraction from two identical rectangular apertures. (Reproduced from Walker, J.
S., *Fourier Analysis*, Oxford University Press, Oxford, 1988. With permission.)**

Section 6

6.31 Using the procedure described in Example 6.8, plot the instrument functions for 31-slit diffraction gratings, using the following functions as unit functions. Use a separation constant of $\delta = 2$.

  (a) $\text{rect}(x)(1 - \text{abs}\,(2x))$

  (b) $\text{rect}(x)(0.5 + 0.5\cos(2\pi x))$

  (c) $\text{rect}(2x)$

  For (a), the grating is called a *triangular grating;* such gratings are used in spectroscopy (usually in reflective gratings).

6.32 Repeat Exercise 6.31, but now use 51-slit gratings and use 8192 points and an interval of $[-64, 64]$.

6.33 Repeat Exercise 6.32, but reduce the separation constant δ to $\delta = 1.5$.

6.34 *Nonuniform illlumination.* Sometimes the light illuminating the diffraction grating is not of uniform intensity. This can be modeled by applying a filter function to the diffraction grating function. Apply the filter function $\exp(-0.4x \wedge 2)$ to the diffraction gratings in Exercise 6.31. What effect does this have on their instrument functions?

Section 7

6.35 Suppose a 31-slit grating, like the one in Example 6.8, is used. If the grating is nonuniformly illuminated, using the function $f(x) = \exp(-0.4x^2)$ as in Exercise 6.34, determine the effect this has on the widths of the spectral lines. What effect does this have on the resolution of distinct spectral lines (based on Rayleigh's criterion)?

6.36 Explain the results of Exercise 6.35 in terms of convolution.

6.37 *Error in slit positions.* Errors can sometimes occur in manufacturing a slit diffraction grating. For example, suppose a 31-slit grating is described by the function

$$f(x) = \text{sumk}(\text{rect}((x - 0.9k)/0.2)(k < 0\,\text{or}\,k > 0))$$

$$+ \text{rect}((x - 0.1)/0.2)\,\backslash k = -15, 15.$$

This represents a grating with a misaligned central slit. How does the instrument function of this grating compare with the instrument function of the grating in Example 6.8? How does the misalignment of the central slit affect the resolution of the grating?

6.38 *Array Theorem.* If MN identical apertures are arranged in N rows and M columns, so that they are evenly spaced within each row and column and nonoverlapping, then the aperture function is

$$A(x, y) = \sum_{m=0}^{M-1} \sum_{n=0}^{N-1} A_0(x - m\delta, y - n\Delta)$$

where δ and Δ are positive constants. Show that

$$I(u, v, D) = I_0(u, v, D)\mathcal{S}_M(u)\mathcal{S}_N(v)$$

where

$$\mathcal{S}_M(u) = \frac{\sin^2(\pi M \delta u/\lambda D)}{\sin^2(\pi \delta u/\lambda D)}, \qquad \mathcal{S}_N(v) = \frac{\sin^2(\pi N \Delta v/\lambda D)}{\sin^2(\pi \Delta v/\lambda D)}.$$

Explain why the diffraction pattern from such an array will consist mainly of high-intensity dots located at the points $(u, v) = (m\lambda D/\delta, n\lambda D/\Delta)$ where m and n are integers, the intensity at these points being $M^2 N^2 I_0(m\lambda D/\delta, n\lambda D/\Delta, D)$.

6.39 *Parallelogram Array.* Suppose that an aperture function $A(\mathbf{x})$ is defined by

$$A(\mathbf{x}) = \sum_{m=0}^{M-1} \sum_{n=0}^{N-1} A_0(\mathbf{x} - m\mathbf{a} - n\mathbf{c})$$

where $\mathbf{a} = (a, b)$ and $\mathbf{c} = (c, d)$ are two nonzero, nonparallel vectors in $\mathbf{R}^2$. Show that

$$I(\mathbf{u}, D) = I_0(\mathbf{u}, D)\mathcal{S}_M^{\mathbf{a}}(\mathbf{u})\mathcal{S}_N^{\mathbf{c}}(\mathbf{u}).$$

where

$$\mathcal{S}_M^{\mathbf{a}}(\mathbf{u}) = \frac{\sin^2(\pi M \mathbf{a} \cdot \mathbf{u}/\lambda D)}{\sin^2(\pi \mathbf{a} \cdot \mathbf{u}/\lambda D)}, \qquad \mathcal{S}_N^{\mathbf{c}}(\mathbf{u}) = \frac{\sin^2(\pi N \mathbf{c} \cdot \mathbf{u}/\lambda D)}{\sin^2(\pi \mathbf{c} \cdot \mathbf{u}/\lambda D)}.$$

Explain why the diffraction pattern has a form which consists mainly of high-intensity dots located at the points (u, v) which satisfy

$$au + bv = m\lambda D, \qquad cu + dv = n\lambda D \tag{6.212}$$

where m and n are both integers. Also explain why the intensities at these points are equal to $M^2 N^2$ times the intensity $I_0(u, v, D)$. Finally, show that the points satisfying both equations in (6.212) *lie on the intersections of lines perpendicular to the vectors* $\mathbf{a}$ *and* $\mathbf{c}$.

REMARK 6.14:   The results of Exercise 6.39 are important in crystallography. The aperture function $A(\mathbf{x})$ corresponds to a *two-dimensional crystal.*   ▌

6.40 Suppose that a vertical slit grating is constructed so that the separation constant δ equals $3a$, where a is the width of each vertical slit. Show that for all frequencies λ, every spectral line at $3\lambda D/\delta$ is eliminated (since their intensities are all zero).

6.41 Verify Inequality (6.111).

6.42 Using Rayleigh's criterion, derive the relative separation needed for wavelength resolution in the kth-order spectrum (the spectrum generated by spectral lines at $k\lambda D/\delta$) of a slit grating.

6.43 Suppose that the unit function of a grating is $f(x) = \text{rect}(x)(1 - \text{abs}(2x))$ with separation constant $\delta = 2$ (i.e., a triangular grating). Find the relative separation needed for the resolution of wavelengths in the first-order spectrum, in the second-order spectrum, in the kth-order spectrum.

6.44 Using *FAS*, compare the two sides of (6.110), assuming $\lambda D = 1, \alpha = 0.4, \beta = 0.5$, and $N = 100$.

6.45 Using *FAS*, compare the instrument functions from sinusoidal gratings with frequencies of $N = 50$ and $N = 100$, both with $\beta = 0.5, \alpha = 0.4$.

6.46 Suppose a sinusoidal grating (with $N = 100, \alpha = 0.4, \beta = 0.5$) is nonuniformly illuminated (see Exercise 6.34). Use an intensity of illumination of $\exp(-0.4x^2)$. Using Rayleigh's criterion, determine the relative separation needed to resolve wavelengths. How does your result compare with the uniformly illuminated case described by Formula (6.111)?

6.47 Using *FAS*, determine approximately the relative separation needed for resolution of wavelengths in the first-order spectrum (based on Rayleigh's criterion) for the nonuniformly illuminated grating described in Exercise 6.35.

Section 8

6.48 Suppose that a lens is plano-convex. That is, the lens surface at the exit plane is a plane surface, while the lens surface at the entrance plane is a spherical surface bulging outward. Show that the focal length formula is

$$\frac{1}{f} = (\eta - 1)\frac{1}{R} .$$

where R is the radius of curvature of the spherical side.

6.49 *Creation of coherent light.* Consider the setup shown in Figure 6.2. (a) Show that for **x** and **z** in the exit plane of the lens, the coherency function Γ satisfies

$$\Gamma(\mathbf{x}, \mathbf{z}) \approx \left(\frac{P(\mathbf{x})}{|P(\mathbf{x})|}\right)\left(\frac{P(\mathbf{z})}{|P(\mathbf{z})|}\right)^* \tag{6.213}$$

where P is the pupil function of the lens. In particular, if P is the pupil function in (6.130), then the light in the exit plane of the lens is (approximately) coherent provided **x** and **z** are in the lens aperture. [*Hint:* use Formula (6.203) in Exercise 6.3.] (b) Show also that the intensity of the light in the exit plane of the lens is approximately equal to a positive constant.

REMARK 6.15: Exercise 6.49 shows that, as we assumed in Section 6.1, the setup in Figure 6.2 produces coherent light of constant intensity. ∎

6.50 Suppose that the lens is a positive meniscus. That is, the lens surface at the entrance plane bulges outward spherically with a radius of curvature of R_1, while the lens surface at the exit plane bulges inward spherically with a radius of curvature of R_2. Show that the focal length formula is

$$\frac{1}{f} = (\eta - 1)\left[\frac{1}{R_1} - \frac{1}{R_2}\right].$$

6.51 Suppose that an aperture is placed directly in front of a lens, i.e., in Figure 6.22 the distance D equals 0. Assuming that the lens is double convex with spherical surfaces, derive the following formula:

$$\psi(\mathbf{X}, \Delta + \epsilon, t) \approx \frac{e^{i\delta}}{\lambda\Delta} \int_{\mathbf{R}^2} A(\mathbf{u})\psi(\mathbf{u}, 0, t) P(\mathbf{u}) e^{\frac{i\pi}{\lambda\Delta}|\mathbf{X}-\mathbf{u}|^2} e^{\frac{-i\pi}{\lambda f}|\mathbf{u}|^2} d\mathbf{u}. \quad (6.214)$$

6.52 Show that, under the assumption of coherent illumination, Formula (6.214) in Exercise 6.51 leads to

$$I(\mathbf{X}, \Delta + \epsilon) \approx \left| \frac{1}{\lambda\Delta} \int_{\mathbf{R}^2} A(\mathbf{u}) P(\mathbf{u}) e^{\frac{i\pi}{\lambda\Delta}|\mathbf{X}-\mathbf{u}|^2} e^{\frac{-i\pi}{\lambda f}|\mathbf{u}|^2} d\mathbf{u} \right|^2. \quad (6.215)$$

And that, *if the aperture defined by $A(\mathbf{u})$ fits inside the lens aperture*, then [assuming $P(\mathbf{u})$ is described by (6.130)]

$$I(\mathbf{X}, \Delta + \epsilon) \approx \left| \frac{1}{\lambda\Delta} \int_{\mathbf{R}^2} A(\mathbf{u}) e^{\frac{i\pi}{\lambda\Delta}|\mathbf{X}-\mathbf{u}|^2} e^{\frac{-i\pi}{\lambda f}|\mathbf{u}|^2} d\mathbf{u} \right|^2. \quad (6.216)$$

6.53 Suppose that $\Delta = f$. Show that Formula (6.216) in Exercise 6.52 becomes

$$I(\mathbf{X}, f + \epsilon) \approx \left| \frac{1}{\lambda f} \hat{A}\left(\frac{\mathbf{X}}{\lambda f}\right) \right|^2. \quad (6.217)$$

Formula (6.217) shows that *the intensity in the back focal plane of the lens is equal to the modulus-squared of the (scaled) Fourier transform of the aperture function.* [Compare with Formula (6.60).]

6.54 Show that if $\Delta = f$, then Formula (6.214) in Exercise 6.51 becomes

$$\psi(\mathbf{X}, f + \epsilon, t) \approx \frac{e^{i\delta}}{\lambda f} e^{\frac{i\pi}{\lambda f}|\mathbf{X}|^2} \int_{\mathbf{R}^2} A(\mathbf{u})\psi(\mathbf{u}, 0, t) P(\mathbf{u}) e^{-i\frac{2\pi}{\lambda f}\mathbf{X}\cdot\mathbf{u}} d\mathbf{u}. \quad (6.218)$$

6.55 Explain why, if a properly selected lens is placed directly behind the observation plane in Figure 6.22 (so that its entrance plane coincides with the observation plane), then Formula (6.218) in Exercise 6.54 can be replaced by

$$\psi(\mathbf{X}, f + \epsilon', t) \approx \frac{e^{i\delta'}}{\lambda f} \mathcal{P}(\mathbf{X}) \int_{\mathbf{R}^2} A(\mathbf{u})\psi(\mathbf{u}, 0, t) P(\mathbf{u}) e^{-i\frac{2\pi}{\lambda f}\mathbf{X}\cdot\mathbf{u}} d\mathbf{u} \quad (6.219)$$

where $\mathcal{P}(\mathbf{X})$ is the pupil function for this second lens. (ϵ' and δ' are positive constants resulting from the thickness of the second lens.)

Section 9

6.56 Suppose that

$$\frac{1}{D} + \frac{1}{\Delta} - \frac{1}{f} = \pm\delta$$

where δ is a small positive constant. (This is known as *defocus.*) Show that the pupil function $P(\mathbf{u})$ in (6.142) must be replaced by $P(\mathbf{u})e^{\pm i\pi\delta|\mathbf{u}|^2/\lambda}$.

6.57 Suppose fourth-power terms are included in the approximations in (6.121). [This would be more necessary the larger the lens aperture.] Show that the pupil function in (6.142) must then be replaced by $P(\mathbf{u})e^{i\pi\sigma|\mathbf{u}|^4}$ where σ is a constant. (This is known as *spherical aberration,* the constant σ is the *spherical aberration constant.*)

6.58 For a double convex lens, describe what shape of lens surfaces would be needed to ensure that no spherical aberration occurs (of the type described in Exercise 6.57).

Section 10

6.59 Repeat Example 6.9, but now use the following wavelengths: $\lambda = 4 \times 10^{-4}$ mm, 6×10^{-4} mm, and 8×10^{-4} mm. What is the effect of varying the wavelength on the edge image?

6.60 Suppose that we have a one-dimensional pupil function $P(\lambda\Delta x) = \text{rect}(x/40)$. Using an initial interval of $[-512, 512]$ and 2048 points, Fourier transform this pupil function to get a *PSF* for coherently imaging the following object:

$$A(x) = \exp(-(x/0.8) \wedge 20) - \text{rect}(100x), \qquad -1 \le x \le 1.$$

You should observe a pronounced effect in the image due to the spreading out of the point-like part of the object, $-\text{rect}(100x)$, which represents a dust particle scattering light.

6.61 Repeat Exercise 6.60, but now use the object function:

$$A(x) = \exp(-(x/0.8) \wedge 20) - \text{rect}(100(x - .04)) - \text{rect}(100(x + .04)).$$

6.62 Suppose that instead of an edge, the object in Example 6.9 is changed to

$$A(x, y) = \begin{cases} 0 & \text{if } x < -2 \\ (x+2)/2 & \text{if } -2 \le x \le 0 \\ 1 & \text{if } x > 0. \end{cases}$$

Keeping all other parameter values the same as in that example, graph the resulting image. Compare your results with those in Example 6.9. What happens if the image detector (film, or charge coupled device) records $I(X, Y)^{\frac{1}{2}}$ instead of $I(X, Y)$?

6.63 Suppose that a *Gaussian filter* is placed over the lens aperture in Example 6.9, so that the pupil function is now

$$P(x, y) = e^{-0.1(x^2+y^2)} \, \text{rect}\left(\frac{x}{40}\right) \text{rect}\left(\frac{y}{40}\right)$$

$$= e^{-0.1x^2} \, \text{rect}\left(\frac{x}{40}\right) e^{-0.1y^2} \, \text{rect}\left(\frac{y}{40}\right). \qquad (6.220)$$

Keeping all other parameter values the same as in Example 6.9, graph the image of an edge. Compare your results with those in Example 6.9. Is there a bright band (Gibbs' effect) along the image edge when this Gaussian pupil is used? Is there extraneous oscillation away from the edge? What happens with other Gaussian pupils (wider or narrower ones)? In which case is the imaged edge closer to the true edge? (*Note:* using a pupil function like (6.220) is described in detail in [Mi-T].)

6.64 Repeat Example 6.12, but now use

$$A(x) = \exp(-(x/3) \wedge 10)[1 + \beta \cos(12\pi x)]$$

for $\beta = 0.04, 0.08, 0.16, 0.2$, and 0.25. Estimate the improvement in contrast over the bright field image for each value of β. At what value of β do false details begin to appear?

6.65 Repeat Exercise 6.64, but change the gray field pupil to

$$P_{GF}(\lambda \Delta x) = \begin{cases} 0 & \text{if } 40 < |x| \\ 10 & \text{if } 3 < |x| < 40 \\ 2 & \text{if } |x| < 3. \end{cases}$$

Describe the effects of raising the value of P_{GF} for $|x| < 3$.

6.66 The *Schlieren method* of imaging consists of placing an opaque material over a half-plane immediately behind the lens aperture. This exercise describes a one-dimensional model of Schlieren imaging. Assume that the pupil function is

$$P(\lambda \Delta x) = \begin{cases} 1 & \text{if } 0 < x < 40 \\ 0 & \text{if } x < 0 \text{ or } x > 40 \end{cases}$$

and transform this function over $[-128, 128]$ using 2048 points to create a *complex PSF*. Compare Schlieren imaging with gray field imaging of the function $A(x)$ in (6.158).

Section 11

6.67 Suppose that a second double convex lens, having focal length F, is placed at a distance F in back of the observation plane (so that its distance behind the first lens is $f + F$ and it is aligned parallel with the first lens along a common optic axis). Using Formula (6.181), show that in the back focal plane of this second lens a magnified, inverted image of $A(\mathbf{x})$ is formed. Find the magnification factor of this image and an expression for the *PSF*.

6.68 Why is there no Fourier transforming property (at the back focal plane, or any plane) when incoherent illumination is used?

Section 12

6.69 Graph the *PSF* of the square lens aperture from Example 6.10, assuming all parameter values are the same, but under incoherent illumination. What advantages does the *PSF* for incoherent illumination have over the coherent one? What would the image of several tiny point-like objects (like dust particles) look like for these two types of illumination?

6.70 Suppose that we have a one-dimensional pupil function $P(\lambda \Delta x) = \text{rect}(x/40)$. Using an initial interval of $[-512, 512]$ and 2048 points, compute the power spectrum of this pupil function to get a *PSF* for incoherently imaging the following object:

$$A(x) = \exp(-(x/0.8) \wedge 20) - \text{rect}(100x), \qquad -1 \le x \le 1.$$

Compare your result with the coherent image obtained for Exercise 6.60. Which image seems to be a better representation of the object?

6.71 Repeat Exercise 6.70, but now use the object function:

$$A(x) = \exp(-(x/0.8) \wedge 20) - \text{rect}(100(x - .04)) - \text{rect}(100(x + .04)).$$

Compare your result with the coherent image obtained for Exercise 6.61. Which image seems to be a better representation of the object?

6.72 Given that

$$P(\lambda \Delta x) = \begin{cases} 1 & \text{if } |x| < 10 \\ 0 & \text{if } |x| > 10 \end{cases}$$

compare the one-dimensional images of an object whose *intensity function* is

$$\exp(-(x/3) \wedge 10)[1 + \cos(22\pi x)]$$

under coherent and incoherent illumination. *Note:* to obtain the *PSFs* you should compute the Fourier transform and the power spectrum of $P(\lambda \Delta x)$ over $[-128, 128]$ using 2048 points.

6.73 Repeat Exercise 6.72, but use the following function for the object *intensities:*

$$\exp(-(x/3) \wedge 10)[1 + \cos(c\pi x)]$$

where $c = 17, 19, 21, 23$, and 25. What advantages or disadvantages do you see in using coherent or incoherent illumination, especially in regard to *relative contrast* [see (6.161)] of the oscillations near 0? There is some advantage for coherent illumination when $c = 17$ and 19; what is it?

REMARK 6.16: In studying incoherent imaging it is a standard practice to divide the *PSF* by the value of its integral over $\mathbf{R}^2$. *This forces its integral over $\mathbf{R}^2$ to be* 1. This new function is called the *normalized PSF.* By Parseval's equality, the integral of the *PSF* over $\mathbf{R}^2$ is equal to the square of the 2-Norm of $P(\lambda \Delta \mathbf{x})$. Therefore, the normalized *PSF* can be found by computing the power spectrum of

$$P(\lambda \Delta \mathbf{x})/\|P(\lambda \Delta \mathbf{x})\|_2. \tag{6.221}$$

The function in (6.221) is called the *normalized pupil function.* When using *FAS* it is easy to do the operation in (6.221). You just have to create $P(\lambda \Delta \mathbf{x})$, then find its 2-Norm, and then go back and create $P(\lambda \Delta x)$ *divided by this 2-Norm.* (See the User's Manual in Appendix A for further information about how to compute the 2-Norm of a function.) ∎

6.74 Show that the integral of the normalized *PSF* over $\mathbf{R}^2$ is 1. And, show that the power spectrum of the function in (6.221) is indeed the same as the normalized *PSF.*

6.75 Consider the image of an edge discussed in Example 6.13. Show that the normalized *PSF along the X-direction* is given by $40 \text{ sinc}^2(40X)$. Using this result, obtain the image of the edge along the X-direction by convolving the edge function with $40 \text{ sinc}^2(40X)$. (Use the edge function given in Example 6.13, and convolve over $[-32, 32]$, using 4096 points.) Your result should be a close approximation to the original edge.

6.76 Repeat Exercise 6.73, but now use the normalized pupil function for incoherent imaging (follow the instructions at the end of Remark 6.16). Your comparison of coherent and incoherent imaging should now be done *directly* without appeal to the relative contrast formula.

6.77 Compute the inverse Fourier transform of the coherent *PSF* and the normalized incoherent *PSF* from the previous exercise. Explain the results of that exercise in terms of these inverse tranforms. [*Hint:* use the *convolution theorem;* the images are the Fourier transforms of the products of the object transforms times the inverse transforms of the *PSFs.*]

REMARK 6.17: The inverse Fourier transform of the coherent *PSF* is called the *coherent transfer function.* And the inverse Fourier transform of the normalized incoherent *PSF* is called the *optical transfer function.* For further discussion, see [Go, Chapter 6]. ∎

6.78 *Blurring.* This exercise describes a one-dimensional model of image blurring. For the normalized *PSF* use the function $4 \operatorname{rect}(4x)$. This function models a *blurred point:* Using *FAS*, convolve this normalized *PSF* with the following object intensity functions over $[-1, 1]$ using 1024 points.

  (a) $\exp(-(x/0.8) \wedge 10)(1 + 0.5\cos(4\pi x))$

  (b) $\exp(-(x/0.8) \wedge 10)(1 + 0.5\cos(8\pi x))$

  (c) $\exp(-(x/0.8) \wedge 10)(1 + 0.5\cos(12\pi x))$

  (d) $\exp(-(x/0.8) \wedge 10)(1 + 0.5\cos(16\pi x))$

  (e) $\exp(-(x/0.8) \wedge 10)(1 + 0.5\cos(20\pi x))$

The resulting images are blurred. Notice, in particular, that in the image for (c) the oscillations are *out of phase* with the oscillations of the original object by about $90°$. Explain this effect, as well as the *absence* of oscillations in images (b) and (d), by comparing the optical transfer function with the transforms of the objects (a) to (e).

A

User's Manual for Fourier Analysis Software

In this User's Manual, we describe all the aspects of using Fourier Analysis Software (*FAS*). *FAS* can be found on the disk under the file name FAS.EXE. At the DOS prompt, entering FAS or FAS.EXE will load the program. If you have a PC with a 286 processor or higher operating under DOS 5 or higher, and you have VGA video, then *FAS* should run properly. It will run best with at least 540KB of free RAM, and at least 1 MB of extended memory (XMS memory, loaded using HIMEM.SYS, which is fairly standard with DOS 5 or higher). For instructions on how to install all the program files to another drive or directory, see the section **Installing the Program** at the end of this appendix.

Display menu

FAS is a menu-driven program. Each menu is displayed at the bottom of the screen. When you begin the program, the following menu appears:

Next   X-int   Y-int   Graph   Cap \Calc   Print   Save   Rem   Trace   Options

This menu is called the **display menu.** It is the main menu for the program. You can select a choice from any menu by either pressing the first letter of the selection, or using the Spacebar or Backspace keys to move the cursor to the selection, and then pressing the *Enter* key, or, if you have a mouse installed, double clicking on the menu item that you want to select. You may back up from any menu to a previous menu by pressing the *Esc* key. If you are at a step requiring input, you can still back up by pressing the *Esc* key. We will now describe what happens with each choice in the display menu.

NEXT

This choice leads you to the main procedure of *FAS*. It is here that you tell the program what type of Fourier analysis you want to perform. First, the following menu appears:

Series   Transforms   Convolutions   Graphs   Erase screen   Quit

This menu is called the **Fourier analysis menu.** We will now explain each of the choices on the Fourier analysis menu.

Initial Setup

FAS can be used to calculate Fourier series, Fourier cosine series, Fourier sine series, sampling (interpolation) series, Fourier transforms, Fourier cosine transforms, Fourier sine transforms, convolutions, and for general graphing purposes. Before discussing each of these procedures, however, we must first discuss the common sequence of steps that are performed at the beginning of each of these procedures. These first three steps are

1) *Choice of number of points.* You are presented with a menu of the form

    1) 128   2) 256   3) 512   4) 1024   5) 2048   6) 4096   7) 8192

from which you can select the number of points you want to work with. Choosing 128, 256, or 512 points gives very short processing time in exchange for low resolution, while choosing 2048, 4096, or 8192 points gives higher resolution in exchange for longer processing time. The choice of 1024 points gives a useful mean between the two, hence it is the default choice.

2) *Choice of interval type,* $[-L, L]$ or $[0, L]$. *FAS* graphs functions over two types of intervals, either an interval of the form $[-L, L]$ or of the form $[0, L]$. After choosing which interval type you want, you are then asked for the value of L. You may enter a positive real number (double precision) or a number of the form $\#\pi$ where $\#$ represents a real number. This format allows for multiples of π to be used. To enter the symbol π, hold down the *Alt key* and press *p*. (*Note:* do not use commas in your numbers. A number like $20,000$ should be entered as either 20000 or 20 000.)

3) *Choice of function.* After specifying the interval, you are presented with the following **function menu:**

Formula   Diffeq   Recursive   Get data   Quit

Or, if you have already graphed a function and *are using the same number of points,* then the function menu looks like

Formula   Diffeq   Recursive   Get data   Select graph   Quit

These choices are described below.

FORMULA

This choice allows you to compute functions of x. The prompt:

$$f(x) =$$

followed by a blinking cursor, will appear. You may then type in up to 3 lines of text for your formula.

FAS contains a fairly sophisticated function compiler. This compiler allows you to write function formulas using elementary functions, or summations, or products, or logical operators, or comparison operators, or iteration. The functions that are allowed in formulas are

sin( )	[sine function]
cos( )	[cosine function]
tan( )	[tangent function]
exp( ) or e∧( )	[exponential function]
sqr( )	[positive square root function]
abs( )	[absolute value function]
log( ) or ln( )	[natural logarithm function]
atn( )	[arctangent function]
sinc( )	[sinc function]
sinh( )	[hyperbolic sine function]
cosh( )	[hyperbolic cosine function]

tanh( )   [hyperbolic tangent function]
rect( )   [rectangle function]
gri( )   [greatest integer function]
sgn( )   [sign function]
δ( )   [delta function]
ran( )   [random number function]
fac( )   [factorial function]

The arctangent function is, of course, also known as the inverse tangent function. The sinc and rect functions are defined by

$$\text{sinc}\,(x) = \begin{cases} 1 & \text{for } x = 0 \\ \dfrac{\sin \pi x}{\pi x} & \text{for } x \neq 0 \end{cases} \quad , \quad \text{rect}\,(x) = \begin{cases} 1 & \text{for } |x| < 0.5 \\ 0.5 & \text{for } |x| = 0.5 \\ 0 & \text{for } |x| > 0.5. \end{cases}$$

The sgn function is defined by

$$\text{sgn}\,(x) = \begin{cases} 1 & \text{for } x > 0 \\ 0 & \text{for } x = 0 \\ -1 & \text{for } x < 0. \end{cases}$$

The δ–function, δ( ), which can also be specified by del( ), is defined by

$$\delta\,(x) = \begin{cases} 1 & \text{for } x = 0 \\ 0 & \text{for } x \neq 0. \end{cases}$$

The character δ is entered by holding down the *Alt* key and pressing the letter *d*. The random number function, ran( ), returns a random number between -1 and 1 each time it is called (in particular, ran(0) is not a constant). Like all random number generators, the one that *FAS* uses is not truly random. It uses an arithmetic method to generate numbers that provide a good simulation of randomness. (There are, however, correlations that exist within large sets of these numbers which would not exist if the numbers were truly random.) See [PFTV, Chapter 7] for a good introduction to random number generators. The factorial function, fac( ), will give the factorial value of a non-negative integer. That is, fac $(n) = n!$ whenever n is a non-negative integer. The factorial function will only give correct values for non-negative integers.

You may combine these functions using composition and the 5 binary operations, $\wedge$ (raise to a power), / (divide), $*$ (multiplication), $-$ (subtraction), and $+$ (addition). Multiplication may also be done using *juxtaposition*. The order of operations is the familiar one from elementary algebra: raising to a power is first, then division, then multiplication, then subtraction, and addition is last.

For example, the following are all valid formulas:

$$2 \sin(3x) - 4 \cos(5x)$$

$$\exp[\sin(3x)] - 4 \cos(3\pi x)\mathrm{atn}(4x)$$

$$3 \exp\{-(2x - 1)(3x - 4)\}$$

$$x - x \wedge 3/3! + x \wedge 5/5! - x \wedge 7/7!$$

Notice that brackets, [ ], or curly brackets, { }, are allowed instead of parentheses. And the factorial symbol, !, is allowed, too. The factorial symbol will only give the correct values if non-negative integer constants are used before it. As explained above, π is entered by holding down the *Alt* key and pressing *p*.

The editor used for typing formulas will respond to the *Home* key, the *End* key, and the arrow keys. *Home* or *End* will take you to the beginning or end of a line, respectively. The arrow keys allow you to move about the formula while typing. If you press the *Del* key, the letter you are on will be deleted. Pressing the *Ins* key allows you to toggle between typing over previous letters (normal mode) and inserting new letters (insert mode).

If a formula has been previously entered, then that formula will appear again after the "$f(x) =$" prompt. This allows for simple modifications for creating related formulas. To clear the formula completely, press the *F9* key. *FAS* keeps a record of the last 150 formulas entered. You can retrieve an old formula by pressing either the *PgUp* key or the *PgDn* key. *PgUp* will go back in the list, then to move forward press *PgDn*. To recall a formula that was previously saved (see *Save*), press the *F3* key. A list of formula file names will appear in the help box, and you can select a formula.

There is also an extensive amount of on-line help available; just press the F1 key when you are at the "$f(x) =$" prompt. In addition to information on how to create formulas, *the on-line help contains every formula used in this book* (not including the exercises) *which you can enter into FAS by clicking with the mouse* (assuming your computer has a mouse installed).

When you have finished entering a formula, then press the *Enter* key and the program will graph the function described by that formula.

If the formula does not have the proper syntax, e.g., too many left or right parentheses, a beeping sound will result and the following error message will appear:

Invalid formula. Press any key to continue.

When you press a key, your formula will reappear and you may correct it. Similarly, if the formula is correct, but an overflow occurs during computation or the domain is too large (which can happen with sqr and log), then the program will beep and the following error message will appear:

Run Time Error. Press any key to continue.

Again, after you press a key, you can modify your formula to handle the problem.

To back out of the *Formula* procedure, just press the *Esc* key and you will return to the function menu.

In addition to the elementary types of formulas described above, *FAS* also allows several other kinds of formulas which we will now describe.

Sums and Products

Formulas can include the following four sum operators: sumi(), sumj(), sumk(), or sumn(); and the following two product operators: prodp(), or prodq(). A couple of examples should make it clear how to use these operators. Consider the following formula:

$$f(x) = 1 + 2 \operatorname{sumj}[\cos(jx)].$$

If you enter this formula, then you get the following message prompting you for input:

Enter lower, upper limits.

j:

You are being asked for the lower and upper limits for the sum over j indicated in the formula above. For example, if you enter 1, 15 then *FAS* will compute the following function:

$$f(x) = 1 + 2 \sum_{j=1}^{15} \cos(jx).$$

You can also specify this same sum using the following formula:

$$f(x) = 1 + 2 \operatorname{sumj}[\cos(jx)] \setminus j{=}1{,}15.$$

In this formula we have used a *command string*. Any expression that is preceded by a backslash, \, is a command string. In this case the command string is \j=1,15. This tells *FAS* to use the values 1 through 15 for the index j in the sum. Consequently, no prompt for the lower and upper limits of j will appear with this last formula. The advantage of using command strings is that when you save a formula, then you also save the information in the command string. *Command strings must always appear at the end of a formula.*

You can also use command strings to include a comment about your formula. For example, the previous formula could be modified to include a comment as follows:

$$f(x) = 1 + 2 \operatorname{sumj}(\cos(jx)) \setminus j{=}1{,}15 \setminus \text{Dirichlet kernel, 15 harmonics.}$$

Here is an example of a product operator. If you enter the following formula:

$$f(x) = \text{prodp}(x - p)$$

then you will receive the following prompt for input:

Enter lower, upper limits.
p:

You are being asked for the lower and upper limits for the product over p indicated in the formula above. For example, if you enter -4, 4 then *FAS* will compute the following function:

$$f(x) = \prod_{p=-4}^{4} (x - p)$$

which is the same function as the more complicated formula

$$f(x) = (x + 4)(x + 3)(x + 2)(x + 1)x(x - 1)(x - 2)(x - 3)(x - 4).$$

You could also use a command string by entering the following formula:

$$f(x) = \text{prodp}(x - p) \backslash p = -4, 4.$$

The major limitation to this method of specifying sums and products is that *you may not repeat any operator more than once in a single formula.* For example, the following formula is not allowed:

$$f(x) = \text{sumk}(x + k) - \text{sumk}(x - k)$$

because the operator sumk is used twice. In this case, however, the formula can easily be rewritten as

$$f(x) = \text{sumi}(x + i) - \text{sumk}(x - k)$$

Although this restriction limits the number of possible formulas, it is not a very severe limitation. Most formulas of interest can be described with this many sum and product operators.

Parameters
Formulas can include parameters. The following variables can be used as parameters: a, b, c, m, n, t, u, v, w, and z. (The variable y can also be used as a

parameter, unless the formula is specifying a differential equation or a recursion equation.) For example, if you enter the following formula:

$$f(x) = a\cos(vx) + b\sin(wx)$$

then you will be prompted successively for the values of the parameters a, b, v, and w. If you enter, say, $a = 2$, $b = -3$, $v = 4$, and $w = 6$, then *FAS* will graph the function

$$f(x) = 2\cos(4x) - 3\sin(6x).$$

You can also graph this last function using command strings as follows:

$$f(x) = a\cos(vx) + b\sin(wx) \setminus a{=}2 \setminus b{=}{-}3 \setminus v{=}4 \setminus w{=}6.$$

Substitutions

Parameters can also be used to perform substitutions. For example, suppose you want to graph the following function:

$$f(x) = \cos(\mathrm{sqr}(47)x) - \sin(\mathrm{sqr}(47)x)/\mathrm{sqr}(47).$$

Then instead of entering the formula as written above, you can enter the simpler formula

$$f(x) = \cos(bx) - \sin(bx)/b \setminus b = \mathrm{sqr}(47)$$

and let *FAS* do the substitution of sqr(47) in place of b. (If you do not specify b with a command string, then you will be prompted for it.)

Functions of Graphs

A powerful feature of *FAS* is its ability to handle functions of graphs. Suppose, for example, you have graphed two functions on the screen. If you enter the following formula

$$f(x) = \mathrm{sqr}\,(\mathrm{g1}\,(x) \wedge 2 + \mathrm{g2}\,(x) \wedge 2)$$

then *FAS* will compute the square root of the sum of the squares of the values of the two functions already on the screen. Here, g1 specifies the first function you plotted, and g2 specifies the second function you plotted. If you have a color monitor, then the first function will be plotted in green and the second in light blue.

Or, if you have three graphs on the screen, then

$$f(x) = 2\mathrm{g1}\,(x) - 4\mathrm{g3}\,(x)$$

will tell *FAS* to compute twice the values of the first function that you plotted minus four times the values of the third function that you plotted.

There are two reservations to keep in mind when you are using graph variables. First, any graph on screen is assigned values of 0 outside of the interval over which it was originally defined. For example, if you enter the formula $f(x) = g1(2x)$, then for those values of $2x$ that lie outside the interval for which $g1$ was graphed $f(x)$ will be given the value of 0. Second, you cannot use a graph variable of greater index than the number of graphs displayed. For example, if two graphs are displayed and you use either $g3$, $g4$, or $g5$ in a formula, then you will get an error message.

Comparison Operations

FAS allows for the following six operations of comparison in formulas:

>	[greater than]
<	[less than]
≥	[greater than or equal]
≤	[less than or equal]
=	[equal]
<>	[not equal]

The symbol $\geq$ is entered by holding down the *Alt* key and pressing the key with > on it (you do not have to hold down the *Shift* key). Likewise, the symbol $\leq$ is entered by holding down the *Alt* key and pressing the key with < on it. You can also specify $\geq$ by entering >=, and $\leq$ by entering <=.

All of these operations return a value of 1 if they are true and a value of 0 if they are false. For example, if you enter

$$f(x) = \cos(x) > 0.5$$

then the function that is graphed will equal 1 when $\cos(x) > 0.5$ and 0 when $\cos(x) \leq 0.5$. If you enter

$$f(x) = (\cos(x) > 0.5)\cos(x) + (\cos(x) \leq 0.5)0.5$$

then the function that is graphed will equal $\cos(x)$ when $\cos(x) > 0.5$ and 0.5 when $\cos(x) \leq 0.5$. Or, suppose you want to graph the function

$$f(x) = \begin{cases} 0 & \text{if } x = 0 \\ x\sin(1/x) & \text{if } x \neq 0. \end{cases}$$

Then, you could enter the following formula:

$$f(x) = (x <> 0)x\sin(1/[x + \delta(x)])$$

Notice that when $x \neq 0$ then $\delta(x) = 0$, so, in this case, $f(x) = x \sin(1/x)$. While if $x = 0$, then $(x <> 0) = 0$ so $f(x) = 0$. You may wonder why we did not recommend $f(x) = (x <> 0)x \sin(1/x)$. The reason is that either possible interval of x-values, $[-L, L]$ or $[0, L]$, contains $x = 0$. *FAS* will therefore try to substitute $x = 0$ into $1/x$, resulting in an error. The expression $x + \delta(x)$, however, equals 1 when $x = 0$; hence this error is avoided with the recommended formula.

Logical Operations

FAS allows for the following three logical operations in formulas:

&	[logical and]
or	[logical or]
not( )	[logical not]

These are logical functions, which are meant to be used in combination with functions which return the following two values: 1 representing TRUE, and 0 representing FALSE. For example, if you want to graph the following function:

$$f(x) = \begin{cases} \sin(x) & \text{for } -5 \le x < -2 \\ \cos(x) & \text{for } -2 \le x < 3 \\ \sin(x) & \text{for } 3 \le x \le 5 \end{cases}$$

then you could enter the following formula:

$$[(-5 \le x < -2) \text{ or } (3 \le x \le 5)] \sin(x) + (-2 \le x < 3) \cos(x).$$

Notice that we used expressions such as $(-2 \le x < 3)$. *FAS* will interpret this last expression as an abbreviation for $(-2 \le x \text{ and } x < 3)$.

Iteration

FAS also allows for iteration of functions. For example, suppose you want to repeatedly compose $f(x) = \cos(x)$ with itself, say 10 times. Here is the way to do it. First, graph $f(x) = x$, and then enter the following formula:

$$f(x) = \cos(f).$$

You will get the following prompting for input:

Iterations:

You then enter 10 for the number of iterations. The graph which results is the composition of $\cos(x)$ with itself 10 times (also called the 10th iterate of cos). If you now enter this last formula again, and enter 1 for the number of iterations, then the function cos is applied to the last function graphed. The resulting function is therefore the 11th iterate of cos. In this way you can graph successive iterates of cos.

You can also specify the number of iterations using a command string. For instance, the previous example could have been done by graphing $f(x) = x$ and then entering the formula

$$f(x) = \cos(f) \setminus \text{Iter}=10.$$

The command string, $\setminus$Iter=10, tells *FAS* to perform 10 iterations.

Here is another example. Suppose you choose $[0, 4]$ for the interval and 4096 points. Then, you graph $f(x) = 0.5$. Now, suppose you enter the function

$$f(x) = xf(1 - f)$$

and perform, say, 50 iterations followed by 1 more iteration four times in a row (so that you are graphing 50, 51, 52, 53, and 54 iterates). If you display all the graphs simultaneously you get a figure that approximates the bifurcation diagram for the chaotic dynamical system generated by the family of functions $f(x) = cx(1 - x)$, where c is a parameter ranging over $[0, 4]$.

This concludes our discussion of the *Formula* procedure on the function menu. We now turn to the remaining procedures listed on the function menu.

DIFFEQ

This procedure can be used for plotting approximations of solutions to initial-value problems for differential equations. It begins with the following prompt,

$$f(x) =$$

followed by a blinking cursor. You then enter a formula that describes your differential equation. Here are a few examples.

Example A.1:
Suppose that you want to approximate the solution to the following initial-value problem:

$$y' + 4y = -x, \quad y(0) = 2.$$

Then, you enter the following formula:

$$f(x) = -4y - x.$$

This specifies the differential equation, after solving for y'. After entering this formula, the following message will appear:

The order of the D.E. is: 1

If the differential equation is of a higher order than 1 (as in our next example), then you can change the order at this point. Since, in this case, the differential equation is of order 1, you just press the *Enter* key to continue. Then, the following prompt appears:

$$Y0:$$

You then enter the initial value of 2.

You can also tell *FAS* which initial value to use by including a command string, as follows:

$$f(x) = -4y - x \, \backslash \text{Diff} \backslash y0 = 2$$

Notice that this formula contains the command string \Diff. This tells *FAS* to treat the formula as specifying a differential equation whether you selected the choice *Diffeq* or not. This is useful to remember, because the *Formula* procedure will not treat $-4y - x$ as specifying a differential equation unless you explicitly tell it to do so with a command string.

After you have specified the initial value, *FAS* will plot an approximation to the solution of the initial-value problem over the interval [0, 10], using the Runge-Kutta algorithm. The following caption will appear below the graph:

Max.Tr.Err.=9.71238576472716D−07, Cum.Tr.Err.=5.87396570754148D−06

These are *truncation errors,* which give estimates of how accurate a solution has been obtained. In this case, the *maximum truncation error* (Max.Tr.Err.) is about 9.7×10^{-7}. This is an estimate of the largest (relative) error occurring at all of the points (it does not take into account rounding error). The *cumulative truncation error* (Cum.Tr.Err.) is about 5.87×10^{-6}. The cumulative truncation error is another estimate of error; it is obtained by summing the truncation errors at all the points. Of these two estimates, the cumulative truncation error is the most reliable, but it is often overly conservative. In this case, since the exact solution is $y = -x/4 + 1/16 + (31/16)e^{-4x}$, we can determine the exact error. The worst error at any one point is no more than about 1.43×10^{-8}. In this case, the Max.Tr.Err. gave the better estimate (this is not true in all cases).

Example A.2:

Suppose that you want to graph a solution of

$$y'' + 36y = \cos(6x), \quad y(0) = 1, \quad y'(0) = 0.$$

Then, you could enter the formula

$$f(x) = -36y + \cos(6x).$$

After entering this formula, the same initial prompt appears as in the previous example. This time, however, you enter the number 2 in order to tell *FAS* that your equation is of order 2. You then enter $y(0)$ and $y'(0)$ as follows (the underlined numbers are the data that you enter):

$$Y0: \underline{1}$$

$$Y1: \underline{0}$$

This tells *FAS* to use 1 for $y(0)$ and 0 for $y'(0)$.

You can also use command strings for this example. If you enter the following formula:

$$f(x) = -36y + \cos(6x) \setminus \text{Diff} \setminus \text{Order=2} \setminus y0=1 \setminus y1=0$$

then *FAS* will plot the same graph as above, but without the prompting. The command string, \Order=2, tells *FAS* to treat the differential equation as an order 2 equation.

Example A.3:

Suppose that you want to graph a solution to

$$y^{(4)} + x(y^{(3)})^4 - \sin(x)y^{(1)} - y^6 = e^{x/5}$$

$$y(0) = 1, \ y^{(1)}(0) = -2, \ y^{(2)}(0) = 3, \ y^{(3)}(0) = -4$$

where $y^{(k)}$ denotes the kth derivative of y. Then, you could enter the formula

$$f(x) = -x(y3 \wedge 4) + \sin(x)y1 + y \wedge 6 + \exp(x/5).$$

The following prompt would appear:

The order of the D.E. is: 4.

Since the differential equation does have order 4, you press the *Enter* key. Then you enter the following data:

$$Y0 : \underline{1}$$

$$Y1 : \underline{-2}$$

$$Y2 : \underline{3}$$

$$Y3 : \underline{-4}$$

in order to tell *FAS* which initial values to use. You can also tell *FAS* this same information using command strings as follows:

$$f(x) = -x(y3 \wedge 4) + \sin(x)y1 + y \wedge 6 + \exp(x/5) \setminus y0 = 1 \setminus y1 = -2 \setminus y2 = 3 \setminus y3 = -4.$$

Limits on Order of Differential Equations

You can enter a differential equation of any order up to 10. You use the variables y, $y1$, $y2$, $y3$, $y4$, ..., $y9$ to refer to the function y and its first 9 derivatives.

RECURSIVE

This procedure allows you to create recursively defined functions, which are solutions of *difference equations*. Before we explain how to plot solutions of difference equations with *FAS*, let's briefly review some basic definitions. A **difference equation of order** k is an expression of the form

$$y_n = F(n, y_{n-1}, y_{n-2}, \ldots, y_{n-k})$$

$$y_0 = \alpha_0, \quad y_1 = \alpha_1, \ldots, \quad y_{k-1} = \alpha_{k-1}.$$

The first equation shows that y_n is a function, F, of the *index* n and of k previous y-values (y_{n-1}, ..., y_{n-k}). The second set of equations specify the following *initial values:* $y_0 = \alpha_0$, ..., $y_{k-1} = \alpha_{k-1}$. These initial values are used to get the first equation started.

For example, the following is a difference equation of order 2:

$$y_n = y_{n-1} + y_{n-2}$$

$$y_0 = 1, \quad y_1 = 2$$

If $n = 2$, then

$$y_2 = y_1 + y_0 = 3.$$

If $n = 3$, then

$$y_3 = y_2 + y_1 = 5$$

and so on.

Notice that when we put n equal to the order 2, we got

$$y_2 = y_1 + y_0 , \qquad (n = 2).$$

In *FAS* you enter the following formula in order to plot the values of y_n:

$$f(x) = y1 + y0.$$

The right side of this formula, $y1 + y0$, is a direct translation of the right side of the previous equation, $y_1 + y_0$. Once you enter the formula above, you will be prompted for the initial values of $y0$ and $y1$, respectively. If you enter 1 and 2, then you are specifying the correct initial values for this example. After you enter these initial values *FAS* will define $f(0) = 1$ and $f(1) = 2$, and then all subsequent values of f will be computed recursively by (if you chose, say, 256 points)

$$f(n) = f(n - 1) + f(n - 2), \quad n = 2, 3, \ldots, 256.$$

Using this method you can specify any difference equation up to order 10, using the variables $y0, y1, \ldots, y9$.

You can also use command strings to specify some of the information in this last example. For example, you could enter the formula

$$f(x) = y1 + y0 \setminus \text{Rec} \setminus y0=1 \setminus y1=2$$

The command string, \Rec, tells *FAS* to switch to *Recursive* mode, while the other two command strings assign the initial values 1 and 2 to $y0$ and $y1$.

Here is another example. Suppose you want to plot the sequence $\{y_n\}$ generated by the difference equation

$$y_n = -0.5 y_{n-3}, \quad y_0 = 1, \ y_1 = 2, \ y_2 = 6$$

using, say, 512 points. After choosing 512 points, you could enter the following formula:

$$f(x) = -0.5 y0 \setminus \text{Rec} \setminus \text{Order}=3 \setminus y0=1 \setminus y1=2 \setminus y2=6.$$

Or, you could also enter

$$f(x) = -0.5y \setminus \text{Rec} \setminus \text{Order}=3 \setminus y=1 \setminus y1=2 \setminus y2=6$$

since $y0$ can also be specified by just the character y alone. We used command strings for this example. If you leave them off, then you will be prompted for the necessary information. It is important to recognize that the difference equation above has order 3, which is why we used the command string, $\setminus \text{Order}=3$, to tell *FAS* that the order is 3 (otherwise, *FAS* will think that the order is 1 and graph the solution to $y_n = -0.5y_{n-1}$, $y_0 = 1$).

Sometimes difference equations are specified as in the following equation:

$$y_{n+2} = 4y_{n+1} - 3y_n, \quad y_0 = 2, \ y_1 = -3.$$

This difference equation is not in standard form, so it is necessary for you to convert it into standard form by using an index shift. The equivalent difference equation is

$$y_n = 4y_{n-1} - 3y_{n-2}, \quad y_0 = 2, \ y_1 = -3$$

which can be entered into *FAS* as

$$f(x) = 4y1 - 3y0 \setminus \text{Rec} \setminus y0=2 \setminus y1=-3.$$

We now discuss how to handle the case when n appears as a variable, as well as an index, in a difference equation. For example, suppose you want to plot the difference equation

$$y_n = y_{n-3}/n + y_{n-1}, \quad y_0 = 1, \ y_1 = -2, \ y_2 = 4$$

for $n = 0$ to 128. You begin by choosing 128 points and specifying the interval [0, 128]. The variable n, which appears as a divisor of y_{n-3}, *must be written as* x (as described above, in *FAS*, n denotes an index for the sumn( ) operation, not a variable). Thus, we would enter the difference equation above as

$$f(x) = y0/x + y2 \setminus \text{Rec} \setminus y0=1 \setminus y1=-2 \setminus y2=4.$$

Here is another example. Suppose the difference equation is

$$y_{n+3} = \frac{y_{n+1} - y_n}{(n+6)(n+8)}, \quad y_0 = 1, \ y_1 = 0, \ y_2 = -4.$$

First, you make an index shift to express the equation in the form

$$y_n = \frac{y_{n-2} - y_{n-3}}{(n+3)(n+5)}, \quad y_0 = 1, \ y_1 = 0, \ y_2 = -4.$$

Then, the variables $n+3$ and $n+5$ are replaced by $x+3$ and $x+5$ and you enter either the formula

$$f(x) = (y1 - y0)/[(x+3)(x+5)] \setminus \text{Rec} \setminus \text{Order}{=}3 \setminus y0{=}1 \setminus y1{=}0 \setminus y2{=}{-}4$$

or the formula

$$f(x) = (y1 - y)/[(x+3)(x+5)] \setminus \text{Rec} \setminus \text{Order}{=}3 \setminus y{=}1 \setminus y1{=}0 \setminus y2{=}{-}4.$$

GET DATA

This choice allows you to load numerical data files that were previously saved (using *Save* on the display menu). When you make this choice a list of previously saved numerical data file names will appear on the right side of the screen. Using the spacebar and backspace, or the up and down arrow keys, you can place the cursor on the file you want and then press *Enter*. (If there are more file names than the Help box can hold, then you will scroll through the list of file names once you reach the end of the first list.)

Once you select it, a file will either load and a graph will appear on the screen, or an error message will appear at the bottom of the Help box if the number of points or interval are inconsistent with present values.

The files listed on the right side of the screen were retrieved using the path listed at the bottom left corner of the screen. To change the path, you go through the following steps. If you press the *F5* key, then the following prompt will appear (followed by a blinking cursor):

$$path >$$

where *path* represents the path *FAS* is using to retrieve files. You may then enter either a new drive using the format

$$* :$$

where $*$ represents the letter of the new drive, or change directories using the format

$$cd\backslash pathname$$

where *pathname* stands for any valid DOS path designation. For example, if you enter

$$cd\backslash data$$

then you will change to the directory DATA on the current drive. If you enter

$$cd\ a:\backslash data$$

then the default directory on drive A will be changed to DATA. If A is the current drive, then you will change to this directory. If A is not the current drive, then you will *not* change over to drive A. To do that, you have to enter

$$a:$$

Since the default directory on drive A is DATA, you will now change to drive A and directory DATA. In other words, the files that are listed now will be obtained using the path A:\DATA (and this path will appear at the bottom left corner of the screen).

You may also load external data files if they are in the following format. They must have a name with format $*$.FDT where $*$ represents a valid 8-character DOS file name. These external files must be ASCII sequential files. The first data value must be the number of points (either 128, 256, 512, 1024, 2048, 4096, or 8192). The next entry must be the left endpoint of the interval; it must be either 0 or the negative of the right endpoint of the interval. The entry after that must be the right endpoint of the interval; it must be positive. The remaining entries are the function data values, either 129, 257, 513, 1025, 2049, 4097, or 8193 values. Notice that each entry in this list is *one* more than the corresponding entry in the list of numbers of points. This has to do with the way *FAS* processes the other function choices, and this choice is forced to be consistent with that method. All these numerical values must be in IEEE double precision format. *Note:* many devices for obtaining data collect some power of 2 number of data values. There is then no unique way to assign the last data entry. (One possible way is to give it the value of the first entry. This ensures that transforms are calculated correctly. For Fourier series, however, a better method would be to assign to it some average of the last few data values.)

A good way to see what the function data file format looks like is to plot a function and then save its values in ASCII format (using the choice *Ascii function data* from the *Save* menu). Using any word processor, or the text editor supplied with DOS, you can then examine the format of this function data file.

SELECT GRAPH

If you have already graphed a function and you are using the same number of points, then this choice also appears on the function menu. If you make this choice, then you can choose which function values of the graph are to be used for whatever function you are creating (if only one graph is displayed, then this choice is made automatically; if more than one graph is displayed, then you are asked to supply a graph number). *Note:* after you specify a graph a caption will appear on the caption line verifying this choice; however, *no new graph is drawn.* Although no new graph is drawn, the values of the specified graph will be supplied as function values to whatever procedure you are presently in.

This completes our discussion of the initial setup of function data. We will now discuss the various procedures for doing Fourier analysis with *FAS.*

The first menu you encounter after selecting *Next* from the display menu is the Fourier analysis menu:

Series   Transforms   Convolutions   Graphs   Erase screen   Quit

We will now discuss each of these choices.

SERIES

This choice allows you to calculate Fourier series, Fourier sine series, Fourier cosine series, or sampling (interpolation) series. You are presented with the following menu:

Fourier   Sine   Cosine   Interpolation (sampling)

Choose *Fourier* if you want to do Fourier series, choose *Sine* if you want to do Fourier sine series, choose *Cosine* if you want to do Fourier cosine series, or choose *Interpolation (sampling)* if you want to do sampling (interpolation) series. We now describe each of these procedures.

FOURIER SERIES

Once you have completed the initial setup, then this choice allows you to compute Fourier series partial sums for your function data. The first question you are asked is

$$\text{How many harmonics?} \tag{A.1}$$

The number of harmonics is the number of complex exponential terms (counting positively and negatively indexed harmonics together, and excluding the constant term). For example, the function $f(x) = x$ has the following complex Fourier series over the interval $[0, 2\pi]$:

$$\pi + \sum_{k=-\infty}^{\infty}{}' \frac{i}{k} e^{ikx} \tag{A.2}$$

(where the prime on the summation sign means the $k = 0$ term is omitted). If you choose to graph, say, 15 harmonics from this Fourier series, then *FAS* will compute (via an FFT algorithm) the partial sum

$$\pi + \sum_{k=1}^{15} \left(\frac{i}{k} e^{ikx} - \frac{i}{k} e^{-ikx} \right). \tag{A.3}$$

If you enter f instead of entering a number of harmonics, then *FAS* will graph the original function.

After you choose the number of harmonics you want (except if you enter f), then you are asked the following question:

$$\text{Do you want to apply a filter? (Press y or n)} \tag{A.4}$$

If you press n, then the partial sum in (A.3) is graphed. If you press y, then a menu of filters appears. After selecting a filter, the sum in (A.3) is modified by multiplying by constants $\{F_k\}$ as follows:

$$F_0 \pi + \sum_{k=1}^{15} F_k \left(\frac{i}{k} e^{ikx} - \frac{i}{k} e^{-ikx} \right) \tag{A.5}$$

The nature of the constants $\{F_k\}$ depends on which filter you chose. Some information on this is given in a Help file; just press $F1$ when the cursor is on the particular filter you want information about. Further discussion of these filters can be found in Chapter 4. After the graph is displayed, if you select *Next,* from the display menu, then you will be sent to the next part of the *Fourier series* routine.

After graphing a partial sum, the following menu appears:

$$\text{Partial sum} \qquad \text{New function} \qquad \text{Done} \qquad\qquad (A.6)$$

If you choose *Partial sum,* then you are sent back to the question about the number of harmonics and the procedure begins all over again. If you choose *New function,* then you are sent back to the function menu, where you can create a new function and do the *Fourier series* procedure again. If you choose *Done,* then you are sent back to the display menu (where you can choose *Next* to select another procedure).

For each of the questions in the *Fourier series* procedure, if you press the *Esc* key, then you are sent back to the previous question.

SINE SERIES

The structure of the *Sine series* procedure is essentially the same as the *Fourier series* procedure. The only difference is that instead of computing a Fourier series you are computing a Fourier sine series. For example, if you choose to compute 15 harmonics of the Fourier sine series for $f(x) = x - \pi$ over the interval $[0, 2\pi]$, then *FAS* will graph

$$\sum_{n=1}^{15} \frac{-2}{n} \sin nx. \qquad\qquad (A.7)$$

Other than what is being graphed [the sine series in (A.7) instead of the Fourier series in (A.3)], the questions you are asked and how *FAS* responds to your answers is *identical* to what is described above for the *Fourier series* procedure.

COSINE SERIES

The structure of the *Cosine series* procedure is essentially the same as the *Fourier series* procedure. The only difference is that instead of computing a Fourier series you are computing a Fourier cosine series. For example, if you choose to compute 15 harmonics of the Fourier cosine series for $f(x) = x^2/4$ over the interval $[0, 2\pi]$, then *FAS* will graph

$$\frac{\pi^2}{3} + \sum_{n=1}^{15} \frac{1}{n^2} \cos nx. \qquad\qquad (A.8)$$

Other than what is being graphed [the cosine series in (A.8) instead of the Fourier series in (A.3)], the questions you are asked and how *FAS* responds to your answers is identical to what is described above for the *Fourier series* procedure.

INTERPOLATION (SAMPLING) SERIES

Once you have completed the initial setup, then this choice allows you to compute (periodic) sampling series partial sums for your function data. The following menu appears:

Choose function (reconstruction kernel).
  Formula   Diffeq   Recursive   Get data   (A.9)

You then create the reconstruction kernel to be used in the sampling series. For example, suppose that you plot the function $f(x) = \text{sinc}(4x)$ over the interval $[-8, 8]$ using 512 points. Then *FAS* graphs the function shown in Figure A.1(a).

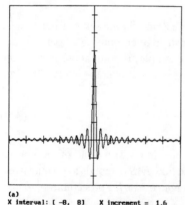
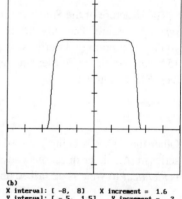

```
(a)
X interval: [ -8,  8]    X increment =  1.6
Y interval: [ -.5,  1.5]    Y increment =  .2
```
```
(b)
X interval: [ -8,  8]    X increment =  1.6
Y interval: [ -.5,  1.5]    Y increment =  .2
```

FIGURE A.1
(a) Reconstruction kernel. (b) User-defined function.

After choosing *Next* from the display menu, the following menu appears:

Choose function.
  Formula   Diffeq   Recursive   Get data   Select graph   (A.10)

You can now create the function that you want to approximate with a sampling series. Suppose you create the function

$$f(x) = \exp[-(x/4) \wedge 20]$$ (A.11)

Then *FAS* produces the graph shown in Figure A.1(b). After pressing *Next* on the display menu, the following request for input is displayed (the underlined number is entered data):

Sampling interval is a multiple of the increment: .03125

Multiples of basic increment (no more than 256): 8

The increment is equal to the interval length divided by the number of points (in this example, $16/512 = .03125$). This increment represents the separation between the points that *FAS* is using for plotting graphs. The number of multiples (in this case we have specified 8) of the increment gives the sampling spacing, which we will call T (in this case $T = .25$). This sampling spacing, T, determines the maximum number of terms allowed in the following interpolation (sampling) series

$$\sum_{k=-K}^{K} f(kT)\text{sinc}\,[4(x - kT)] \qquad (A.12)$$

The function f, in this example, is the one given above in equation (A.11). A limitation on the size of K, in this case it is 32, is needed to ensure that kT does not leave the interval $[-L, L]$ ($[-8, 8]$ in this example). Finally, you are asked for the value of K. The following prompt appears:

Sampling interval is: .25

Order of partial sum (no more than 32): 12

FAS will now plot *a periodic approximation* to the sampling series in (A.12). It does this using two FFTs (see Exercises 5.76 to 5.78 for more details). For many functions and reconstruction kernels, this will result in a close approximation to the series in (A.12). Or, more importantly, it will result in a close approximation to the function f (especially if the maximum value for multiples of the basic increment is used). For this example, we chose $K = 12$ (instead of 32) and the graph shown in Figure A.2(a) results. If instead we choose $K = 32$, then the graph shown in Figure A.2(b) results.

After you have graphed a partial sum and selected *Next* from the display menu, then the following menu appears:

Partial sum   *New function*   *Reconstruction kernel*   *Done*   (A.13)

If you choose *Partial sum,* then you are sent back to the request for the order of the partial sum, and you can graph another approximation to the partial sum

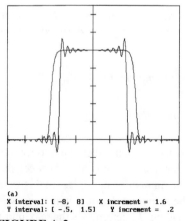
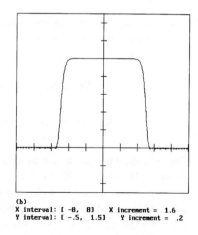

(a)
X interval: [-8, 8]   X increment = 1.6
Y interval: [-.5, 1.5]   Y increment = .2

(b)
X interval: [-8, 8]   X increment = 1.6
Y interval: [-.5, 1.5]   Y increment = .2

FIGURE A.2
(a) User-defined function and its order 12 sampling series. (b) Same user-defined function and its order 32 sampling series.

in (A.12). If you choose *New function,* then you are sent back to menu (A.10). You can then create a new function to be approximated using the same reconstruction kernel. To change the reconstruction kernel, you choose *Reconstruction kernel* from menu (A.13). You are then sent back to menu (A.9) and the *Interpolation (sampling)* procedure begins anew. If you choose *Done,* then you are sent to the display menu, where you can select *Next* to perform a new procedure.

One final remark about sampling series. As pointed out above, the *Interpolation (sampling)* procedure produces a *periodic approximation* to a function f using a subset, $\{f(kT)\}$, of its values. It is also possible to compute a sum like (A.12) using a sum operator. You could begin by choosing the *Draw graph* choice described below. Then, the sum in (A.12) could be computed (after first graphing the function f that you want to approximate) using the formula:

$$\text{sumk } \{g1(kt)\text{sinc}[4(x - kt)]\}$$

where $g1$ produces the values of the function that you plotted first. You will be prompted for the value of the parameter t and for the lower, upper limits of the sum variable k.

TRANSFORMS

This choice allows you to calculate Fourier transforms, Fourier sine transforms, and Fourier cosine transforms. You are presented with the following menu:

Choose transform type.
Fourier   Sine   Cosine

Choose *Fourier* if you want to do Fourier transforms, choose *Sine* if you want to do Fourier sine transforms, choose *Cosine* if you want to do Fourier cosine transforms. We now describe each of these procedures.

FOURIER TRANSFORMS

Fourier transforms of either real or complex valued functions can be computed. You are presented with the following menu:

Real   Complex   (A.14)

If you want to transform a real valued function, then select *Real.* Otherwise, if you want to transform a complex valued function, then select *Complex.*

If you select *Real,* then the initial setup is similar to the one we described above. You begin by choosing a number of points. The selection of interval type, however, is replaced by the following menu:

Choose which interval pair you want for your Fourier transformation.
1) $[-L, L] \rightarrow [-\Omega, \Omega]$ *2)* $[0, L] \rightarrow [-\Omega, \Omega]$
3) $[-L, L] \rightarrow [0, \Omega]$ *4)* $[0, L] \rightarrow [0, \Omega]$

The choice 1) $[-L, L] \rightarrow [-\Omega, \Omega]$ is used when an integral over $[-L, L]$ is used to approximate the Fourier transform integral over $(-\infty, \infty)$. The Fourier transform that *FAS* produces is graphed over an interval of the form $[-\Omega, \Omega]$. This choice was made the first choice because all the transforms produced in this book were obtained using it. The other choices are included, however, because they arise frequently in applications. For example, if the initial function is a function of time, then this function is often defined over an interval of the form $[0, L]$, in which case, either choice 2) $[0, L] \rightarrow [-\Omega, \Omega]$ or choice 4) $[0, L] \rightarrow [0, \Omega]$ might be used. If 2) $[0, L] \rightarrow [-\Omega, \Omega]$ is chosen, then *FAS* graphs the Fourier transform over an interval of the form $[-\Omega, \Omega]$. Consequently, *to invert this Fourier transform,* the choice 3) $[-L, L] \rightarrow [0, \Omega]$ is necessary. The fourth choice 4) $[0, L] \rightarrow [0, \Omega]$ is included because it allows the computation of FFTs of the type described in many digital signal processing books. (For further details about these four choices you should make use of the help in *FAS.*) After specifying the interval pair, you then create your function. When you choose *Complex,* however, you must create two functions. These functions will be the real and imaginary parts of the complex valued function that you are transforming. You will be asked to supply the real part as the first function and the imaginary part as the second function.

Once you have finished entering the function data, then *FAS* will begin approximating Fourier transforms. First, you are presented with the following menu of two choices:

$$\textit{Four Tr} \qquad \textit{Power Spectrum} \qquad\qquad (A.15)$$

If you select *Four Tr,* then the following menu appears:

$$\textit{Real/imaginary} \qquad \textit{Amplitude/phase} \qquad\qquad (A.16)$$

If you choose *Real/imaginary,* then *FAS* will graph the real and imaginary parts of the Fourier transform. If you choose *Amplitude/phase,* then *FAS* will graph the amplitude, $|\hat{f}|$, of the Fourier transform $\hat{f}$, and the phase of the transform (the phase being the angles of the complex values of $\hat{f}$). After making your choice, if you chose *Four Tr* at menu (A.15), then you are asked the following question:

$$\textit{Positive transform exponent } (y = \textit{pos., } n = \textit{neg.)? (Press y or n)} \qquad (A.17)$$

If you press y, then *FAS* will approximate the following Fourier transform,

$$\tilde{f}(x) = \int_{-\infty}^{\infty} f(s) e^{i2\pi x s}\, ds. \qquad\qquad (A.18)$$

while, if you press n, then *FAS* will approximate the following Fourier transform:

$$\hat{f}(x) = \int_{-\infty}^{\infty} f(s) e^{-i2\pi x s}\, ds. \qquad\qquad (A.19)$$

FAS will then ask you the following question:

$$\textit{Do you want to apply a filter? (Press y or n)} \qquad\qquad (A.20)$$

If you press y, then a menu of filter choices will appear. If you choose a filter, then

FAS will approximate the following filtered transform,

$$\int_{-\infty}^{\infty} F(s) f(s) e^{\pm i 2\pi x s} \, ds$$

where the sign in the exponent depends upon how you answered question (A.17). The purposes of the filtering process are described in Chapter 5.

Depending on what choice you made at menu (A.16), *FAS* will either first display the real part or the amplitude of the (filtered or unfiltered) transform. If you choose *Next* from the display menu, then *FAS* will either graph the imaginary part or the phase of the (filtered or unfiltered) transform.

After finishing the display portion of the *Four Tr* procedure, then the following menu will appear:

Transform   New function   Done

If you choose *Transform,* then you are returned to the menu in (A.15) and the *Four Tr* procedure begins again *using the same function you initially chose to transform.* You can perform a transform with a different filter (or without a filter) or you can choose to do a *Power Spectrum.* If you choose *New function,* then you will be returned to menu (A.14). Choosing *Done* will send you to the display menu, where you can choose *Next* to start a new procedure.

If you select *Power Spectrum* from menu (A.15), then *FAS* will approximate the function $|\hat{f}(x)|^2$. If you chose *Real* in (A.14), then it does not matter whether Formula (A.18) or (A.19) is used to compute $\hat{f}(x)$, the result for $|\hat{f}(x)|^2$ is the same. Consequently, you are not asked Question (A.17) about what sign exponent you want (*FAS* just chooses positive automatically). If you chose *Complex,* however, it does matter whether Formula (A.18) or (A.19) is used, so you will be asked Question (A.17).

The structure of the Power Spectrum procedure is the same as what we just described above for the *Four Tr* subprocedure.

SINE TRANSFORM

The structure of the *Sine transform* procedure is almost the same as the *Four Tr* procedure described above for real valued functions. The only difference is that with the *Sine transform* procedure, *FAS* approximates the following sine transform:

$$\hat{f}_S(x) = 2 \int_0^{\infty} f(s) \sin(2\pi x s) \, ds \tag{A.21}$$

Since there is no exponent in (A.21), you are not asked Question (A.17) about the sign of the exponent, as in the *Four Tr* procedure. Furthermore, an interval of the type $[0, L]$ is always used in the initial setup. Other than that, the questions you are asked are the same as in the *Four Tr* procedure.

COSINE TRANSFORMS

The structure of the *Cosine transform* procedure is almost the same as the *Four Tr* procedure described above for real valued functions. The only difference is that with the *Cosine transform* procedure, *FAS* approximates the following cosine transform:

$$\hat{f}_C(x) = 2 \int_0^\infty f(s) \cos(2\pi xs) \, ds \qquad (A.22)$$

Since there is no exponent in (A.22), you are not asked Question (A.17) about the sign of the exponent, as in the *Four Tr* procedure. Furthermore, an interval of the type $[0, L]$ is always used in the initial setup. Other than that, the questions you are asked are the same as in the *Four Tr* procedure.

CONVOLUTION

This choice lets you compute either a convolution, or an autocorrelation, or a pair correlation. After initial setup (choosing the number of points and specifying the interval) you are presented with the following menu:

$$\textit{Convolution} \qquad \textit{Autocorrelation} \qquad \textit{Pair correlation} \qquad (A.23)$$

If you choose *Convolution,* then the following menu appears:

> *Choose first function.*
> *Formula*   *Diffeq*   *Recursive*   *Get data*   (A.24)

from which you then proceed to create your first function, call it $f(x)$. After selecting *Next* from the display menu, *FAS* then asks you to create a second function, call it $g(x)$. The following menu appears:

> *Choose second function.*
> *Formula*   *Diffeq*   *Recursive*   *Get data*   (A.25)

After this second function is graphed, and you have selected *Next* from the display menu, then *FAS* approximates the convolution of f and g. There are several kinds of convolution. You are first asked the following question:

$$\text{Divide by interval length? (Press y or n)} \qquad \text{(A.26)}$$

If you press *y,* and the interval you chose was $[-L, L]$, then *FAS* will approximate the following integral,

$$\frac{1}{2L} \int_{-L}^{L} f(s)g(x - s)\, ds \qquad \text{(A.27)}$$

which is the convolution of f and g over the finite interval $[-L, L]$. If the interval you chose was $[0, L]$, then *FAS* will approximate the following integral,

$$\frac{1}{L} \int_{0}^{L} f(s)g(x - s)\, ds \qquad \text{(A.28)}$$

which is the convolution of f and g over the interval $[0, L]$. See Chapters 4 and 5 for further details about the theory of convolution.

On the other hand, suppose you press *n* in response to Question (A.26). Then, when the interval you chose was $[-L, L]$, *FAS* will compute

$$\int_{-L}^{L} f(s)g(x - s)\, ds. \qquad \text{(A.29)}$$

For certain kinds of functions, f and g, the integral in (A.29) will be a good approximation to the following integral,

$$\int_{-\infty}^{\infty} f(s)g(x - s)\, ds \qquad \text{(A.30)}$$

which is the convolution of f and g over $(-\infty, \infty)$. If the interval you chose was $[0, L]$, then *FAS* will compute

$$\int_{0}^{L} f(s)g(x - s)\, ds. \qquad \text{(A.31)}$$

For certain kinds of functions, f and g, the integral in (A.31) will be a good approximation of

$$\int_{0}^{x} f(s)g(x - s)\, ds \qquad \text{(A.32)}$$

which is the convolution of f and g over $[0, \infty)$.

If you choose *Autocorrelation* from the menu in (A.23), then you will only be asked to create one function. The following menu appears:

> Choose a function.
> *Formula*   *Diffeq*   *Recursive*   *Get data*   (A.33)

After you have graphed this function and selected *Next* from the display menu, then you will be asked Question (A.26). If you press *n*, then the integral

$$\int_{-L}^{L} f(s)f(s-x)\,ds \qquad (A.34)$$

will be computed if you chose the interval $[-L, L]$. If you chose the interval $[0, L]$, then the integral

$$\int_{0}^{L} f(s)f(s-x)\,ds \qquad (A.35)$$

will be computed. If you press *y* in response to Question (A.26), then the integral in (A.34) will be divided by $2L$ while the integral in (A.35) will be divided by L. These integrals are approximations to autocorrelations over finite and infinite intervals.

The *Pair correlation* procedure works exactly like the *Convolution* procedure. The only difference is that the integrands involved are changed to $f(-s)g(x-s)$. Or, by change of variables, $f(s)g(x+s)$. (The one difficulty to keep in mind with pair correlation is that it is not commutative with respect to f and g.)

When each of these procedures is finished, you are then asked the following question:

> *Do another convolution operation? (Press y or n)*   (A.36)

If you press *y*, then you are sent back to the menu in (A.23). If you press *n*, then you are returned to the display menu where you can choose a new procedure.

GRAPHS

This procedure begins by displaying the following menu:

> *Draw graphs*   *View graphbook*   (A.37)

Here is a description of these two choices.

Draw Graphs

This procedure allows you to draw graphs without doing any Fourier analysis. It is useful for overlaying, for comparison purposes, previously saved graphs created by other procedures. For example, you might wish to compare different convolutions, or Fourier transforms of different functions, or certain Fourier series with Fourier transforms. This procedure can also be used as a graphing program for other purposes, such as comparing functions with their sampling series partial sums or their power series partial sums.

After initial setup, the procedure will graph the first function you choose to graph. After you have graphed a function and selected *Next* from the display menu, then you are asked the following question:

$$\text{Graph another function? (Press y or n)} \tag{A.38}$$

If you press *y*, then you are sent to the initial function menu. This gives you the opportunity to overlay graphs for comparison purposes. If you press *n*, then you are returned to the display menu where you can choose to do another procedure.

View Graphbook

This procedure allows you to view a *Graphbook*. A Graphbook is a series of screen images of graphs and remarks about them. (See the discussion below of the *Make GBK* procedure for an explanation of how a Graphbook is created.) The Graphbook file names appear on the right side of the screen. Using the spacebar and backspace (or the up and down arrow keys) to place the cursor on your choice, you then press *Enter* to select the Graphbook that you want to view. The first screen image in the Graphbook will then appear. This will happen quickly if you are using a hard disk; it will take a little longer if you are retrieving the file from a floppy disk. To view successive screen images just press the *PgDn* key to go forward and the *PgUp* key to go backward.

When a screen image is displayed you will see the full display menu. You can use this menu to make modifications of the screen image, such as changing the x or y intervals, or adding remarks. A modified screen image can be saved as part of the Graphbook. First, you select *Save* from the display menu and the following menu appears:

$$\text{Update screen file} \qquad \text{Binary function data} \qquad \text{Ascii function data} \tag{A.39}$$

By selecting *Update screen file,* a modified screen image is automatically saved as part of the Graphbook that you are viewing.

If you select *Binary function data* from menu (A.39), then you can save one of the graphs from the screen image. This graph will then be accessible from the *Get data* choice on the function menu. The choice *Ascii function data* allows you to save one of the graphs from the screen image into an ASCII text file. This ASCII file is readable by other software. Graphs saved in this format are also accessible using the *Get data* choice on the function menu.

When you have finished viewing all the screen images in a Graphbook, then the following question appears:

Are you done? (Press y or n)

If you press *n,* then you are returned to the display menu, and you are viewing the last image (you can then view a previous image by pressing the *Page Up* key). If you press *y,* then you are returned to the display menu (as it appeared when you started the program), and you can choose another procedure.

ERASE SCREEN

Make this choice if you want to clear the screen of all its graphs. Once you clear the screen, all function data are cleared from memory (and are lost unless you have previously saved them). Consequently, you will be prompted for confirmation before the screen is erased.

QUIT

When you select *Quit* you will exit the program. Before doing so, however, *FAS* will ask you if you are sure that you want to quit the program (since all graphs and formulas will be lost unless you have previously saved them).

You can also quit the program as follows. Whenever you are at a menu line or at a question, you can hold down the *Ctrl* key and press the *End* key. *FAS* will again ask you if you are sure that you want to quit.

This completes our discussion of all the procedures that are accessible starting with *Next* on the display menu. We shall now discuss each of the other choices on the display menu.

X-int

Select this choice in order to change the x-interval. You will be presented with the following message

$$\textit{Enter X-interval (left endpt., right endpt.):} \qquad \text{(A.40a)}$$

followed by a blinking cursor. You then type in an interval in the form of two real numbers separated by a comma (you need not include parentheses or brackets around the numbers). The first number will be the new left endpoint and the second number will be the new right endpoint of the x-interval. If the right endpoint is not greater than the left endpoint, then *FAS* will beep and you will be asked to redo your input.

After entering an x-interval, you will be asked the following question:

$$\textit{Change Y-interval? (Press y or n)}$$

Press n if you only want to change the x-interval. If, however, you also want to change the y-interval, then press y and the following request for input will appear:

$$\textit{Enter Y-interval (left endpt., right endpt.):} \qquad \text{(A.40b)}$$

You then type in an interval in the form of two real numbers separated by a comma. The first number will be the new left endpoint and the second number will be the new right endpoint of the y-interval. If the right endpoint is not greater than the left endpoint, then *FAS* will beep and you will be asked to redo your input.

Y-int

Select this choice in order to change the y-interval. You will be presented with the following message:

$$\textit{Enter Y-interval (left endpt., right endpt.):} \qquad \text{(A.41a)}$$

followed by a blinking cursor. You then type in an interval in the form of two real numbers separated by a comma (you need not include parentheses or brackets around the numbers). The first number will be the new left endpoint and the second number will be the new right endpoint of the y-interval. If the right endpoint is not greater than the left endpoint, then *FAS* will beep and you will be asked to redo your input.

After entering a y-interval, you will be asked the following question:

Change X-interval? (Press y or n)

Press *n* if you only want to change the y-interval. If, however, you also want to change the x-interval, then press *y* and the following request for input will appear:

Enter X-interval (left endpt., right endpt.):   (A.41b)

You then type in an interval in the form of two real numbers separated by a comma. The first number will be the new left endpoint and the second number will be the new right endpoint of the x-interval. If the right endpoint is not greater than the left endpoint, then *FAS* will beep and you will be asked to redo your input.

GRAPH

This procedure allows you to perform various operations involving graphs. The following menu will appear:

Integrals   Norm diff   Function   Deriv   Remove   Make GBK   (A.42)

Here is a description of each choice.

Integrals

With this choice you can approximate the p-Norm or the Sup-Norm, or the integral of a graph on the screen. When you make this selection the following menu will appear:

Function norm   Integral

If you choose *Function norm,* then you can approximate either the p-Norm or the Sup-Norm of a graph on the screen. The p-Norm of a function $f(x)$ over an interval $[a, b]$ is defined to be

$$\left[\int_a^b |f(x)|^p \, dx \right]^{\frac{1}{p}}$$

                                                                (A.43a)

while the Sup-Norm is defined by

$$\sup_{a \leq x \leq b} |f(x)| \tag{A.43b}$$

where sup stands for *supremum* and means the maximum of $|f(x)|$ over the interval $[a, b]$. The following request for input appears:

$$\textit{Enter a Power, or } s \textit{ for sup:} \tag{A.44}$$

followed by a blinking cursor. If you enter a real number greater than or equal to 1, then that real number is used for the power p in computing the p-Norm in (A.43a). If you enter a number less than 1, then you will be told to redo your input. By entering the letter s in response to (A.44), you will tell *FAS* to use (A.43b) to compute a Sup-Norm.

If, on the other hand, you choose *Integral,* then the integral $\int_a^b f(x)\,dx$ is approximated (using the trapezoid rule) and the result is displayed on the caption line below the graphs.

Note: these norms and integrals are computed for $[a, b]$ equal to the present x-interval. If this interval is not the original x-interval over which the function was graphed, then you will have to use X-int to change back to that original interval in order to approximate norms or integrals over it.

Norm diff

With this choice you can approximate either the p-Norm difference or the Sup-Norm difference between two graphs on the screen. The following menu will appear:

$$\textit{Absolute} \qquad \textit{Relative}$$

If you choose *Absolute,* then, depending on your response to (A.44), either Formula (A.43a) or Formula (A.43b) is used for $f(x) = h(x) - g(x)$, where $h(x)$ and $g(x)$ are the two functions whose norm difference is being calculated. If there are just two graphs on the screen, then the norm difference between those two graphs is computed and the result is displayed in the caption below the graphs. If there are more than two graphs, then you are presented with the following request for input

$$\textit{Numbers of graphs (separate numbers by a comma):} \tag{A.45}$$

followed by a blinking cursor. You may then enter the numbers for which you want the norm difference calculated (1,2 for graph numbers 1 and 2, etc.). The norm difference is then approximated and the result displayed in the caption line below the graphs.

On the other hand, if you select *Relative,* then the norm of $f = h - g$ is divided by the norm of h, where h stands for the function whose graph is specified by the first number that you enter in response to (A.45). Consequently, entering 1, 2 is *not* the same as entering 2, 1 when you have chosen to do a relative norm difference.

Note: these norm differences are computed for [a, b] equal to the present x-interval. If this interval is not the original x-interval over which the functions were graphed, then you will have to use X-int to change back to that original interval in order to approximate norm differences over it.

Function

This choice allows you to graph new functions on the screen. After you make this choice, the formula menu will appear:

Formula   Diffeq   Recursive   Get data

These choices were discussed above.

Deriv

This choice allows you to compute a numerical approximation to the first derivative of any displayed graph.

Remove

This choice allows you to remove one of the graphs displayed on the screen. (If only one graph is displayed, then this choice does not appear.) When you select this choice the following request for input appears:

Number of graph to be removed:

followed by a blinking cursor. After entering the number of the graph you want removed that graph will be deleted from the screen. *Be careful using this choice!* The data from the deleted graph will be lost unless you have previously saved them.

Make GBK

This choice is used to create a Graphbook, a recording of the screens that you create while using *FAS*. These screens are stored in a file that you can view later using the choice *View graphbook* described above. When you make this choice, the following message appears prompting you for a file name for your Graphbook:

Enter graphbook name:

You then enter a valid DOS file name (without extension) for your Graphbook. If your file name is acceptable (i.e., does not conflict with any previous file names), then you are returned to the display menu. From then on, whenever you choose *Next* after you have produced a new graph, you will be asked the following question:

Save the screen in your graphbook? (Press y or n)

If the screen presently displayed is one that you wish to save, then you press *y.* You press *n* if you do not want to save the present screen in your Graphbook. *Note:* if there are no graphs displayed, then do **not** press *y.* Saving an empty screen containing no graphs results in a corrupted Graphbook file.

If you wish to stop recording at any time, then after selecting *Graph* from the display menu you will see that the choice *End recording* has replaced *Make GBK.* Selecting *End recording* will close your Graphbook file, and you will no longer be asked if you wish to save your screens.

CAP \ CALC

You make this choice in order to either change the caption underneath the graph display, or to do a numerical calculation. The following menu will appear:

$$\text{New caption} \qquad \text{Calculator} \qquad\qquad (A.46)$$

Choose *New Caption* if you want to change the caption. When you make this choice, the present caption will appear at the bottom of the screen, ready for modification. You may then type in the modifications you want, or, by pressing the *F9* key, erase it and type in a new caption. After you press *Enter* the caption you have created will appear in the caption line. If you press the *Esc* key before you have entered the new caption, then the old caption will be retained and you will be back at the display menu.

If you choose *Calculator,* then a cursor will appear at the lower left corner of the screen and you can type in a simple (one-line) statement for evaluation. For example, if you type in the following statement,

$$\text{sumk}(k+1) \setminus k{=}0{,}9$$

and press *Enter,* then the following caption will appear:

$$\text{sumk}(k+1) \setminus k{=}0{,}9 = 55.$$

PRINT

This choice allows you to feed a screen image to a printer to make a printed copy. After making this choice, the following menu will appear:

> *Hp laserjet comp.*   *Epson comp.*   *Other*

The choice *Hp laserjet comp* allows you to print the screen on any laser printer which is running in a mode compatible with the Hewlett-Packard group of laser printers. If you make this choice, then the following menu appears:

> *Print and eject*   *Multi-print, no eject*   *Eject*   *File script*

If you select *Print and eject,* then the screen will be printed on the page and the page will be ejected from the laser printer. If you select *Multi-print, no eject,* then the screen will be printed and the page will remain in the laser printer. This allows you to print several screen images on a single sheet of paper. Finally, if you select *Eject,* then the page that is presently in the printer will be ejected.

If you choose to either print or multi-print, then the following menu will appear:

> *Large*   *Medium*   *Small*

The choices here control the size of screen image produced by the printer. Technically, *Large* equals 75 DPI resolution, *Medium* equals 100 DPI resolution, and *Small* equals 150 DPI resolution, but the easiest thing to do is just to try some examples to see what the sizes look like. After you choose the size of the screen image from this menu, then you will receive the following request for input:

> *Enter: vertical inches, horizontal inches:*

Here you enter two non-negative numbers separated by a comma. They represent the vertical and horizontal distances of the upper left corner of the screen image from the upper left corner of the print region of the page. For example, if you enter 1.5, 2.2, then the upper left corner of the screen image will be printed 1.5 inches below and 2.2 inches to the right of the upper left corner of the print region of the page. Or, if you enter 0,0, then the upper left corner of the screen image will be printed exactly at the upper left corner of the print region of the page. This allows you to have precise control over where the screen images will print out on the page. (*Note:* sometimes you may wish to print only the graphs, without the caption and interval information. In such a case, press the *Del* key when you are at the *Display* menu. The caption and interval information will be removed, and then you can print the screen. To restore the caption and interval information, press the *Ins* key when you are at the *Display* menu.) After you have printed the screen, *FAS* asks you if you want to save the settings that you have chosen to a file. You can then print another page very quickly and accurately by using these saved settings. You retrieve a file of printer settings by choosing *File script* from the initial menu of the *Hp laserjet* procedure. Several files of printer settings are included with the disk accompanying this book. These files can be used to print, for example, two graphs at the top of the page [use the file *2GR_(A)* to print the left graph, followed by *2GR_(B)* to print the right graph and eject the page] or to print four graphs on a page [use the file *4GR_(A)* to print the top-left graph, followed by *4GR_(B)* to print the top-right graph, followed by *4GR_(C)* to print the bottom-left graph, followed by *4GR_(D)* to print the bottom-right graph and eject the page]. There are also files for printing one graph (at the top-center of a page), or for printing three graphs on a page, or for printing six graphs on a page.

The choice *Epson comp* allows you to print the screen image on any Epson-compatible pin printer, or on some bubble jet printers. The image will print out in Landscape mode.

The last choice, *Other,* allows you to print the screen on other types of printers. To utilize this procedure, you must have loaded a program for graphics dumping *before* loading FAS.EXE. For many printers the DOS command file GRAPHICS.COM will serve this purpose (even for laser printers if you have DOS 5 or 6). It saves a lot of trouble if you write a batch file that loads this graphics program followed by FAS.EXE. Consult your DOS manual for instructions on how to create batch files and for the specific option for your printer when loading GRAPHICS.COM. After you select *Other,* if you have a monochrome video, then the display menu is removed from the screen. If you then press the *Print Screen* button (or *Shift +* *Print Screen* on some PCs) the screen image will be fed to the printer *provided* you have loaded the graphics program prior to loading FAS.EXE. After printing, press any key to continue with *FAS.*

If you have color video, then the following menu appears:

$$\textit{Black \& White} \qquad \textit{Color} \qquad\qquad \text{(A.47)}$$

If you have an ordinary black and white printer, then you should select *Black & White*. The screen converts to a monochrome image. If you then press the *Print Screen* (or *Shift + PrSc* on some PCs) the screen image will be fed to the printer. After printing, press any key to continue with *FAS.* If you have a color printer that can print the colors on the screen, then you may want to choose the *Color* choice on menu (A.47). Then the display menu will vanish and you can print the screen as we described above.

SAVE

This procedure allows you to save graphs or formulas. Normally, the menu that appears is the following:

$$\textit{Binary function data} \quad \textit{Formula} \quad \textit{Ascii function data} \qquad (A.48)$$

The other type of menu is (A.39), which we described above in the *View graphbook* procedure.

If you select *Binary function data* from menu (A.48) or menu (A.39), then (provided there is more than one graph) the following message will appear:

$$\textit{Enter the number of the graph you want saved:} \qquad (A.49)$$

followed by a blinking cursor. After entering the number of the graph that you want saved, then the following request for input appears:

$$\textit{File Name:} \qquad (A.50)$$

followed by a blinking cursor. You may enter any valid DOS file name (without an extension). The extension .FDT is automatically appended to your file name and the result is the function data file name. This function data file will be accessible if you select *Get data* from the function menu.

If you select *Formula* from menu (A.48), then you are asked the following question:

$$\textit{Check formula? (Press y or n)} \qquad (A.51)$$

If you want to check to make sure that you are saving the correct formula, then press *y*. Then the formula that is presently stored in memory will appear. If it is the one you want to save, then just press *Enter*. If you want to change the formula, then you may create a new one and then press *Enter*. This new formula will be the one saved and it becomes a new formula stored in memory. The last step of this procedure is the saving of the formula; the following request for input appears:

$$\textit{File Name:} \qquad (A.52)$$

followed by a blinking cursor. You may enter any valid DOS file name (without an extension). The extension .UFM is automatically appended to the name you select and the result is the formula file name. The resulting formula will be accessible whenever you are creating a formula by pressing the *F3* key to access the list of formula files.

If you select *Ascii function data,* then you are asked the same questions, and file names are assigned in the same way, as if you had chosen *Binary function data.* The only difference is that the data for this choice is saved in an ASCII text format. This ASCII file is readable by other software. The data can also be read by FAS using the *Get data* command, as described above.

REM

This procedure allows you to create a note on the right side of the screen, which can be used to describe the various graphs displayed. You can type in a note using the letters on the keyboard and using the following special keys to do some simple editing:

> *Home* = return to left margin
>
> *End* = move to end of line
>
> *PgUp* = new page (up to 20 pages are allowed)
>
> *PgDn* = back to previous page
>
> *Del* = delete character
>
> *Ins* = toggle between *typeover* and *insert* modes
>
> *Up Arrow* = move cursor up one line
>
> *Down Arrow* = move cursor down one line
>
> *Left Arrow* = move cursor left one space
>
> *Right Arrow* = move cursor right one space
>
> *Enter* = start a new line
>
> *Esc* = end recording of note

If you choose to write a note while one is already displayed, then the following menu appears:

$$\textit{Modify} \qquad \textit{Create} \qquad\qquad\qquad (A.53)$$

If you choose *Modify* from menu (A.53), then you can modify the note that is presently displayed. If you choose *Create*, then the present note is erased (it is *lost* unless you have saved it as part of a Graphbook). You are then free to create a new note using the procedure described above.

Since the *PgDn* key can be used to create new pages in a note (up to a maximum of 20 pages), this same key is used when you return to the display menu to view these pages. You may also use *PgUp* to view a previous page.

TRACE

This choice allows you to display the function values from graphs on the screen. When you select *Trace,* a readout of all the values of the functions displayed appears on the right side of the screen. Also, a small cross appears on each graph. The following set of choices appears:

Esc = exit   *Z = Zoom*   *E = Expand*   *I = Incr index*   *D = Decr index*
L = Loop   *S = Start*   *F = Finish*   *C = Center*   →   ←

You can choose any item by pressing the appropriate key on the keyboard (pressing *Esc* will return you to the display menu), or by clicking on an item with the mouse pointer. If you select *Zoom,* then you can zoom in on one of the cross-marked points on one of the graphs. Or, if you select *Expand,* you can enlarge the view around one of the cross-marked graph points. By selecting *Incr index* or *Decr index,* you will make all of the cross markers move simultaneously (through either increasing *x*-values or decreasing *x*-values). If you select *Loop,* then whenever a cross marker reaches the largest *x*-value it will automatically jump back to the starting *x*-value (to undo this choice you select *No loop,* which will appear when *Loop* is selected). The choices *Start, Center,* and *Finish* will position the cross markers at the start of the *x*-values, or the midpoint (center) of the *x*-values, or at the end (finish) of the *x*-values. The arrow keys, ← and →, can be selected by either pressing the corresponding arrow keys on the keyboard, or by clicking on them with the mouse pointer. These arrow selections move the cross markers on the graphs either one step back along the *x*-axis (use ←) or one step forward along the *x*-axis (use →).

OPTIONS

This choice contains some miscellaneous procedures. The following menu will appear:

$$Dots \qquad Axes \qquad Redraw \qquad Command\ prompt \qquad (A.54)$$

If you choose *Dots,* then *FAS* will only plot computed function values point-by-point and will *not* connect these points by line segments. Once you choose *Dots,* then the first item in menu (A.54) will change to *Lines,* which you will want to select if you wish to change back to the first method of graphing (connecting plotted points by line segments). The choice *Axes* is for toggling between displaying and not displaying coordinate axes. If you wish to redraw the graphs on the screen, then select *Redraw.* To temporarily suspend *FAS* and access the DOS command line, select *Command prompt.*

Alt\Anim

When viewing a Graphbook, the *Options* choice on the display menu is replaced by this choice. If you select *Alt\Anim,* then the following menu appears

> *Dots*   *Axes*   *Redraw*   *Graphbook animation*

The choice *Graphbook animation* replaces the choice *Command prompt* in menu (A.54). By selecting *Graphbook animation* you can view a rapid succession of screens, creating an illusion of motion. The Graphbooks *TEMP_CH, VIBR_STR,* and *WHIP_STR,* which are on the disk accompanying this book, are examples of animations.

Help

There is on-line help available to you at any point. All you have to do is press the *F1* key. If you are at a menu, then you can obtain more information about the choice marked by the cursor by pressing the *F1* key (or, if a mouse is installed, by clicking on a choice using the *right* mouse button). If you are being asked a question by *FAS* (requiring you to press either *y* or *n*), then by pressing the *F1* key you can get information about what will happen if you press *y* or *n*. Finally, if you are asked by *FAS* to input some data, you can press the *F1* key to get some information about what to enter. To obtain more information on a particular topic, you should consult either this User's Manual or Chapters 1 to 5.

You must keep the help files (FAS.HLP, FASA.HLP, FASB.HLP, and FASC.HLP) in the same directory as FAS.EXE in order for the help to be accessible. If these files are not in the same directory as FAS.EXE, then an error message will appear.

Special Characters

There are some special characters that can be entered easily from the keyboard. To enter π, you just hold down the *Alt* key and press *p*. To enter δ, which is used for specifying the delta function $\delta(\ )$ in formulas, hold down the *Alt* key and press *d*. To enter |, which is used for indicating absolute values in captions and notes, hold down the *Alt* key and press the key with the backslash character, \.

It is important to remember that the vertical bar, |, is not allowed in formulas (the function abs ( ) is used for computing absolute values in formulas). If you use this character in a formula, then *FAS* will beep and the following error message will appear:

$$| \text{ or } \backslash \text{ not allowed in formulas. Press any key to continue.} \qquad (A.55)$$

After pressing a key, your formula will reappear and you can correct it.

Installing the Program

One way to install *FAS* is to copy all the files from the disk accompanying this book into the directory from which you intend to run *FAS*.

Another way to install *FAS* is to use the batch file *install.bat* which is on the disk accompanying this text. For example, suppose the disk is in drive B and you want to install *FAS* into the directory FOURIER on drive C. You would then enter the following command at the DOS prompt (from WINDOWS you would first have to get to the DOS prompt by selecting the MS–DOS icon):

    b:install b c fourier

The spaces between the letters "b" and "c" and the directory name "fourier" cannot be omitted.

Software Updates

Future versions of *FAS* can be obtained from the following web site: http://www.crcpress.com.

B

Some Computer Programs

In this appendix we include listings of some computer programs for algorithms discussed in Chapter 3. These programs consist of the following procedures.

1. SUB BITREV—This procedure performs bit reversal permutations of complex function data.

2. SUB SINESTANS—This procedure generates the sines and tangents necessary for performing the FFTs in the procedures FTBFLY, REALFFT, and InvRFFT.

3. SUB FTBFLY—This procedure performs the butterflies needed for performing a complex FFT with weight $e^{i2\pi/N}$. In order for it to perform properly, BITREV must first be called in order to permute your data into bit reversed form. Also, FTBFLY uses the sines and tangents generated by the procedure SINESTANS, so SINESTANS must also be called before FTBFLY.

4. SUB REALFFT—This procedure performs an FFT, with weight $e^{i2\pi/N}$, on real function data. Just like FTBFLY, it requires that SINESTANS be called first in order to generate the necessary sines and tangents.

5. SUB InvRFFT—This procedure performs the inverse of an FFT obtained from real function data (using the procedure REALFFT). Just like FTBFLY, it requires that SINESTANS be called first in order to generate the necessary sines and tangents.

6. SUB COSTRAN—This procedure performs a fast cosine transform of real function data. It requires that SINESTANS be called first in order to generate the necessary sines and tangents.

7. SUB SINETRAN—This procedure performs a fast sine transform of real function data. It requires that SINESTANS be called first in order to generate the necessary sines and tangents.

These programs are written in the QuickBASIC programming language (a product of Microsoft Corporation). Some of the nice features of this language are that there are no line numbers, variables are local to the procedures, and the programs are meant to be *compiled*. Indeed, these procedures behave pretty much like procedures in a structured language like PASCAL. Unfortunately, the author has not made an effort to write these programs in a highly structured format. One unintended benefit of this, however, is that the reader might feel the need to write his or her own programs for FFTs (there is no better way to learn about FFT programming!).

When these programs were compiled and a 1024-point complex FFT was computed on a PC, operating at 100 MHz with a built-in coprocessor, the execution time was about 129 milliseconds. For a 1024-point real FFT, the execution time was about 78 milliseconds.

For the reader's convenience, these programs are also available in an ASCII file (readable by standard programming editors) labeled FFT.BAS on the disk accompanying this book.

Procedure Declarations

Here are the procedure declarations. These declaration statements belong at the very beginning of any program which utilizes the procedures. (See *Sample Programs* below.)

DECLARE SUB BITREV (F#( ), G#( ), M%, R%)

DECLARE SUB SINESTANS (N%, R%, S#( ), T#( ), Z#)

DECLARE SUB FTBFLY (F1S#( ), F2S#( ), SA#( ), TA#( ), N%, C9%)

DECLARE SUB InvRFFT (N%, R%, F#( ), G#( ), SA#( ), TA#( ))

DECLARE SUB REALFFT (N%, F1#( ), F2#( ), S#( ), T#( ), R%)

DECLARE SUB SINETRAN (N%, F#( ), FH#( ), SA#( ), TA#( ), R%, ZF#)

DECLARE SUB COSTRAN (N%, F#( ), FH#( ), SA#( ), TA#( ), R%, ZF#)

Procedures

In each of these procedures, all of the arrays are assumed to have been declared in such a way that their indices have initial value 0. This is the convention, for example, in the REDIM statement in each of the sample programs listed below.

The Procedure BITREV

This procedure performs bit reversal permutations of the double precision arrays F( ) and G( ). The number of points in each of these arrays is $M = 2^R$. The positive integers M and R are passed to the procedure when it is called, as are the arrays F( ) and G( ).

The first part of this procedure calculates bit reversal numbers for the integer array J( ), using the method of Buneman. The second part of the procedure (after the DO WHILE loop begins) calculates the indices for swapping and performs the swaps, according to the second algorithm described in Section 3.2 of Chapter 3. To increase the speed of the program, the equations described in Formulas (3.21) and (3.25) of that section have been programmed so that successive additions are performed instead of multiplications.

```
DEFINT A–D, H–N, P–R, Y
DEFDBL E–G, O, S–X, Z
SUB BITREV (F( ), G( ), M, R)
N2 = M \ 2: N = M: M7 = 0: R2 = R \ 2: N1 = N: Y = 0: C = 1
IF R MOD 2 = 1 THEN
N1 = N2: R2 = (R − 1) \ 2: C = 0
END IF
N9 = SQR(N1)
IF R MOD 2 = 1 THEN N8 = N9 + N9 ELSE N8 = N9
'$STATIC
DIM J(64)
J(0) = 0: J(1) = 1: L9 = 2
FOR I9 = 2 TO R2
FOR J9 = 0 TO L9 − 1
J(J9) = J(J9) + J(J9): J(J9 + L9) = J(J9) + 1
NEXT J9
L9 = L9 + L9
NEXT I9
DO WHILE C < 2
FOR L = 1 TO N9 − 1
M7 = M7 + N8: N6 = J(L) + Y
T = G(M7): G(M7) = G(N6): G(N6) = T
T = F(M7): F(M7) = F(N6): F(N6) = T
FOR K = 1 TO L − 1
M6 = M7 + J(K): N6 = N6 + N8
T = G(M6): G(M6) = G(N6): G(N6) = T
T = F(M6): F(M6) = F(N6): F(N6) = T
NEXT K
NEXT L
Y = N9: M7 = N9: C = C + 1
LOOP
END SUB
```

The Procedure SINESTANS

This procedure calculates the values of the sines and tangents that are needed for FFTs and stores them in the double precision arrays S( ) and T( ). The integer R has the value of the power of 2 used for FFTs, and the integer N is the number of points ($N = 2^R$). The program uses Buneman's recursion procedure for computing sines for FFTs (see Section 3.4 of Chapter 3). The double precision variable Z retains the last value of Z in that recursion procedure; it is needed if COSTRAN or SINETRAN is called.

```
DEFINT A–D, H–N, P–R, Y
DEFDBL E–G, O, S–X, Z
SUB SINESTANS (N, R, S( ), T( ), Z)
Z = 0: N2 = N \ 2: H = N2 \ 2: S(0) = 0: S(H) = 1: D = 1
FOR J = 0 TO R – 3
Z = SQR(2 + Z): H9 = H: H = H \ 2: L = H
FOR K = 1 TO D
S(L) = (S(L + H) + S(L – H)) / Z
L = H9 + L
NEXT K
D = D + D
NEXT J
N4 = N \ 4
T(0) = 0: T(N4) = 1
FOR J = 1 TO N4 – 1
T(J) = (1 – S(N4 – J)) / S(J)
NEXT J
END SUB
```

The Procedure FTBFLY

This procedure computes a complex FFT using weight $e^{i2\pi/N}$. The double precision arrays F1S( ) and F2S( ) are the real and imaginary parts of the complex function data. The double precision arrays SA( ) and TA( ) hold the sines and tangents generated by the SINESTANS procedure. The integer N is the number of points. The integer C9 is the step size for walking through angles (retrieving their sines and tangents). For a complex FFT, C9 should be given the value 1 (see the sample programs below). For a real FFT, C9 is given the value 2 (see the calling of FTBFLY in the procedure REALFFT).

The first part of the program (up to the beginning of the DO WHILE loop) executes the first stage of the FFT, which consists of $N/2 (= N2)$ 2-point DFTs. Those DFTs only involve additions and subtractions, so in order to save time they are programmed separately from the other DFTs in the FFT algorithm.

The second part of the program (the DO WHILE loop) performs the higher order DFTs in the remaining $\log_2 N - 1$ stages of the FFT. The first C-loop contains the butterflies which only involve multiplication by i or $-i$. *In terms of real and imaginary parts*, these are not really multiplications. Therefore, for greater speed they are programmed separately. The rest of the program performs the butterflies for which Buneman's algorithm, described in Section 3.3 of Chapter 3, is needed. The real and imaginary parts of the FFT are returned in the arrays F1S( ) and F2S( ), respectively.

Note: there is a simple algorithm for inverting this FFT; we described it in Exercise 3.14 of Chapter 3. (See Sample Program 2 below.)

```
DEFINT A–D, H–N, P–R, Y
DEFDBL E–G, O, S–X, Z
SUB FTBFLY (F1S( ), F2S( ), SA( ), TA( ), N, C9)
N2 = N \ 2: L = 0: M2 = (C9 * N) \ 2: M4 = M2 \ 2
FOR K = 0 TO N2 − 1
T1 = F1S(L) + F1S(L + 1): T2 = F2S(L) + F2S(L + 1)
F1S(L + 1) = F1S(L) − F1S(L + 1): F2S(L + 1) = F2S(L) − F2S(L + 1)
F1S(L) = T1: F2S(L) = T2
L = L + 2
NEXT K
Q = C9 * N2: D = 2: Q2 = N2
DO WHILE D < N
Q2 = Q2 \ 2: Q = Q \ 2: D2 = D: D = D2 + D2
A1 = 0: L = 0: A = A1: D4 = D2 \ 2
FOR C = 1 TO Q2
B1 = A + D2
T = F1S(B1): S = F2S(B1): F1S(B1) = F1S(A) − T: F2S(B1) = F2S(A) − S
F1S(A) = F1S(A) + T: F2S(A) = F2S(A) + S
AV = A + D4: BV = B1 + D4
T = F1S(BV): S = F2S(BV)
F1S(BV) = F1S(AV) + S: F2S(BV) = F2S(AV) − T
F1S(AV) = F1S(AV) − S: F2S(AV) = F2S(AV) + T: A = A + D
NEXT C
L = L + Q: A1 = A1 + 1: A = A1
FOR K = 2 TO D4
SL = SA(L): TL = TA(L)
FOR C = 1 TO Q2
B1 = A + D2
V = F2S(B1) + TL * F1S(B1): T3 = F1S(B1) − V * SL
```

```
T4 = T3 * TL + V
F1S(B1) = F1S(A) − T3: F2S(B1) = F2S(A) − T4
F1S(A) = F1S(A) + T3: F2S(A) = F2S(A) + T4: AV = A + D4
BV = B1 + D4
V = F1S(BV) − TL * F2S(BV): T3 = −F2S(BV) − V * SL
T4 = T3 * TL + V
F1S(BV) = F1S(AV) − T3: F2S(BV) = F2S(AV) − T4
F1S(AV) = F1S(AV) + T3: F2S(AV) = F2S(AV) + T4
A = A + D
NEXT C
L = L + Q: A1 = A1 + 1: A = A1
NEXT K
LOOP
END SUB
```

The Procedure REALFFT

This procedure produces the FFT of the real data contained in the double precision array F1( ). The double precision arrays S( ) and T( ) hold the sines and tangents created by the SINESTANS procedure. And, the integer R is the power of 2.

The first part of the program creates a new set of data by putting the odd indexed values of F1( ) into the first half of F2( ) and putting the even-indexed values of F1( ) into the first half of F1( ). Then, an $N/2$-point FFT is performed by calling BITREV and FTBFLY (note that C9 is fed the value 2).

The remainder of the program performs the butterflies needed for the last stage of the real FFT algorithm and completes the FFT calculation according to Equations (3.63) to (3.65) of Chapter 3. The real and imaginary parts of the FFT are returned in the arrays F1( ) and F2( ), respectively.

```
DEFINT A–D, H–N, P–R, Y
DEFDBL E–G, O, S–X, Z
SUB REALFFT (N, F1( ), F2( ), S( ), T( ), R)
N2 = N \ 2: K = 0: N4 = N2 \ 2
FOR I = 0 TO N − 1 STEP 2
F1(K) = F1(I): F2(K) = F1(I + 1)
K = K + 1
NEXT I
CALL BITREV(F1( ), F2( ), N2, R−1)
CALL FTBFLY(F1( ), F2( ), S( ), T( ), N2, 2)
FOR L = 1 TO N4 − 1
```

```
SL = S(L): TL = T(L)
T1 = F1(L) − F1(N2 − L): T2 = F2(L) + F2(N2 − L)
V = T2 + TL * T1: T3 = T1 − V * SL: T4 = T3 * TL + V
F1(N − L) = (F1(L) + F1(N2 − L) + T4) / 2
F2(N − L) = (−F2(L) + F2(N2 − L) + T3) / 2
LJ = N4 + L
T1 = F1(LJ) − F1(N2 − LJ): T2 = F2(LJ) + F2(N2 − LJ)
V = T1 − TL * T2: T3 = −T2 − V * SL: T4 = T3 * TL + V
F1(N − LJ) = (F1(LJ) + F1(N2 − LJ) + T4) / 2
F2(N − LJ) = (−F2(LJ) + F2(N2 − LJ) + T3) / 2
NEXT L
F1(N − N4) = F1(N4): F2(N − N4) = −F2(N4)
FOR L = 1 TO N2 − 1
F1(L) = F1(N − L): F2(L) = −F2(N − L)
NEXT L
F1(N2) = F1(0) − F2(0): F1(0) = F1(0) + F2(0)
F2(0) = 0: F2(N2) = 0
END SUB
```

The Procedure InvRFFT

This procedure inverts a real FFT (obtained from the procedure REALFFT). The integer N is the number of points. The integer R is the power of 2, the double precision arrays F( ) and G( ) are the real and imaginary parts of the complex function data. And, the double precision arrays SA( ) and TA( ) contain the sines and tangents generated by the SINESTANS procedure.

The first part of the program (up to the calling of BITREV and FTBFLY) involves the creation of the auxiliary arrays for doing Fast Sine and Fast Cosine Transforms (see Sections 3.9 and 3.7 of Chapter 3). Also, the starting values for the recursions in those algorithms are computed in this part of the procedure.

The second part of the procedure involves the simultaneous calculation of the Fast Sine and Cosine Transforms (of order $N/2$). Both involve doing an FFT on the auxiliary arrays (which are *real*). Therefore, these two real arrays are FFT'd simultaneously. This is performed by calling BITREV and FTBFLY. The K-loop that follows performs the separation of the two FFTs (as described in Section 3.5 of Chapter 3). The remainder of the program implements the two recursion formulas for Fast Sine and Cosine Transforms, and then reconstructs the inverse FFT from them (as described in Section 3.9 of Chapter 3).

The output of this procedure should be thought of as the array F( ). It will be the same (except for rounding errors) as the real array F1( ) which was initially input

for the procedure REALFFT. There is no guarantee, however, that the output G( )
will be identically zero.

```
DEFINT A–D, H–N, P–R, Y
DEFDBL E–G, O, S–X, Z
SUB InvRFFT (N, R, F( ), G( ), SA( ), TA( ))
N2 = N \ 2: N4 = N \ 4: SPV2 = F(N2): SPV1 = −F(0)
J = 1: SPV3 = 0: SPV4 = 0
FOR K = 1 TO N2 − 1
J = −J
SPV3 = SPV3 + J * F(K): SPV4 = SPV4 + F(K)
NEXT K
SPV3 = SPV3 + SPV3: SPV4 = SPV4 + SPV4
SPV3 = F(0) + F(N2) + SPV3: SPV4 = SPV4 + SPV2 − SPV1
F1 = F(0): G(0) = 0: G1 = 0
FOR K = 1 TO N4 − 1
F1 = F1 + SA(K) * (F(N4 − K) − F(N4 + K))
NEXT K
FOR K = 1 TO N4 − 1
T = SA(K)
S = (F(K) + F(N2 − K)) / 2: U = F(K) − F(N2 − K)
F(K) = S − T * U: F(N2 − K) = S + T * U
S1 = (G(K) − G(N2 − K)) / 2: U1 = T * (G(K) + G(N2 − K))
G(K) = S1 + U1: G1 = G1 + U1: G(N2 − K) = U1 − S1
NEXT K
U = G(N4): G1 = G1 + U: G(N4) = U + U
CALL BITREV(F( ), G( ), N2, R − 1)
CALL FTBFLY(F( ), G( ), SA( ), TA( ), N2, 2)
FOR K = 1 TO N4 − 1
F(N2 + K) = G(K) − G(N2 − K): F(N − K) = G(N2 − K) + G(K)
NEXT K
G(1) = F1 + F1
FOR K = 1 TO N4 − 1
G(K + K) = F(N2 − K) + F(K)
G(K + K + 1) = G(K + K − 1) + F(N2 + K)
F(N2 + K) = F(N2 − K) − F(K)
NEXT K
F(1) = G1 + G1
FOR K = 1 TO N4 − 1
F(K + K) = F(N2 + K)
F(K + K + 1) = F(K + K − 1) + F(N − K)
NEXT K
F(0) = SPV4
FOR K = 1 TO N2 − 1
```

```
SPV2 = −SPV2: U = SPV1 + SPV2 + G(K)
F(N − K) = U − F(K): F(K) = U + F(K)
NEXT K
F(N2) = SPV3
END SUB
```

The Procedure COSTRAN

This procedure computes a Fast Cosine Transform of the real data contained in the double precision array F( ). The other double precision array FH( ) is auxiliary to the procedure (it should have all values equal to 0 when it is input to this procedure). The double precision arrays SA( ) and TA( ) hold the sines and tangents generated by the SINESTANS procedure. The integer R is the power of 2. The double precision variable ZF is to be given the value of Z obtained from the SINESTANS procedure.

The first part of the program forms the auxiliary array and computes a starting value for the recursion used in the Fast Cosine Transform algorithm described in Section 3.7 of Chapter 3. Then, a real FFT is done on this auxiliary array. After REALFFT is called, the program finishes by recursively computing the Fast Cosine Transform. The values of this Fast Cosine Transform are then contained in the array F( ).

```
DEFINT A–D, H–N, P–R, Y
DEFDBL E–G, O, S–X, Z
SUB COSTRAN (N, F( ), FH( ), SA( ), TA( ), R, ZF)
N2 = N \ 2: F1 = F(0): V = 1 / SQR(2 + ZF)
FOR K = 1 TO N2 − 1
IF K MOD 2 = 1 THEN
T = (SA((K + 1) \ 2) + SA((K − 1) \ 2)) * (−V)
ELSE
T = SA(K \ 2)
END IF
F1 = F1 + T * (F(N2 − K) − F(N2 + K))
NEXT K
FOR K = 1 TO N2 − 1
IF K MOD 2 = 1 THEN
T = (SA((K + 1) \ 2) + SA((K − 1) \ 2)) * V
ELSE
T = SA(K \ 2)
END IF
S = (F(K) + F(N − K)) / 2: U = F(K) − F(N − K)
```

```
F(K) = S − T * U: F(N − K) = S + T * U
NEXT K
CALL REALFFT(N, F( ), FH( ), SA( ), TA( ), R)
F(N − 1) = F1
FOR K = N2 − 1 TO 1 STEP −1
F(K + K) = F(K)
F(K + K − 1) = F(K + K + 1) − FH(K)
NEXT K
END SUB
```

The Procedure SINETRAN

This procedure computes a Fast Sine Transform of the real data contained in the double precision array F( ). The other double precision array FH( ) is auxiliary to the procedure (it should have all values equal to 0 when it is input to this procedure). The double precision arrays SA( ) and TA( ) hold the sines and tangents generated by the SINESTANS procedure. The integer R is the power of 2. The double precision variable ZF is to be given the value of Z obtained from the SINESTANS procedure.

The first part of the program forms the auxiliary array and computes a starting value for the recursion used in the Fast Sine Transform algorithm described in Section 3.7 of Chapter 3. Then, a real FFT is done on this auxiliary array. After REALFFT is called, the program finishes by recursively computing the Fast Sine Transform. The values of this Fast Sine Transform are then contained in the array F( ).

```
DEFINT A–D, H–N, P–R, Y
DEFDBL E–G, O, S–X, Z
SUB SINETRAN (N, F( ), FH( ), SA( ), TA( ), R, ZF)
F(0) = 0: N2 = N \ 2: V = 1 / SQR(2 + ZF): J = −1
FOR K = 1 TO N2 − 1
IF K MOD 2 = 1 THEN
T = (SA((K + 1) \ 2) + SA((K − 1) \ 2)) * V
ELSE
T = SA(K \ 2)
END IF
S = (F(K) − F(N − K)) / 2: U = T * (F(K) + F(N − K)) J = −J
F(K) = S + U: F1 = F1 + J * U: F(N − K) = U − S
NEXT K
U = F(N2): F1 = F1 − U: F(N2) = U + U
CALL REALFFT(N, F( ), FH( ), SA( ), TA( ), R)
```

```
F(N − 1) = F1: F(0) = 0: F(N) = 0
FOR K = N2 − 1 TO 1 STEP −1
F(K + K − 1) = F(K + K + 1) − F(K)
F(K + K) = FH(K)
NEXT K
END SUB
```

Sample Programs

Here are two sample programs which utilize the procedures described above. These programs are not included in the file FFT.BAS.

Sample program 1

This program checks that the complex FFT and the real FFT of a sequence of real data produce the same results. The author used a modified version of it to obtain the FFT execution times given in the introduction to this appendix.

```
DEFINT A–D, H–N, P–R, Y
DEFDBL E–G, O, S–X, Z
DECLARE SUB BITREV (F#( ), G#( ), M%, R%)
DECLARE SUB SINESTANS (N%, R%, S#( ), T#( ), Z#)
DECLARE SUB FTBFLY (F1S#( ), F2S#( ), SA#( ), TA#( ), N%, C9%)
DECLARE SUB REALFFT (N%, F1#( ), F2#( ), S#( ), T#( ), R%)
CLS
INPUT "Enter an integer from 3 to 12: ", IR
CLS
M = 2 ∧ IR
REDIM FR(M), FI(M), SN(M / 4 + 1), TN(M / 4 + 1)
FOR J = 0 TO M − 1
FR(J) = EXP(−J): FI(J) = 0
NEXT J
T = TIMER
CALL SINESTANS(M, IR, SN( ), TN( ), ZI)
CALL BITREV(FR( ), FI( ), M, IR)
CALL FTBFLY(FR( ), FI( ), SN( ), TN( ), M, 1)
T = TIMER − T
PRINT "Complex FFT, "; M; "Points. Elapsed time "; T; "seconds"
PRINT
PRINT "Index", "Real Part", "Imaginary Part"
FOR J = 0 TO M − 1
PRINT J, CSNG(FR(J)), CSNG(FI(J))
```

```
NEXT J
REDIM FR(M), FI(M), SN(M / 4 + 1), TN(M / 4 + 1)
FOR J = 0 TO M − 1
FR(J) = EXP(−J): FI(J) = 0
NEXT J
T = TIMER
CALL SINESTANS(M, IR, SN( ), TN( ), ZI)
CALL REALFFT(M, FR( ), FI( ), SN( ), TN( ), IR)
T = TIMER − T
PRINT "Real FFT, "; M; "Points. Elapsed time "; T; "seconds"
PRINT
PRINT "Index", "Real Part", "Imaginary Part"
FOR J = 0 TO M − 1
PRINT J, CSNG(FR(J)), CSNG(FI(J))
NEXT J
END
```

Sample program 2

This program shows how to invert a complex FFT without generating any new sines or tangents. It implements the method described in Exercise 3.14 of Chapter 3.

```
DEFINT A–D, H–N, P–R, Y
DEFDBL E–G, O, S–X, Z
DECLARE SUB BITREV (F#( ), G#( ), M%, R%)
DECLARE SUB SINESTANS (N%, R%, S#( ), T#( ), Z#)
DECLARE SUB FTBFLY (F1S#( ), F2S#( ), SA#( ), TA#( ), N%, C9%)
CLS
INPUT "Enter an integer from 3 to 12: ", IR
CLS
M = 2 ∧ IR
REDIM FR(M), FI(M), SN(M / 4 + 1), TN(M / 4 + 1)
FOR J = 0 TO M − 1
FR(J) = J: FI(J) = 2 ∗ J
NEXT J
CALL SINESTANS(M, IR, SN( ), TN( ), ZI)
CALL BITREV(FR( ), FI( ), M, IR)
CALL FTBFLY(FR( ), FI( ), SN( ), TN( ), M, 1)
PRINT "Complex FFT, "; M; "Points"
PRINT
PRINT "Index", "Real Part", "Imaginary Part"
FOR J = 0 TO M − 1
PRINT J, CSNG(FR(J)), CSNG(FI(J))
NEXT J
FOR J = 0 TO M − 1
```

```
FI(J) = −FI(J)
NEXT J
CALL BITREV(FR( ), FI( ), M, IR)
CALL FTBFLY(FR( ), FI( ), SN( ), TN( ), M, 1)
FOR J = 0 TO M − 1
FR(J) = FR(J) / M: FI(J) = −FI(J) / M
NEXT J
PRINT
PRINT "Inverse Complex FFT"; M; "Points"
PRINT
PRINT "Index", "Real Part", "Imaginary Part"
FOR J = 0 TO M − 1
PRINT J, CSNG(FR(J)), CSNG(FI(J))
NEXT J
END
```

C

The Schwarz Inequality

In this appendix we will prove the Schwarz inequality, which we made use of a few times in the text. After proving this inequality, we will show how it applies to Inequality (5.158) in Chapter 5.

THEOREM C.1: THE SCHWARZ INEQUALITY
If f and g are two functions (real or complex valued) defined over the interval [a, b], then

$$\left| \int_a^b f(t)g(t)\,dt \right| \le \sqrt{\int_a^b |f(t)|^2\,dt} \sqrt{\int_a^b |g(t)|^2\,dt}. \qquad \text{(C.1)}$$

[We assume that all integrals in (C.1) are defined and finite.]

PROOF Since $|f(t)g^*(s) - f(s)g^*(t)|^2 \ge 0$ for all s and t in $[a, b]$, we have

$$\int_a^b \int_a^b |f(t)g^*(s) - f(s)g^*(t)|^2\,dt\,ds \ge 0. \qquad \text{(C.2)}$$

Multiplying out the integrand in (C.2) and integrating the separate terms, we obtain

$$0 \le \int_a^b |f(t)|^2\,dt \int_a^b |g(s)|^2\,ds + \int_a^b |f(s)|^2\,ds \int_a^b |g(t)|^2\,dt$$

$$- \int_a^b f(t)g(t)\,dt \int_a^b f^*(s)g^*(s)\,ds - \int_a^b f^*(t)g^*(t)\,dt \int_a^b f(s)g(s)\,ds.$$

Replacing the dummy variable s by the dummy variable t in each integral above, and bringing the complex conjugates outside of each integral (this is allowed since *t* is real), we get

$$0 \le 2 \int_a^b |f(t)|^2 \, dt \int_a^b |g(t)|^2 \, dt - 2 \left| \int_a^b f(t)g(t) \, dt \right|^2 .$$

Hence,

$$\left| \int_a^b f(t)g(t) \, dt \right|^2 \le \int_a^b |f(t)|^2 \, dt \int_a^b |g(t)|^2 \, dt. \tag{C.3}$$

Taking positive square roots of both sides of (C.3), we obtain (C.1). ∎

We made an application of the Schwarz inequality in Chapter 5, when we discussed the proof of the sampling theorem [see (5.158) in Chapter 5]. We used the inequality

$$\int_{-L}^L \left| \hat{f}(u) - S_M^{\hat{f}}(u) \right| |W(u)| \, du \le \left\| \hat{f} - S_M^{\hat{f}} \right\|_2 \|W\|_2 \tag{C.4}$$

where the notation $\|.\|_2$ is defined by

$$\|g\|_2 = \sqrt{\int_{-L}^L |g(u)|^2 \, du}.$$

We will now show that (C.4) follows from the Schwarz inequality.

Since the two functions $|\hat{f}(u) - S_M^{\hat{f}}(u)|$ and $|W(u)|$ on the left side of (C.4) are both non-negative, we have

$$\int_{-L}^L \left| \hat{f}(u) - S_M^{\hat{f}}(u) \right| |W(u)| \, du = \left| \int_{-L}^L \left| \hat{f}(u) - S_M^{\hat{f}}(u) \right| |W(u)| \, du \right|$$

$$\le \sqrt{\int_{-L}^L \left| \hat{f}(u) - S_M^{\hat{f}}(u) \right|^2 \, du} \sqrt{\int_{-L}^L |W(u)|^2 \, du}$$

where the inequality is an application of the Schwarz inequality. By definition of $\|.\|_2$ we then have (C.4).

D

Solutions to Odd-Numbered Exercises

CHAPTER 1

Section 1

1.1 We have

$$e^{i\phi} = \sum_{n=0}^{\infty} \frac{(i\phi)^n}{n!} = \sum_{k=0}^{\infty} \frac{(i\phi)^{2k}}{(2k)!} + \sum_{k=0}^{\infty} \frac{(i\phi)^{2k+1}}{(2k+1)!}$$

$$= \sum_{k=0}^{\infty} \frac{(-1)^k \phi^{2k}}{(2k)!} + i \sum_{k=0}^{\infty} \frac{(-1)^k \phi^{2k+1}}{(2k+1)!} = \cos\phi + i\sin\phi.$$

1.3 See Figure D.1.[1]

1.5 (a) See Figure D.2.

1.7 We have

$$\int_{-P/2}^{P/2} h(x)\,dx = \int_{0}^{P/2} h(x)\,dx + \int_{-P/2}^{0} h(x)\,dx.$$

Replace x by $x + P$ in the last integral and use $h(x + P) = h(x)$ to obtain $\int_{P/2}^{P} h(x)\,dx$. This results in $\int_{-P/2}^{P/2} h(x)\,dx = \int_{0}^{P} h(x)\,dx$. The identity in Remark 1.1(c) follows by replacing $h(x)$ by $(1/P)g(x)e^{-i2\pi nx/P}$ for each n.

---

[1]Figures appear at the end of this appendix.

Section 2

1.9

(a) $\sum \dfrac{2\cos(n\pi/2)}{\pi(1-4n^2)}e^{i2nx}$

(b) $3 + \sum' \left(\dfrac{9}{2\pi^2 n^2} + i\dfrac{9}{2\pi n}\right)e^{i2\pi nx/3}$

(c) $\sum \dfrac{(-1)^n[e^1 - e^{-1}]}{2(1-n\pi i)}e^{in\pi x}$

(d) $\sum \dfrac{i}{n\pi}[\cos(n\pi) - 1]e^{i2\pi nx}$

(e) $\dfrac{1}{2} + \sum' \dfrac{i}{2n\pi}(e^{-in\pi} - 1)e^{inx}$

1.11 (a) See Figure D.3.   (b) See Figure D.4.   (d) See Figure D.5.

Section 3

1.13 View the GraphBook CH1SEC3 in *FAS* .

Section 4

1.15 (a) For $0 < x < 1$, the F. series converges to e^x. For $x = 0$, the F. series converges to $[e+1]/2$. See Figure D.6.   (c) For $x = 0$, the F. series converges to 1, for $0 < x < 1$ the F. series converges to 0, for $x = 1$ the F. series converges to $1/2$, for $1 < x < 2$ the F. series converges to 1, for $x = 2$ the F. series converges to $3/2$, for $2 < x < 3$ the F. series converges to 2, and for $x = 3$ the F. series converges to 1. See Figure D.7.   (e) For $-\pi/2 < x < \pi/2$ the F. series converges to $\sin x$, for $x = \pm\pi/2$ the F. series converges to 0. See Figure D.8.

1.17 On $[-0.1, 0.1]$, 41 harmonics. On $[0.1, 0.3]$, 9 harmonics. On $[0.3, 0.5]$, 7 harmonics.

Section 5

1.19 View the GraphBook CH1SEC5 in *FAS* .

1.21 The periodic extension of $|x|$ is continuous and has a piecewise continuous derivative [equal to -1 on the interval $(-1, 0)$ and equal to 1 on the interval $(0, 1)$] hence Theorem 1.3 is applicable. The F. series for $|x|$ on $[-1, 1]$ is $0.5 - (4/\pi^2)\sum_{k=0}^{\infty}[\cos(2k+1)\pi x]/(2k+1)^2$. Consequently,

$$\left||x| - S_{2M+1}(x)\right| = \left|\dfrac{4}{\pi^2}\sum_{k=M+1}^{\infty}\dfrac{\cos(2k+1)\pi x}{(2k+1)^2}\right| \le \dfrac{4}{\pi^2}\sum_{k=M+1}^{\infty}\dfrac{1}{(2k+1)^2}$$

$$\le \dfrac{4}{\pi^2}\int_{M}^{\infty}\dfrac{1}{(2x+1)^2}\,dx = \dfrac{2}{\pi^2}\dfrac{1}{2M+1}$$

It follows that $\sup \left||x| - S_{2M+1}(x)\right| \le .01$ if $2M + 1 \ge 21$. Thus, 21 harmonics are needed. Using *FAS* the Sup-Norm difference between abs (x) and a 21-harmonic Fourier series partial sum is found to be about 9×10^{-3}, which confirms this estimate.

A simple explanation for why $S_M(x)$ is furthest from $g(x)$ at $x = 0$ is that $g'(0)$ fails to exist, but $S_M(x)$ is infinitely differentiable at $x = 0$.

Section 6

1.23

(a) cosine series: $\dfrac{1}{3}\sin 3 + \displaystyle\sum_{n=1}^{\infty} \dfrac{(-1)^{n+1}6\sin 3}{n^2\pi^2 - 9} \cos \dfrac{n\pi x}{3}$

sine series: $\displaystyle\sum_{n=1}^{\infty} \dfrac{2n\pi(1 - (-1)^n \cos 3)}{n^2\pi^2 - 9} \sin \dfrac{n\pi x}{3}$

(c) cosine series: $\dfrac{1}{3} + \displaystyle\sum_{n=1}^{\infty} \dfrac{4(1 + 3(-1)^n)}{n^2\pi^2} \cos \dfrac{n\pi x}{2}$

sine series: $\displaystyle\sum_{n=1}^{\infty} \left[\dfrac{4n^2\pi^2(-1)^{n+1} + 16((-1)^n - 1)}{n^3\pi^3} \right] \sin \dfrac{n\pi x}{2}$

(e) cosine series: $\dfrac{1}{3} + \displaystyle\sum_{n=1}^{\infty} \dfrac{2}{n\pi}(\sin \dfrac{2n\pi}{3} - \sin \dfrac{n\pi}{3}) \cos \dfrac{n\pi x}{3}$

sine series: $\displaystyle\sum_{n=1}^{\infty} \dfrac{2}{n\pi}(\cos \dfrac{n\pi}{3} - \cos \dfrac{2n\pi}{3}) \sin \dfrac{n\pi x}{3}$

1.25 If $g(x) = \sum_{n=1}^{\infty} B_n \sin(n\pi x/L)$, then

$$\dfrac{2}{L}\int_0^L g(x)\sin\dfrac{n\pi x}{L}\,dx = \dfrac{2}{L}\int_0^L \sum_{m=1}^{\infty} B_m \sin\dfrac{m\pi x}{L}\sin\dfrac{n\pi x}{L}\,dx$$

$$= \sum_{m=1}^{\infty} B_m \dfrac{2}{L}\int_0^L \sin\dfrac{m\pi x}{L}\sin\dfrac{n\pi x}{L}\,dx = B_n.$$

1.27 If $g(x) = A_0/2 + \sum_{n=1}^{\infty} A_n \cos(n\pi x/L)$, then

$$\dfrac{2}{L}\int_0^L g(x)\,dx = \dfrac{2}{L}\int_0^L \dfrac{1}{2}A_0\,dx + \sum_{n=1}^{\infty} A_n \dfrac{2}{L}\int_0^L \cos\dfrac{n\pi x}{L}\,dx = A_0$$

and

$$\frac{2}{L} \int_0^L g(x) \cos \frac{n\pi x}{L} \, dx = \frac{A_0}{L} \int_0^L \cos \frac{n\pi x}{L} \, dx$$

$$+ \sum_{m=1}^{\infty} A_m \frac{2}{L} \int_0^L \cos \frac{m\pi x}{L} \cos \frac{n\pi x}{L} \, dx = A_n$$

1.29 (a) See Figure D.9. (c) See Figure D.10.

Section 7

1.31 View the GraphBook CH1SEC7 in *FAS* .

CHAPTER 2

Section 1

2.1 If $f_j = 1$ for $j = 0, \ldots, N-1$, then $F_k = N$ if k is divisible by N and $F_k = 0$ if k is not divisible by N (in particular, $F_0 = N$ and $F_k = 0$ for $k = 1, \ldots, N-1$).

2.3 $F_k = 1 + 2 \sum_{j=1}^{3} \cos(\pi j k / 8)$

2.5 Using the result of Exercise 2.4, we have

$$c_k \approx \frac{1}{N} \left[g(P) + \sum_{j=1}^{N-1} g\left(j\frac{P}{N}\right) e^{-i2\pi jk/N} \right] \tag{D.1}$$

hence by averaging (D.1) and Formula (2.3) we obtain

$$c_k \approx \frac{1}{N} \left[\frac{1}{2}[g(0) + g(P)] + \sum_{j=1}^{N-1} g\left(j\frac{P}{N}\right) e^{-i2\pi jk/N} \right]. \tag{D.2}$$

The expression in brackets in (D.2) is the DFT of $\{\,[g(0) + g(P)]/2, g(P/N), \ldots,$ $g[(N-1)P/N]\}$.

Note: Formula (D.2) can also be obtained by using the *Trapezoidal Rule* to approximate the integral for c_k in Formula (2.1).

Section 2

2.7 $G_{2k} = \sum_{j=0}^{2N-1} g_j e^{-i2\pi j(2k)/(2N)} = \sum_{j=0}^{N-1} h_j e^{-i2\pi jk/N} = H_k$.

Section 3

2.9 Aliasing begins with $k = 512$ (to see this, it helps to look at the case of $k = 513$).

Section 4

2.11 Sup-Norm difference = 7.196×10^{-5}.

2.13 For explicit partial sum use $f(x) = 0.5 + (2/\pi)\text{sumk}\,(\sin((2k+1)x)/(2k+1))\,\backslash k=0,17$. The Sup-Norm difference is 3.516×10^{-2}.

Section 5

2.15 For explicit partial sum use $f(x) = -32\text{sumk}[\sin(a\pi x/2)/(a\pi) \wedge 3] \backslash a=2k+1 \backslash k=0,20$. The Sup-Norm difference is 1.467×10^{-10}.

2.17 For explicit partial sum use $f(x) = (-4/\pi)\text{sumk}((-1) \wedge k \, \sin(k\pi x/2)/k) \backslash k=1,41$. The Sup-Norm difference is 7.5×10^{-4}.

2.19 For explicit partial sum use $f(x) = (2/\pi)\text{sumk}((1 - \cos(2k\pi/3)) \sin(k\pi x/3)/k) \backslash k=1,41$. The Sup-Norm difference is 6.8×10^{-3}.

2.21 For (2.17) the Sup-Norm difference is 4.7×10^{-5}. For (2.19) the Sup-Norm difference is 1.71×10^{-3}.

CHAPTER 3

Section 1

3.1 The 16-point FFT is as follows (*Note:* $W^8 = -1$ and $W^4 = i$):

$$
\begin{array}{llllll}
h_0 & \to h_0 + h_8 & \to h_0 + h_8 + h_4 + h_{12} & = g_0 & \to g_0 + g_4 & = q_0 \to \\
h_8 & \to h_0 - h_8 & \to h_0 - h_8 + i(h_4 - h_{12}) & = g_1 & \to g_1 + W^2 g_5 & = q_1 \to \\
h_4 & \to h_4 + h_{12} & \to h_0 + h_8 - h_4 - h_{12} & = g_2 & \to g_2 + ig_6 & = q_2 \to \\
h_{12} & \to h_4 - h_{12} & \to h_0 - h_8 - i(h_4 - h_{12}) & = g_3 & \to g_3 + W^6 g_7 & = q_3 \to \\
h_2 & \to h_2 + h_{10} & \to h_2 + h_{10} + h_6 + h_{14} & = g_4 & \to g_0 - g_4 & = q_4 \to \\
h_{10} & \to h_2 - h_{10} & \to h_2 - h_{10} + i(h_6 - h_{14}) & = g_5 & \to g_1 - W^2 g_5 & = q_5 \to \\
h_6 & \to h_6 + h_{14} & \to h_2 + h_{10} - h_6 - h_{14} & = g_6 & \to g_2 - ig_6 & = q_6 \to \\
h_{14} & \to h_6 - h_{14} & \to h_2 - h_{10} - i(h_6 - h_{14}) & = g_7 & \to g_3 - W^6 g_7 & = q_7 \to \\
h_1 & \to h_1 + h_9 & \to h_1 + h_9 + h_5 + h_{13} & = g_8 & \to g_8 + g_{12} & = q_8 \to \\
h_9 & \to h_1 - h_9 & \to h_1 - h_9 + i(h_5 - h_{13}) & = g_9 & \to g_9 + W^2 g_{13} & = q_9 \to \\
h_5 & \to h_5 + h_{13} & \to h_1 + h_9 - h_5 - h_{13} & = g_{10} & \to g_{10} + ig_{14} & = q_{10} \to \\
h_{13} & \to h_5 - h_{13} & \to h_1 - h_9 - i(h_5 - h_{13}) & = g_{11} & \to g_{11} + W^6 g_{15} & = q_{11} \to \\
h_3 & \to h_3 + h_{11} & \to h_3 + h_{11} + h_7 + h_{15} & = g_{12} & \to g_8 - g_{12} & = q_{12} \to \\
h_{11} & \to h_3 - h_{11} & \to h_3 - h_{11} + i(h_7 + h_{15}) & = g_{13} & \to g_9 - W^2 g_{13} & = q_{13} \to \\
h_7 & \to h_7 + h_{15} & \to h_3 + h_{11} + h_7 + h_{15} & = g_{14} & \to g_{10} - ig_{14} & = q_{14} \to \\
h_{15} & \to h_7 - h_{15} & \to h_3 - h_{11} - i(h_7 - h_{15}) & = g_{15} & \to g_{11} - W^6 g_{15} & = q_{15} \to
\end{array}
$$

$$
\begin{array}{ll}
q_0 + q_8 & = H_0 \\
q_1 + W q_9 & = H_1 \\
q_2 + W^2 q_{10} & = H_2 \\
q_3 + W^3 q_{11} & = H_3 \\
q_4 + i q_{12} & = H_4 \\
q_5 + W^5 q_{13} & = H_5 \\
q_6 + W^6 q_{14} & = H_6 \\
q_7 + W^7 q_{15} & = H_7 \\
q_0 - q_8 & = H_8 \\
q_1 - W q_9 & = H_9 \\
q_2 - W^2 q_{10} & = H_{10} \\
q_3 - W^3 q_{11} & = H_{11} \\
q_4 - i q_{12} & = H_{12} \\
q_5 - W^5 q_{13} & = H_{13} \\
q_6 - W^6 q_{14} & = H_{14} \\
q_7 - W^7 q_{15} & = H_{15}
\end{array}
$$

Note: the scheme above requires 10 multiplications (since multiplications by i can be programmed without using multiplications). This is a significant improvement over the $(1/2)16 \log_2 16 = 32$ multiplications described in the text. The programs for FFTs in Appendix B take advantage of this increased efficiency. If $N = 1024$, then this more efficient FFT requires only 3586 multiplications, a savings by a factor of almost 300 over a direct DFT.

3.3 For $N = 4^R$, the N-point DFT $H_k = \sum_{j=0}^{N-1} h_j W^{jk}$ decomposes into

$$H_k = H_k^0 + H_k^1 W^k + H_k^2 W^{2k} + H_k^3 W^{3k} \tag{D.3}$$

where

$$H_k^0 = \sum_{j=0}^{\frac{1}{4}N-1} h_{4j}(W^4)^{jk}, \qquad H_k^1 = \sum_{j=0}^{\frac{1}{4}N-1} h_{4j+1}(W^4)^{jk}$$

$$H_k^2 = \sum_{j=0}^{\frac{1}{4}N-1} h_{4j+2}(W^4)^{jk}, \qquad H_k^3 = \sum_{j=0}^{\frac{1}{4}N-1} h_{4j+3}(W^4)^{jk}$$

are the $N/4$-point DFTs of the subsequences of $\{h_j\}$ whose base 4 expansions of their indices end in 0, 1, 2, and 3, respectively. Based on (D.3) and the fact that these $N/4$-point DFTs each have period $N/4$, and using the relations $W^{N/4} = i$, $W^{N/2} = -1$, and $W^{3N/4} = -i$ (when $W = e^{i2\pi/N}$), we obtain (for $k = 0, 1, \ldots, N/4 - 1$):

$$H_k = H_k^0 + (H_k^1 W^k) + (H_k^2 W^{2k}) + (H_k^3 W^{3k})$$

$$H_{k+\frac{1}{4}N} = H_k^0 + i(H_k^1 W^k) - (H_k^2 W^{2k}) - i(H_k^3 W^{3k})$$

$$H_{k+\frac{1}{2}N} = H_k^0 - (H_k^1 W^k) + (H_k^2 W^{2k}) - (H_k^3 W^{3k})$$

$$H_{k+\frac{3}{4}N} = H_k^0 - i(H_k^1 W^k) - (H_k^2 W^{2k}) + i(H_k^3 W^{3k}). \tag{D.4}$$

The computations in (D.4) constitute the basic computations that are needed for a 4^R-point FFT (they are analogous to the butterflies used for 2^R-point FFTs). The 4^R-point FFT is implemented by continuing the decomposition described above. It requires $R = \log_4 N$ stages. Since only the quantities $(H_k^1 W^k)$, $(H_k^2 W^{2k})$, and $(H_k^3 W^{3k})$ for $k = 1, 2, \ldots, N/4 - 1$ in the first line of (D.4) require complex multiplications [their values are stored and retrieved from memory in order to perform the computations in the second, third, and fourth lines of (D.4)], each stage of the 4^R-point FFT requires $3N/4 - 3$ complex multiplications. Actually, to be more precise, since the first stage of the FFT uses weight $W^{N/4} = i$, it does not require multiplications (see the 16-point FFT diagrammed below). Thus a 4^R-point FFT requires $(3N/4 - 3)(\log_4 N - 1)$ multiplications. For example, if $N = 1024 = 4^5$, then 3060 complex multiplications are needed [instead of $1,046,529$ multiplications for a direct DFT, or 3586 multiplications for a power of 2 FFT (see the solution of Exercise 3.1 above)]. *Note: FAS* does not have code for radix 4 FFTs, since that would require two separate codings for $N = 4^R$ and $N = 2^P$ (when P is odd).

To illustrate the radix 4 FFT described above, here is how it works for the case of $N = 16 = 4^2$. The left column below is the *base* 4 *digit reversal indexing* of $\{h_j\}_{j=0}^{15}$. For checking these computations the reader should use the relations $W^4 = i$, $W^8 = -1$, and $W^{12} = -i$.

$$
\begin{aligned}
h_0 &\to h_0 + h_4 + h_8 + h_{12} &= g_0 &\to g_0 + g_4 + g_8 + g_{12} &= H_0 \\
h_4 &\to h_0 + ih_4 - h_8 - ih_{12} &= g_1 &\to g_1 + g_5 W + g_9 W^2 + g_{13} W^3 &= H_1 \\
h_8 &\to h_0 - h_4 + h_8 - h_{12} &= g_2 &\to g_2 + g_6 W^2 + g_{10} W^4 + g_{14} W^6 &= H_2 \\
h_{12} &\to h_0 - ih_4 - h_8 + ih_{12} &= g_3 &\to g_3 + g_7 W^3 + g_{11} W^6 + g_{15} W^9 &= H_3 \\
h_1 &\to h_1 + h_5 + h_9 + h_{13} &= g_4 &\to g_0 + ig_4 - g_8 - ig_{12} &= H_4 \\
h_5 &\to h_1 + ih_5 - h_9 - ih_{13} &= g_5 &\to g_1 + ig_5 W - g_9 W^2 - ig_{13} W^3 &= H_5 \\
h_9 &\to h_1 - h_5 + h_9 - h_{13} &= g_6 &\to g_2 + ig_6 W^2 - g_{10} W^4 - ig_{14} W^6 &= H_6 \\
h_{13} &\to h_1 - ih_5 - h_9 + ih_{13} &= g_7 &\to g_3 + ig_7 W^3 - g_{11} W^6 - ig_{15} W^9 &= H_7 \\
h_2 &\to h_2 + h_6 + h_{10} + h_{14} &= g_8 &\to g_0 - g_4 + g_8 - g_{12} &= H_8 \\
h_6 &\to h_2 + ih_6 - h_{10} - ih_{14} &= g_9 &\to g_1 - g_5 W + g_9 W^2 - g_{13} W^3 &= H_9 \\
h_{10} &\to h_2 - h_6 + h_{10} - h_{14} &= g_{10} &\to g_2 - g_6 W^2 + g_{10} W^4 - g_{14} W^6 &= H_{10} \\
h_{14} &\to h_2 - ih_6 - h_{10} + ih_{14} &= g_{11} &\to g_3 - g_7 W^3 + g_{11} W^6 - g_{15} W^9 &= H_{11} \\
h_3 &\to h_3 + h_7 + h_{11} + h_{15} &= g_{12} &\to g_0 - ig_4 - g_8 + ig_{12} &= H_{12} \\
h_7 &\to h_3 + ih_7 - h_{11} - ih_{15} &= g_{13} &\to g_1 - ig_5 W - g_9 W^2 + ig_{13} W^3 &= H_{13} \\
h_{11} &\to h_3 - h_7 + h_{10} - h_{15} &= g_{14} &\to g_2 - ig_6 W^2 - g_{10} W^4 + ig_{14} W^6 &= H_{14} \\
h_{15} &\to h_3 - ih_7 - h_{10} + ih_{15} &= g_{15} &\to g_3 - ig_7 W^3 - g_{11} W^6 + ig_{15} W^9 &= H_{15}
\end{aligned}
$$

This scheme requires 9 multiplications, which is a slight improvement over the 10 multiplications needed for the radix 2 FFT diagrammed in the solution of Exercise 3.1.

Section 2

3.5 If $N = 4^R$, there are R stages. The first stage begins with the numbers 0, 1, 2, and 3. To proceed from one stage to the next, you multiply the preceding numbers by 4 to get the first quarter numbers of the next stage, then you add 1 to the first quarter numbers to get the second quarter numbers, then add 2 to the first quarter numbers to get the third quarter numbers, then add 3 to the first quarter numbers to get the fourth quarter numbers. After $R = \log_4 N$ stages, this process generates the digit reversal permutation $n \to P_N(n)$. You then swap each h_n with $h_{P_N(n)}$.

3.7 If $N = 4^R$ where R is even, then Equation (3.21) remains the same (only P_M is the base 4 digit reversal permutation for $M = \sqrt{N}$). If R is odd, then (3.25) is replaced by

$$n = P_M(K)(4M) + L \quad \longrightarrow \quad P_N(n) = P_M(L)(4M) + K$$

$$n + M \quad \longrightarrow \quad P_N(n + M) = P_N(n) + M$$

$$n + 2M \quad \longrightarrow \quad P_N(n + 2M) = P_N(n) + 2M$$

$$n + 3M \quad \longrightarrow \quad P_N(n + 3M) = P_N(n) + 3M$$

for $K = 1, 2, \ldots, M - 1$ and $L = 0, 1, \ldots, K - 1$, and $M = \sqrt{N/4}$.

Section 3

3.9 $S(N/2 - m) = \sin[2\pi(N/2 - m)/N] = \sin(\pi - 2\pi m/N) = \sin(2\pi m/N) = S(m)$

3.11 $\{S(n)\}_{n=0}^{N/4-1}$ and $\{T(n)\}_{n=0}^{N/4-1}$ (if $W^{N/4} = i$ is used to reduce multiplications by W^k, for $N/4 \le k \le N/2 - 1$, to multiplications by $i\,W^{k-N/4}$).

Section 4

3.13 Define z_4 by $z_4 = (2 + z_3)^{1/2}$, then

$$S(\frac{1}{64}N) = \frac{S(\frac{1}{32}N) + S(0)}{z_4}, \qquad S(\frac{3}{64}N) = \frac{S(\frac{1}{16}N) + S(\frac{1}{32}N)}{z_4}$$

$$S(\frac{5}{64}N) = \frac{S(\frac{3}{32}N) + S(\frac{1}{16}N)}{z_4}, \qquad S(\frac{7}{64}N) = \frac{S(\frac{1}{8}N) + S(\frac{3}{32}N)}{z_4}$$

$$S(\frac{9}{64}N) = \frac{S(\frac{5}{32}N) + S(\frac{1}{8}N)}{z_4}, \qquad S(\frac{11}{64}N) = \frac{S(\frac{3}{16}N) + S(\frac{5}{32}N)}{z_4}$$

$$S(\frac{13}{64}N) = \frac{S(\frac{7}{32}N) + S(\frac{3}{16}N)}{z_4}, \qquad S(\frac{15}{64}N) = \frac{S(\frac{1}{4}N) + S(\frac{7}{32}N)}{z_4}.$$

All other sine values have been previously computed.

3.15 The FFT, $\sum_{k=0}^{N-1} F_k^* W^{jk}$, requires only the sines $\{S(j)\}_{j=0}^{N/4}$ and the tangents $\{T(j)\}_{j=0}^{N/4}$. Therefore, the computation required by formula (3.106) needs only these same values in order to invert an FFT.

3.17 We have

$$R_k = h_0 + \sum_{j=1}^{N-1} s_j \cos \frac{2\pi jk}{N}, \qquad \text{since } \{a_j \cos \frac{2\pi jk}{N}\} \text{ is odd}$$

$$= h_0 + \sum_{j=1}^{N-1} \frac{1}{2} h_j \cos \frac{2\pi jk}{N} + \sum_{j=1}^{N-1} \frac{1}{2} h_{N-j} \cos \frac{2\pi jk}{N}$$

$$= \sum_{j=0}^{N-1} h_j \cos \frac{2\pi jk}{N} = H_{2k}^C.$$

Thus $H_{2k}^C = R_k$. And

$$I_k = \sum_{j=1}^{N-1} a_j \sin \frac{2\pi jk}{N}, \qquad \text{since } \{s_j \sin \frac{2\pi jk}{N}\} \text{ is odd}$$

$$= \sum_{j=1}^{N-1} h_{N-j} \sin \frac{j\pi}{N} \sin \frac{2\pi jk}{N} - \sum_{j=1}^{N-1} h_j \sin \frac{j\pi}{N} \sin \frac{2\pi jk}{N}$$

$$= -\sum_{j=1}^{N-1} 2h_j \sin \frac{j\pi}{N} \sin \frac{2\pi jk}{N}, \qquad \text{since } \{\sin \frac{j\pi}{N} \sin \frac{2\pi jk}{N}\} \text{ is odd}$$

$$= \sum_{j=1}^{N-1} h_j \left\{ \cos \left(\frac{2k+1}{N} j\pi \right) - \cos \left(\frac{2k-1}{N} j\pi \right) \right\} = H_{2k+1}^C - H_{2k-1}^C.$$

Hence $H_{2k+1}^C = H_{2k-1}^C + I_k$.

3.19 We have

$$\sum_{n=0}^{N-1} \cos \frac{(n+\frac{1}{2})(m+\frac{1}{2})\pi}{N} \cos \frac{(n+\frac{1}{2})(k+\frac{1}{2})\pi}{N}$$

$$= \frac{1}{2} \sum_{n=0}^{N-1} \{ \cos \frac{(n+\frac{1}{2})(m+k+1)\pi}{N}$$

$$+ \cos \frac{(n+\frac{1}{2})(m-k)\pi}{N} \}$$

$$= \frac{1}{2} \mathrm{Re} \{ \sum_{n=0}^{N-1} e^{i(m+k+1)\pi/(2N)} e^{in(m+k+1)\pi/N}$$

$$+ \sum_{n=0}^{N-1} e^{i(m-k)\pi/(2N)} e^{in(m-k)\pi/N} \}$$

$$= \frac{1}{2} \mathrm{Re} \{ e^{i(m+k+1)\pi/(2N)} \frac{1 - e^{i(m+k+1)\pi}}{1 - e^{i(m+k+1)\pi/N}}$$

$$+ e^{i(m-k)\pi/(2N)} \frac{1 - e^{i(m-k)\pi}}{1 - e^{i(m-k)\pi/N}} \}.$$

Now, if m and k have opposite parity (i.e., one is odd and one is even), then $1 - e^{i(m+k+1)\pi} = 0$. On the other hand, if they have the same parity, then $1 - e^{i(m-k)\pi} = 0$. And, if m and k have the same parity, then $m + k + 1$ is odd. Let's say $m + k + 1 = 2j + 1$ for some integer j. In that case, $e^{i(m+k+1)\pi} = -1$, so

$$e^{i(m+k+1)\pi/(2N)} \frac{1 - e^{i(m+k+1)\pi}}{1 - e^{i(m+k+1)\pi/N}} = e^{i(2j+1)\pi/(2N)} \frac{2}{1 - e^{i(2j+1)\pi/N}}$$

$$= e^{i(2j+1)\pi/(2N)} \frac{2(1 - e^{-i(2j+1)\pi/N})}{2 - 2\cos[(2j+1)\pi/N]}$$

$$= i \, \frac{2 \sin[(2j + 1)\pi/(2N)]}{1 - \cos[(2j + 1)\pi/N]}.$$

Thus, when m and k have the same parity,

$$\text{Re}\,\{e^{i(m+k+1)\pi/(2N)}\,\frac{1 - e^{i(m+k+1)\pi}}{1 - e^{i(m+k+1)\pi/N}}\} = 0.$$

Similarly, when m and k have opposite parity

$$\text{Re}\,\{e^{i(m-k)\pi/(2N)}\,\frac{1 - e^{i(m-k)\pi}}{1 - e^{i(m-k)\pi/N}}\} = 0.$$

Thus, if $m \neq k$

$$\sum_{n=0}^{N-1} \cos \frac{(n + \frac{1}{2})(m + \frac{1}{2})\pi}{N} \cos \frac{(n + \frac{1}{2})(k + \frac{1}{2})\pi}{N} = 0.$$

While if $m = k$

$$\sum_{n=0}^{N-1} \cos \frac{(n + \frac{1}{2})(m + \frac{1}{2})\pi}{N} \cos \frac{(n + \frac{1}{2})(k + \frac{1}{2})\pi}{N} = \frac{1}{2}\text{Re}\,\{\sum_{n=0}^{N-1} 1\} = \frac{1}{2}N.$$

This demonstrates (3.107). A similar argument can be used to demonstrate that

$$\sum_{n=0}^{N-1} \sin \frac{(n + \frac{1}{2})(m + \frac{1}{2})\pi}{N} \sin \frac{(n + \frac{1}{2})(k + \frac{1}{2})\pi}{N} = \begin{cases} \frac{1}{2}N & \text{if } m = k \\ 0 & \text{if } m \neq k. \end{cases}$$

which is the analog of formula (3.107), when sines are used instead of cosines.

CHAPTER 4

Section 1

4.1 (a) See Figure D.11. (b) See Figure D.12.

4.3 To demonstrate (4.159), we have

$$|b_n| = \left| \frac{2}{L} \int_0^L f(x) \sin \frac{n\pi x}{L} \, dx \right| \leq \frac{2}{L} \int_0^L |f(x)| \, |\sin \frac{n\pi x}{L}| \, dx$$

$$\leq \frac{2}{L} \int_0^L |f(x)| \, dx = \frac{2}{L}\|f\|_1.$$

To demonstrate (4.160), we have

$$\left| u(x,t) - \sum_{n=1}^{M} b_n e^{-(n\pi a/L)^2 t} \sin \frac{n\pi x}{L} \right| = \left| \sum_{n=M+1}^{\infty} b_n e^{-(n\pi a/L)^2 t} \sin \frac{n\pi x}{L} \right|$$

$$\leq \sum_{n=M+1}^{\infty} |b_n| e^{-(n\pi a/L)^2 t}$$

$$\leq \frac{2}{L} \|f\|_1 e^{-(M+1)^2 (\pi a/L)^2 t} \sum_{k=0}^{\infty} e^{-(2M+3)k(\pi a/L)^2 t}$$

$$= \frac{2}{L} \|f\|_1 \frac{e^{-(M+1)^2 (\pi a/L)^2 t}}{1 - e^{-(2M+3)(\pi a/L)^2 t}} \, .$$

4.5 Using inequality (4.160) from Exercise 4.3, we have

$$\left| u(x,t) - b_1 e^{-(a\pi/L)2t} \sin \frac{\pi x}{L} \right| \leq \frac{2}{L} \|f\|_1 \frac{e^{-4(a\pi/L)^2 t}}{1 - e^{-5(a\pi/L)^2 t}} \to 0 \text{ as } t \to \infty.$$

4.7 (a) See Figure D.13.   (d) See Figure D.14.

Section 2

4.9 (a) See Figure D.15. The string is vibrating twice as fast when $c = 200$.

4.11 We have

$$y_{tt} = -\left(\frac{nc\pi}{L}\right)^2 K \cos \frac{nc\pi t}{L} \sin \frac{n\pi x}{L}$$

$$= c^2 \left[-\left(\frac{n\pi}{L}\right)^2 K \cos \frac{nc\pi t}{L} \sin \frac{n\pi x}{L} \right] = c^2 y_{xx}.$$

The rest of the verifications are just as easy. See Figure D.16 for a graph of this solution when $n = 3$, $K = -1$, $L = 10$, $c = 100$, and $t = 1.01$, 1.02, and 1.03.

Section 3

4.15 The initial distributions can be interpreted as probabilities of electrons passing through two slits. Hence the subsequent distributions illustrate the *interference effects* observed in the classic double-slit effect in quantum mechanics. For example, see Figure D.17 (the solution of Exercise 4.14(a) for $t = 0.3$) where interference fringes are clearly visible. As the distance between the rectangles increases, the frequency of the interference fringes undergoes a proportional increase.

4.17 For Example 4.4 and $t = 0.15$, see Figure D.18. For Example 4.5 and $t = 0.15$, see Figure D.19. Letting m_n stand for the mass of a neutron and m_e stand for the mass of an electron, we have $m_n = 1836\,m_e$. Hence, the neutron diffraction pattern for the time $t_n = 1836\,t_e$ will be identical to the electron diffraction pattern at time t_e.

4.19 See the solution of 4.17.

Section 4

4.21 Cesàro: 147 harmonics. dlVP: 29 harmonics. hanning: 59 harmonics. Hamming: 56 harmonics.

Section 5

4.23 For $\omega = 29.99\pi$, see Figure D.20. The graphs have a similar form to the graphs for Exercise 4.11 for $n = 3$, $K = -1$, $L = 10$, and $c = 100$ (see Figure D.16). (The graphs for Exercise 4.11 illustrate the motion of what is called a *fundamental harmonic*. See [Wa, Chapter 3.2] for further discussion of fundamental harmonics.)

4.25 If $\omega \approx \omega_m$ then the filter function $F(x)$ in (4.80) has a sharp peak at $x = m/M$. Thus, the mth harmonic of $y(x, t)$ is amplified the most. Hence, the expression for y in (4.77) becomes

$$y(x, t) \approx \left[\frac{\omega_m \sin \omega t - \omega \sin \omega_m t}{\omega_m (\omega_m^2 - \omega^2)} \right] K_m \sin \frac{m\pi x}{L}.$$

For $\omega \to \omega_m$, L'Hospital's rule yields

$$y(x, t) \approx \left[\frac{\sin \omega_m t - \omega_m t \cos \omega_m t}{2\omega_m^2} \right] K_m \sin \frac{m\pi x}{L}. \tag{D.5}$$

In Exercise 4.23, for example, $\omega_m = \omega_3 = 30\pi$, so the dominant term in (D.5) (when $t \approx 1$) is $-(t/60\pi)(\cos 30\pi t) K_3 \sin(0.3\pi x)$. And this dominant term (again for $t \approx 1$) is similar to a multiple of the fundamental harmonic $\cos(30\pi t) \sin(0.3\pi x)$ (as we saw above in the solution of Exercise 4.23).

4.27 See Figure D.21.

4.29 See Figure D.22.

Section 6

4.31 View the GraphBook CH4SEC6 in *FAS* .

4.33 For hanning filtering and 20 harmonics, the graphs for parts (a) and (b) are both shown in Figure D.23; their Sup-Norm difference is 1.02×10^{-13}. For 40 harmonics, their Sup-Norm difference is 1.57×10^{-13}. For Hamming filtering, using 20 harmonics, the Sup-Norm difference is 1.15×10^{-13}, and using 40 harmonics, the Sup-Norm difference is 1.61×10^{-13}. For dlVP filtering, using 20 harmonics, the Sup-Norm difference is 1.398×10^{-13}, and using 40 harmonics, the Sup-Norm difference is 1.97×10^{-13}.

4.35 We have

$$f * g(x) = \frac{1}{2L} \int_{-L}^{L} f(s)g(x-s)\,ds = \frac{1}{2L} \int_{x+L}^{x-L} f(x-v)g(v)\,(-dv)$$

$$= \frac{1}{2L} \int_{x-L}^{x+L} f(x-v)g(v)\,dv = \frac{1}{2L} \int_{-L}^{L} f(x-v)g(v)\,dv$$

$$= g * f(x)$$

Section 7

4.37 $(9/4)N \log_2 N + 29N/4 - 4$ real multiplications. Here is how this answer was obtained. As shown in Section 3.5 of Chapter 3, the two initial FFTs can be performed simultaneously; this requires $(3/2)N \log_2 N$ multiplications. Multiplying the transforms requires $4N$ real multiplications. If you implement the method of Chapter 3, Section 3.9, to perform the inverse FFT, this requires $2(N/2-1) = N-2$ initial multiplications to produce the two initial sequences. Then, the FFTs can be done simultaneously, requiring $(3/4)N \log_2(N/2) = (3/4)N \log_2 N - 3N/4$ multiplications. Finally, $2(N-1) = 2N-2$ multiplications are needed to produce the two starting values for each recursion. Dividing every computed value by N requires an additional N multiplications. The total is $(9/4)N \log_2 N + 29N/4 - 4$ real multiplications.

When $N = 1024$, this method is 34 times faster than a direct computation (which takes N^2 multiplications).

Section 8

4.39 The graphs for hanning's kernel (8 harmonics) using both methods are shown in Figure D.24. The Sup-Norm difference between the two graphs is 7.84×10^{-13}. For 16 harmonics, the Sup-Norm difference is 2.57×10^{-12}. For 32 harmonics, the Sup-Norm difference is 9.19×10^{-12}.

4.41 The graphs for Cesàro's kernel (8 harmonics) using both methods are shown in Figure D.25. The Sup-Norm difference between the two graphs is 8.37×10^{-13}. For 16 harmonics, the Sup-Norm difference is 2.79×10^{-12}. For 32 harmonics, the Sup-Norm difference is 1.022×10^{-11}.

4.43 View the GraphBook CH4SEC8 in *FAS* .

Section 9

4.45 The graph of $f * I_{16}$ is shown in Figure D.26. The Sup-Norm difference between $f * I_{16}$ and f over the interval $[-0.5, 0.5]$ is 4.307×10^{-2}. The Sup-Norm difference between $f * I_{32}$ and f over $[-0.5, 0.5]$ is 1.196×10^{-2}. The Sup-Norm difference between $f * I_{64}$ and f over $[-0.5, 0.5]$ is 3.092×10^{-3}.

4.47 Since

$$D_M(x) = [\sin(M + 1/2)x][1/\sin(x/2)]$$

and $1/\sin(x/2) \geq 1$ on the intervals $(0, \pi]$ and $[-\pi, 0)$, it follows that $|D_M(x)| \geq 1$ whenever $\sin(M+1/2)x = \pm 1$ (which happens for $x = (2k+1)\pi/(2M+1)$, $k = 0, \pm 1, \pm 2, \ldots, \pm M$). Therefore, property (c) in Definition 4.4 does not hold. It can also be shown that property (b) does not hold (i.e., that $\lim_{M \to \infty} \int_{-\pi}^{\pi} |D_M(x)|\, dx = \infty$), but we will omit the proof.

Section 10

4.49 Using the results in Exercise 4.48, we have $t_{2M} \approx 2\pi/(2M + 1/2)$. Since $(2M + 1/2)/(M + 1/2) \to 2$ as $M \to \infty$, it follows that $t_{2M} \approx x_M$ for large values of M.

4.51 For V_{50}, V_{100}, and V_{200}, we obtain the estimates shown in Table D.1.

Table D.1   Maximum Values of some Filtered Partial Sums

50 harmonics	$x_{50} \approx 8.3602 \times 10^{-2}$	$V_{50}(x_{50}) \approx 1.0715$
100 harmonics	$x_{100} \approx 4.1417 \times 10^{-2}$	$V_{100}(x_{100}) \approx 1.0713$
200 harmonics	$x_{200} \approx 2.0709 \times 10^{-2}$	$V_{200}(x_{200}) \approx 1.0714$

These results indicate that, as $M \to \infty$,

$$x_M \to 0 \quad \text{and} \quad V_M(x_M) \to 1.07\ldots > 1.$$

Thus, the amount of overshooting appears to be approximately 0.07.

CHAPTER 5

Section 1

5.1   (a) $(3/2)\,\text{sinc}\,(u/2)$,   (b) $2\,\text{sinc}\,(u)\cos(8\pi u)$,   (c) $2/[1 + (2\pi u)^2]$,   (d) $4\,\text{sinc}\,[8(u - 1)] + 4\,\text{sinc}\,[8(u + 1)]$,   (e) $-iue^{-\pi u^2}$.

5.3 (a) Exact transform is $(1/3)\,\text{sinc}^2(u/3)$. See Figure D.27 for graphs of the exact and *FAS*-computed transforms. The Sup-Norm difference between the two transforms, over $[-16, 16]$, is 1.34×10^{-3}.   (b) Exact transform is $(1/3)\,\text{sinc}\,(u/3)$. Sup-Norm difference between exact transform and *FAS*-computed transform, over $[-16, 16]$, is 1.66×10^{-2}.   (c) Exact transform is $-i\pi u/(1+\pi^2 u^2)^2$. Sup-Norm difference between exact transform and *FAS*-computed transform, over $[-16, 16]$, is 7.87×10^{-6}.   (d) Exact transform is $-i(\sin 4\pi u)(\,\text{sinc}\,u)$. Sup-Norm difference between exact transform and *FAS*-computed transform is 1.776×10^{-2}.

Section 2

5.5 Applying (5.14) to f'' in place of f, we get $|\widehat{f}''(u)| \leq \|f''\|_1$. And, by Theorem 5.2(a) applied twice, we have $f'' \xrightarrow{\mathcal{F}} (i2\pi u)^2 \hat{f}(u) = -(4\pi^2 u^2)\hat{f}(u)$. Thus $4\pi^2 u^2 |\hat{f}(u)| \leq \|f''\|_1$ and the desired inequality must hold.

5.7 We have

(a) $\quad x^3 e^{-\pi x^2} \xrightarrow{\mathcal{F}} i[u^3 - \dfrac{3}{2\pi}u]e^{-\pi u^2}$

(b) $\quad xe^{-2\pi|x|} \xrightarrow{\mathcal{F}} \dfrac{-iu}{\pi^2(1+u^2)^2}$

(c) $\quad (\sin 8\pi x)\operatorname{rect}(x/6) \xrightarrow{\mathcal{F}} 3i\operatorname{sinc}[6(u+4)] - 3i\operatorname{sinc}[6(u-4)]$

(d) $\quad e^{-|x|}\cos 6\pi x \xrightarrow{\mathcal{F}} \dfrac{1}{1+[2\pi(u-3)^2]} + \dfrac{1}{1+[2\pi(u+3)^2]}$

(e) $\quad e^{-|x-1|} + e^{-|x+1|} \xrightarrow{\mathcal{F}} \dfrac{4\cos(2\pi u)}{1+(2\pi u)^2}$

Section 3

5.9 If f is the function defined in (5.21), then $\|f - S_8^f\|_2 \approx 0.2625$, $\|f - S_{16}^f\|_2 \approx 0.1873$, $\|f - S_{32}^f\|_2 \approx 0.1327$, $\|f - S_{64}^f\|_2 \approx .0939$.

5.11 Letting $x = x_M$ yield the first maximum of $S_M^f(x)$ to the right of $x = 0$, we obtain the results in Table D.2.

Table D.2   Maximum Values of S_M^f.

$M = 8$	$x_8 \approx .0625$	$S_8^f(x_8) \approx 2.678$
$M = 16$	$x_{16} \approx .03125$	$S_{16}^f(x_{16}) \approx 3.14$
$M = 32$	$x_{32} \approx .0156$	$S_{32}^f(x_{32}) \approx 3.41$
$M = 64$	$x_{64} \approx .00781$	$S_{64}^f(x_{64}) \approx 3.55$

These results indicate that, as $M \to \infty$, $S_M^f(x_M)$ overshoots the limiting value of $f(0+) = \pi$. In other words, there is a Gibbs' phenomenon.

Section 4

5.13 If 1024 points are used, over an interval of $[-16, 16]$, then the Sup-Norm difference between the *FAS*-computed transform and the exact transform is 5.57×10^{-4}. Over the interval $[-4, 4]$, the Sup-Norm difference is 3.33×10^{-4}.

5.15 If 1024 points are used, over an interval of $[-16, 16]$, then the Sup-Norm difference between the *FAS*-computed transform and the exact transform is 2.887×10^{-15}. Over the interval $[-4, 4]$, the Sup-Norm difference is 2.887×10^{-15}.

Section 5

5.17 If $x < 0$, then $x - 0.5 < -0.5$ so $\int_{x-1.5}^{x-0.5} \text{rect}(v)\, dv = \int_{x-1.5}^{x-0.5} 0\, dv = 0$. A similar calculation gives 0 if $x > 2$. If $0 \leq x \leq 1$, then $\int_{x-1.5}^{x-0.5} \text{rect}(v)\, dv = \int_{-0.5}^{x-0.5} 1\, dv = x$. If $1 \leq x \leq 2$, then $\int_{x-1.5}^{x-0.5} \text{rect}(v)\, dv = \int_{x-1.5}^{0.5} 1\, dv = 2 - x$.

5.19 (a) We have

$$f *_y P(x) = \int_{-0.5}^{0.5} \frac{1}{\pi} \frac{y}{y^2 + (x-s)^2}\, ds = \int_{x-0.5}^{x+0.5} \frac{1}{\pi} \frac{y}{y^2 + v^2}\, dv$$

$$= \frac{1}{\pi} \text{Tan}^{-1}\left(\frac{v}{y}\right)\Big|_{x-0.5}^{x+0.5}$$

$$= \frac{1}{\pi} \text{Tan}^{-1}\left(\frac{x+0.5}{y}\right) - \frac{1}{\pi} \text{Tan}^{-1}\left(\frac{x-0.5}{y}\right).$$

(b) The Sup-Norm differences are as follows: $y = 0.5$: 1.495×10^{-3}, $y = 1.0$: 9.89×10^{-4}, $y = 2.0$: 5.84×10^{-4}. **(c)** $f *_y P(x) = (1/\pi) \sum_{k=0}^{3} (-1)^k \text{Tan}^{-1}[(x+1.5-k)/y]$ **(d)** The Sup-Norm differences are as follows: $y = 0.5$: 3.51×10^{-3}, $y = 1.0$: 1.626×10^{-3}, $y = 2.0$: 2.336×10^{-3}.

5.21 $\Lambda *_y P$ for $y = 0.25$ is shown in Figure D.28. $f *_y P$ for $y = 0.25$ and $f(x) = 1/(1+x^2)$ is shown in Figure D.29.

5.23 The Fourier transform version of the solution is

$$W(x, t) = \int_{-\infty}^{\infty} \hat{f}(u)(\cos 2\pi ut) e^{i2\pi ux}\, du.$$

Expressing $\cos 2\pi ut$ as $0.5e^{i2\pi ut} + 0.5e^{-i2\pi ut}$, combining exponentials and applying Fourier inversion, we obtain $W(x, t) = (1/2)f(x+t) + (1/2)f(x-t)$ which solves the given problem (provided f is twice differentiable). A graph of $W(x, 1.6)$ for $f(x) = e^{-0.5|x|}$ is shown in Figure D.30.

Section 6

5.25 (b) See Figure D.31. **(d)** See Figure D.32. **(f)** See Figure D.33.

5.27 We have $_t H *_\tau H \xrightarrow{\mathcal{F}} e^{-(2\pi a)^2 u^2 t} e^{-(2\pi a)^2 u^2 \tau} = e^{-(2\pi a)^2 u^2 (t+\tau)}$. Therefore, we have

$$e^{-(2\pi a)^2 u^2 (t+\tau)} \xrightarrow{\mathcal{F}^{-1}} {}_{(t+\tau)} H(x).$$

It follows that $_t H *_\tau H = {}_{(t+\tau)} H$. A similar argument shows that $_y P *_\Upsilon P = {}_{(y+\Upsilon)} P$.

5.29 (a) We have

$$f * {}_tH(x) = \int_{-\infty}^{\infty} 20\,\text{rect}(s/2)\, {}_tH(x-s)\,ds = \int_{-1}^{1} \frac{20}{\sqrt{4\pi t}} e^{-(x-s)^2/4t}\,ds$$

$$= \int_{x-1}^{x+1} \frac{20}{\sqrt{4\pi t}} e^{-v^2/4t}\,dv = \int_{(x-1)/\sqrt{4t}}^{(x+1)/\sqrt{4t}} \frac{20}{\sqrt{\pi}} e^{-s^2}\,ds$$

$$= 10\left[\frac{2}{\sqrt{\pi}} \int_{0}^{(x+1)/\sqrt{4t}} e^{-s^2}\,ds + \frac{2}{\sqrt{\pi}} \int_{(x-1)/\sqrt{4t}}^{0} e^{-s^2}\,ds \right]$$

$$= 10(\,\text{erf}[(x+1)/\sqrt{4t}] - \text{erf}[(x-1)/\sqrt{4t}]\,).$$

(c) For $t = 0.5$, the Sup-Norm difference is 2.518×10^{-4}. For $t = 1.0$, the Sup-Norm difference is 4.469×10^{-4}. For $t = 1.5$, the Sup-Norm difference is 4.56×10^{-4}.

Section 7

5.31 See Figure D.34 for graphs of the *FAS* approximation of $|\psi|^2$ and of

$$\frac{1}{(4\pi\hbar/2m)t} \left| \hat{f}\left(\frac{x}{4\pi\hbar t/2m} \right) \right|^2$$

where $\hat{f}(u) = 0.5\,\text{sinc}\,(u/4)$ and $t = 0.1$. The 1-Norm difference between the graphs is approximately 4×10^{-3}. For $t = 0.15$ the 1-Norm difference is approximately 1.94×10^{-3} and for $t = 0.2$ the 1-Norm difference is approximately 1.22×10^{-2}. (It is best to use 1-Norms since we are comparing p.d.f.s.)

5.33 See Figure D.35 for graphs of the *FAS* approximation of $|\psi|^2$ and of

$$\frac{1}{(4\pi\hbar/2m)t} \left| \hat{f}\left(\frac{x}{4\pi\hbar t/2m} \right) \right|^2$$

where $\hat{f}(u) = \sqrt{2}e^{-2\pi u^2}$ and $t = 1.0$. The 1-Norm difference between them is approximately 3.54×10^{-2}. For $t = 2.0$ the 1-Norm difference is approximately 9.09×10^{-3}.

Section 8

5.35 Noise suppression will work well when Step 5 (multiplying the FFT by a filter) produces a good approximation of the transform of the original signal. The examples in the text are good illustrations of when this occurs. The transform of the noise function is widely spread out and of low amplitude in comparison to the transform of the signal which is very narrow and highly concentrated near one frequency. Hence, when the filter is applied, most of the transform of the signal is retained, while very little of the transform of the noise is allowed through.

5.37 The hanning filtered transform is shown in Figure D.36.

5.39 The frequency is 2. For further details, including a reconstruction of the original signal, view the GraphBook CH5SEC8 in *FAS* .

5.41 *signal3:* 1 0 1 1 1 0 0 1 0 1 0 0 0 1 1 1,   *signal4:* 0 0 0 1 1 0 1 0 0 0 0 1 1 1 0 0,   *signal5:* 1 0 1 0 0 0 1 1 1 0 1 0 1 0 0 1. To see a reconstruction of the pulse train for *signal3*, view GraphBook CH5SEC8 in *FAS* .

5.43 The frequency is 25.

5.45 The frequency is 27. The phase shift is 0.2.

5.47 The frequencies are *signal11:* 28, *signal12:* 39, *signal13:* 14.5, *signal14:* 19, *signal15:* 30.5, *signal16:* 40.

Section 9

5.49 Using geometric series formulas we can rewrite the right side of (5.120) as follows:

$$\sum_{n=-\infty}^{\infty} \pi e^{-2\pi\rho|n|} e^{i2\pi nx} = \sum_{n=0}^{\infty} \pi [e^{-2\pi\rho} e^{i2\pi x}]^n$$

$$+ \sum_{n=0}^{\infty} \pi e^{-2\pi\rho} e^{-i2\pi x} [e^{-2\pi\rho} e^{-i2\pi x}]^n$$

$$= \frac{\pi}{1 - e^{-2\pi\rho} e^{i2\pi x}} + \frac{\pi e^{-2\pi\rho} e^{-i2\pi x}}{1 - e^{-2\pi\rho} e^{-i2\pi x}}$$

$$= \frac{\pi(1 - e^{-4\pi\rho})}{(1 - e^{-2\pi\rho} e^{i2\pi x})(1 - e^{-2\pi\rho} e^{-i2\pi x})}$$

$$= \frac{\pi(1 - e^{-4\pi\rho})}{1 - 2e^{-2\pi\rho} \cos 2\pi x + e^{-4\pi\rho}} .$$

Using this result in (5.120) yields (5.122). Formula (5.123) follows from (5.122) by putting $\rho = 1$ and $x = 0$. Also, using some algebra on (5.123) we obtain

$$\sum_{n=1}^{\infty} \frac{1}{1 + n^2} = \frac{1}{2} \left[\frac{\pi(1 + e^{-2\pi})}{1 - e^{-2\pi}} - 1 \right]$$

and

$$\sum_{n=0}^{\infty} \frac{1}{1+n^2} = \frac{1}{2}\left[\frac{\pi(1+e^{-2\pi})}{1-e^{-2\pi}}+1\right].$$

5.51 See Figure D.37. This is a graph of $S_{64}(x) = \sum_{n=-64}^{64} e^{-2n^2\pi^2(.001)}e^{i2\pi nx}$ which is an excellent approximation of the full series. $S_{64}(x)$ is a 64-harmonic, filtered Fourier series partial sum for the function $f(x) = 1024\delta(x)$ (using 1024 points).

5.53 The sum is

$$\frac{\pi}{2a}\frac{1+e^{-2\pi a}}{1-e^{-2\pi a}} - \frac{1}{2a^2}.$$

Section 10

5.55 $V_M = \sum_{n=-M}^{M} F(n/M)e^{inx}$ where

$$F(x) = \begin{cases} 1 & \text{if } |x| < \frac{1}{2} \\ 2(1-|x|) & \frac{1}{2} \le |x| \le 1 \\ 0 & \text{if } |x| > 1. \end{cases}$$

Clearly, $F(x)$ is even and continuous, and since it is 0 outside of $[-1, 1]$ it satisfies (5.127). Furthermore,

$$\hat{F}(u) = \begin{cases} (\cos\pi u - \cos 2\pi u)/(\pi u)^2 & \text{if } u \ne 0 \\ 3/2 & \text{if } u = 0 \end{cases}$$

It follows, since $\hat{F}$ is continuous at $u = 0$ and $|\hat{F}(u)| \le 2/(\pi u)^2$ for $u \ne 0$, that (5.128) holds. Consequently, by Theorem 5.15, V_M is a summation kernel.

5.57 $\mathcal{P}_M(x) = \sum_{n=-M}^{M} F(n/M)e^{inx}$ where $F(x) = (1-x^2)^\rho\,\text{rect}(x/2)$. Clearly, $F(0) = 1$; also $F(x)$ is continuous and is 0 outside of $[-1, 1]$. When $\rho \ge 2$, we have

$$F'(x) = -2\rho x(1-x^2)^{\rho-1}\,\text{rect}(x/2)$$

$$F''(x) = [-2\rho(1-x^2)^{\rho-1} + \rho(\rho-1)(2x)^2(1-x^2)^{\rho-2}]\,\text{rect}(x/2).$$

Hence F' and F'' are both continuous when $\rho \ge 2$. Thus, F satisfies all the requirements given in Exercise 5.56, so the Riesz kernel is a summation kernel when $\rho \ge 2$.

Section 11

5.59 (a) Nyquist rate: 8. $\|f - S_{32}\|_{\sup} \approx 2.09 \times 10^{-3}$. (b) Nyquist rate: 16. $\|f - S_{16}\|_{\sup} \approx 6.74 \times 10^{-4}$. (c) Nyquist rate: 8. $\|f - S_{32}\|_{\sup} \approx 1.38 \times 10^{-7}$.

5.61 By Parseval's equalities:

$$2L \int_{-\infty}^{\infty} \text{sinc } (2Lx - m) \text{ sinc}(2Lx - n) \, dx$$

$$= 2L \int_{-\infty}^{\infty} \frac{1}{2L} \text{rect} \left(\frac{x}{2L}\right) e^{-i2\pi mx/2L} \frac{1}{2L} \text{rect} \left(\frac{x}{2L}\right) e^{+i2\pi nx/2L} \, dx$$

$$= \frac{1}{2L} \int_{-L}^{L} e^{-i2\pi mx/2L} e^{+i2\pi nx/2L} \, dx = \begin{cases} 1 & \text{if } m = n \\ 0 & \text{if } m \neq n. \end{cases}$$

The last equality holding because of the orthogonality of the complex exponentials $\{e^{i2\pi nx/2L}\}$. If f is band limited with $\hat{f} = 0$ outside of $[-L, L]$, then

$$f(x) = \sum_{m=-\infty}^{\infty} f\left(\frac{m}{2L}\right) \text{sinc}(2Lx - m).$$

Multiplying this equation by $2L \text{ sinc}(2Lx - n)$ and integrating from $-\infty$ to ∞, we obtain

$$2L \int_{-\infty}^{\infty} f(x) \text{ sinc}(2Lx - n) \, dx =$$

$$\sum_{m=-\infty}^{\infty} f\left(\frac{m}{2L}\right) 2L \int_{-\infty}^{\infty} \text{sinc}(2Lx - m) \text{ sinc}(2Lx - n) \, dx = f\left(\frac{n}{2L}\right).$$

5.63 Graphs of both reconstructions are shown in Figure D.38. The Sup-Norm difference between the two graphs is 1.87×10^{-3}.

Section 12

5.65 Graph of S_8 shown in Figure D.39. $\|f - S_8\|_{\sup} \approx 6.39 \times 10^{-3}$, $\|f - S_{16}\|_{\sup} \approx 6.39 \times 10^{-3}$, and $\|f - S_{32}\|_{\sup} \approx 6.39 \times 10^{-3}$.

5.67 For the kernel sinc$(32x)$, the graph of S_{64} (i.e. $L = 16$) is shown in Figure D.40. Also, $\|f - S_{16}\|_{\sup} \approx 8.58 \times 10^{-6}$, $\|f - S_{32}\|_{\sup} \approx 2.09 \times 10^{-6}$, and $\|f - S_{64}\|_{\sup} \approx 5.11 \times 10^{-7}$. For the kernel S we have $\|f - S_{16}\|_{\sup} \approx 1.02 \times 10^{-5}$, $\|f - S_{32}\|_{\sup} \approx 2.72 \times 10^{-6}$, and $\|f - S_{64}\|_{\sup} \approx 6.93 \times 10^{-7}$.

5.69 See Table D.3.

5.71 See Table D.4.

Section 13

5.73 See Table D.5.

Table D.3   Comparison of Two Sampling Series for the
Function f

2L, M	Sup-Norm diff., kernel S_1	Sup-Norm diff., kernel S_2
4, 16	2.16×10^{-4}	2.02×10^{-4}
8, 32	1.31×10^{-4}	1.10×10^{-4}
16, 64	7.08×10^{-5}	5.68×10^{-5}
32, 128	3.68×10^{-5}	2.97×10^{-5}

Table D.4   New Columns for Tables
5.1 and 5.2, Section 12

2L, M	Function f_1	Function f_2
4, 16	1.77×10^{-2}	1.28×10^{-2}
8, 32	8.90×10^{-3}	2.46×10^{-3}
16, 64	4.46×10^{-3}	5.83×10^{-4}
32, 128	2.24×10^{-3}	1.43×10^{-4}

Table D.5   Relative 2-Norm Errors for Three Kernels

Time Span, τ	Kernel S_1	Kernel S_2	Kernel S_3
4.0	3.68×10^{-3}	1.77×10^{-3}	1.72×10^{-4}
3.5	4.69×10^{-3}	2.33×10^{-3}	1.72×10^{-4}
3.0	6.36×10^{-3}	3.10×10^{-3}	1.73×10^{-4}
2.5	8.29×10^{-3}	4.08×10^{-3}	1.79×10^{-4}
2.0	1.16×10^{-2}	5.67×10^{-3}	1.735×10^{-4}
1.5	1.68×10^{-2}	8.25×10^{-3}	1.74×10^{-4}
1.0	2.69×10^{-2}	1.32×10^{-2}	1.89×10^{-4}
0.5	5.57×10^{-2}	2.76×10^{-2}	5.12×10^{-4}

5.75 The entries for column three are as follows: 4.0: 4.79×10^{-4}, 3.5: 4.71×10^{-4}, 3.0: 4.75×10^{-4}, 2.5: 4.81×10^{-4}, 2.0: 5.10×10^{-4}, 1.5: 7.99×10^{-4}, 1.0: 1.88×10^{-3}, 0.5: 1.57×10^{-2}. The kernel S appears to have a slightly longer response time than S_3. It does have a shorter response time than both S_2 and S_1.

5.77 Letting S_M^F stand for the *FAS*-computed partial sum, and S_M stand for the explicitly computed partial sum, the following approximations of relative 2-Norms were obtained:

(a) $\|f - S_{16}^F\|_2 / \|f\|_2 = 2.887 \times 10^{-2}$, $\quad \|S_{16}^F - S_{16}\|_2 / \|S_{16}\|_2 = 1.445 \times 10^{-2}$

(b) $\|f - S_{32}^F\|_2 / \|f\|_2 = 7.438 \times 10^{-3}$, $\quad \|S_{16}^F - S_{32}\|_2 / \|S_{32}\|_2 = 7.050 \times 10^{-3}$

(c) $\|f - S_{32}^F\|_2 / \|f\|_2 = 5.854 \times 10^{-3}$, $\quad \|S_{32}^F - S_{32}\|_2 / \|S_{32}\|_2 = 6.215 \times 10^{-5}$

(d) $\|f - S_4^F\|_2 / \|f\|_2 = 1.578 \times 10^{-1}$, $\quad \|S_{16}^F - S_4\|_2 / \|S_4\|_2 = 5.800 \times 10^{-2}$

(e) $\|f - S_{32}^F\|_2 / \|f\|_2 = 3.360 \times 10^{-4}$, $\quad \|S_{32}^F - S_{32}\|_2 / \|S_{32}\|_2 = 3.005 \times 10^{-5}$.

The two kernels for case (c) and case (e) give acceptable results (the relative 2-Norm error $\|S_{32}^F - S_{32}\|_2 / \|S_{32}\|_2$ is reasonably small for both of these cases). For case (b), the relative 2-Norm error is not as small, but probably acceptable using the rule of thumb described in the text.

5.79 To see that formula (5.209) uses a periodic reconstruction kernel, you should graph the reconstruction kernel over $[-12, 12]$ using the formula

$$f(x) = \text{sinc} (4t) \setminus t = u - w/2 \setminus u = v - w \text{gri}(v/w) \setminus v = x + w/2 \setminus w = 8$$

You will see that this graph consists of three repetitions of the graph of $\text{sinc} (4x)$ (initially defined over $[-4, 4]$). The method derived in Exercise 5.76 uses two FFTs to compute a discrete convolution. As described in Section 4.7 of Chapter 4, this presumes periodicity. In particular, the sequences $f(x_k)$ and $\mathbf{S}(x_k)$ should be periodic with respect to their indices. Therefore, if we use a periodic function $f(x)$ and a periodic kernel $\mathbf{S}(x)$, then the method of evaluating (5.208) using two FFTs produces a sampling series sum with a periodic kernel. Since the sums in Exercise 5.77 only involve terms where the sample points $k/(2L)$ belong to the interval $[-4, 4]$, where $f(x)$ equals its periodic extension, the *FAS*-computed sampling series partial sums must match the partial sums using the periodically extended kernels computed in Exercise 5.78. (The very tiny Sup-Norm differences are attributable to differences in rounding error between the two methods.)

Section 14

5.81 Approximate the integral $\int_0^\infty$ by $\int_0^{\Omega/2}$ for some positive Ω. Then, use Riemann sums to obtain

$$\hat{f}_S(u) \approx \frac{\Omega}{N} \sum_{j=0}^{N-1} f\left(j\frac{\Omega}{2N}\right) \sin\left(\frac{\pi u j \Omega}{N}\right).$$

Drop the $j = 0$ term (since it is 0) and put $u = k/\Omega$, obtaining

$$\hat{f}_S\left(\frac{k}{\Omega}\right) \approx \frac{\Omega}{N} \sum_{j=0}^{N-1} f\left(j\frac{\Omega}{2N}\right) \sin\left(\frac{\pi j k}{N}\right)$$

which shows that $\hat{f}_S(k/\Omega)$ can be approximated by Ω/N times a DST. This DST is then performed using a fast sine transform. Similarly, a fast cosine transform can be used to approximate a cosine transform.

CHAPTER 6

Section 1

6.1 $|\psi(x, 0, t)|^2 = |e^{i2\pi vt}|^2 = 1$, hence $I(x, 0) = 1$. Similarly, $I(z, 0) = 1$. Therefore, $\Gamma(x, z) = (1/T) \int_0^T e^{i2\pi vt} e^{-i2\pi vt} \, dt = (1/T) \int_0^T 1 \, dt = 1$.

6.3 Let $0 = (0, 0)$ and let the point source be located at $(0, -D)$. Then, by Example 6.1, the light at $(x, 0)$ on the screen is given by

$$\frac{\psi(0, -D, t)e^{i\frac{2\pi}{\lambda}[|x|^2+D^2]^{\frac{1}{2}}}}{\lambda\sqrt{|x|^2 + D^2}}.$$

Hence, setting $S(t)$ equal to $\psi(0, -D, t)$, we are done.

6.5 (a) Let $C(x) = 1$ and $C(z) = 1$. (b) For 6.1: $C(x) = 1 = C(z)$. For 6.2: $C(x) = e^{i(2\pi/\lambda)\mathbf{a}\cdot\mathbf{x}}$ and $C(z) = e^{i(2\pi/\lambda)\mathbf{a}\cdot\mathbf{z}}$. For 6.3: $C(x) = e^{i(2\pi/\lambda)[D^2+|x|^2]^{\frac{1}{2}}}$ and $C(z) = e^{i(2\pi/\lambda)[D^2+|z|^2]^{\frac{1}{2}}}$. (c) Replace the integral in brackets in (6.18) by

$$C(x)C^*(z)\sqrt{I(x, 0)}\sqrt{I(z, 0)}$$

and separate integrals.

6.7 By the Schwarz inequality,

$$\left| \int_0^T \psi(x, t)\psi^*(z, t) \, dt \right| \le \sqrt{\int_0^T |\psi(x, t)|^2 \, dt} \sqrt{\int_0^T |\psi(z, t)|^2 \, dt}$$

$$= T\sqrt{I(x, 0)}\sqrt{I(z, 0)}.$$

Therefore, $|\Gamma(x, z)| \le 1$.

Section 2

6.9 The graphs are similar to those shown in Figure 6.31.

6.11 The graph of $I_1(u, 200)$ is shown in Figure D.41.

6.13 The graphs are similar to the one in Figure 6.32. Increasing the wavelength decreases the number of oscillations per unit length.

6.15 (a) The graph of I_1 is similar to the graph in Figure 6.4(d). It was obtained as follows: (1) $F_C^{\lambda D}$ was graphed over the interval $[0, 5]$ by entering the formula

$$f(x) = \cos(\pi x \wedge 2/0.15) \, \backslash \text{Diff} \, \backslash y=0$$

(2) $F_S^{\lambda D}$ was graphed over $[0, 5]$ by entering the formula

$$f(x) = \sin(\pi x \wedge 2/0.15) \backslash \text{Diff} \backslash y=0$$

(3) I_1 was graphed over the interval $[-4, 4]$ by entering the formula:

$$f(x) = (1/0.15) \ \{[\text{sign}(a)g1(\text{abs}(a)) - \text{sign}(b)g1(\text{abs}(b))] \wedge 2$$

$$+ [\text{sign}(a)g2(\text{abs}(a)) - \text{sign}(b)g2(\text{abs}(b))] \wedge 2\}$$

$$\backslash a=x+1 \backslash b=x-1$$

(b) The graph of I_1 using the convolution method is shown in Figure 6.4(d). (c) The Sup-Norm difference between the two graphs of I_1 over $[-2, 2]$ is 1.47×10^{-4}.

Section 3

6.17 $\hat{A}(-\mathbf{u}) = \int_{\mathbf{R}^2} A(\mathbf{x}) e^{i2\pi \mathbf{u} \cdot \mathbf{x}} \, d\mathbf{x} = \left(\int_{\mathbf{R}^2} A(\mathbf{x}) e^{-i2\pi \mathbf{u} \cdot \mathbf{x}} \, d\mathbf{x} \right)^*$. Thus, $\hat{A}(-\mathbf{u}) = [\hat{A}(\mathbf{u})]^*$. It follows that $|\hat{A}(-\mathbf{u})|^2 = |\hat{A}(\mathbf{u})|^2$, hence $I(-\mathbf{u}, D) = I(\mathbf{u}, D)$.

6.19 If an aperture is rotated by an angle θ, then the diffraction pattern must also rotate through the same angle θ. Consequently, if the aperture is unchanged by the rotation, then the diffraction pattern is unchanged as well.

6.21 We have

$$\left(\frac{\partial^2}{\partial x^2} + \frac{\partial^2}{\partial y^2} \right) \left(e^{-i2\pi \mathbf{u} \cdot \mathbf{x}} \right) = -(2\pi)^2 |\mathbf{u}|^2 e^{-i2\pi \mathbf{u} \cdot \mathbf{x}}.$$

Therefore, if $\mathbf{u} \neq (0, 0)$, we obtain

$$\int_{\mathcal{R}} e^{-i2\pi \mathbf{u} \cdot \mathbf{x}} \, d\mathbf{x} = \int_{\mathcal{R}} \left(\frac{\partial^2}{\partial x^2} + \frac{\partial^2}{\partial y^2} \right) \left[\frac{-1}{(2\pi)^2 |\mathbf{u}|^2} e^{-i2\pi \mathbf{u} \cdot \mathbf{x}} \right] d\mathbf{x}.$$

Applying Green's identity to the second integral, we obtain

$$\int_{\mathcal{R}} e^{-i2\pi \mathbf{u} \cdot \mathbf{x}} \, d\mathbf{x} = \int_{\partial \mathcal{R}} \frac{\partial}{\partial \eta} \left[\frac{-1}{(2\pi)^2 |\mathbf{u}|^2} e^{-i2\pi \mathbf{u} \cdot \mathbf{x}} \right] ds$$

$$= \frac{-1}{(2\pi)^2 |\mathbf{u}|^2} \int_{\partial \mathcal{R}} \frac{\partial}{\partial \eta} \left[e^{-i2\pi \mathbf{u} \cdot \mathbf{x}} \right] ds$$

which proves formula (6.210).

Section 4

6.23 The first three zeroes of $\hat{a}(\rho)$ are $\rho = 0.61$, 1.12, and 1.62. The first two local extrema to the right of the origin are $\hat{a}(\rho) = -0.416$ at $\rho = 0.82$ (a local min.), and $\hat{a}(\rho) = 0.202$ at $\rho = 1.34$ (a local max.).

6.25 The graph of $\hat{A}(\rho)$ for $R_1 = 1$ and $R_2 = 2$ is shown in Figure D.42. The graph of the intensities $I_1(\mathbf{u}, D)$ have radial forms that are similar to the graph shown in Figure 6.9.

Section 5

6.27   When the mica is placed over the right aperture, the aperture function is $A(x, y) = A_0(x + \delta/2, y) + i A_0(x - \delta/2, y)$. Therefore, $\hat{A}(u, v) = \hat{A}_0(u, v)[e^{i\pi\delta u} + i e^{-i\pi\delta u}]$. Hence,

$$I(u, v, D) = I_0(u, v, D)\left[2 + 2\sin\left(\frac{2\pi\delta u}{\lambda D}\right)\right].$$

It follows that the diffraction pattern looks like the pattern for a single aperture, but with dark vertical interference fringes at $u = (\lambda D/\delta)(-1/4 + k)$ for each integer k. Notice that the interference fringes are displaced slightly to the right of the origin. By a similar analysis, if the mica is placed over the right aperture, then the dark interference fringes are located at $u = (\lambda D/\delta)[1/4 + k]$ for each integer k. Hence, the fringes are displaced slightly to the left of the origin. Thus, we can see that the diffraction patterns differ, depending on whether the mica is placed over the left or right aperture.

6.29 $A(\mathbf{x}) \xrightarrow{\mathcal{F}} \hat{A}_0(\mathbf{u})\left[e^{-i2\pi\mathbf{c}\cdot\mathbf{u}} + e^{i2\pi\mathbf{c}\cdot\mathbf{u}}\right]$, so $I(\mathbf{u}, D) = I_0(\mathbf{u}, D)\,4\cos^2(2\pi\mathbf{c}\cdot\mathbf{u}/\lambda D)$. Since $\mathbf{c}\cdot\mathbf{u} = au + bv$, this gives the desired intensity function. Since $\cos\theta = 0$ when $\theta = \pm(2k+1)\pi/2$ for each integer k, putting $\theta = 2\pi\mathbf{c}\cdot\mathbf{u}/\lambda D$ leads to $\mathbf{c}\cdot\mathbf{u} = \pm\lambda D(2k+1)/4$. Since $\mathbf{c}$ is a normal vector for this last set of equations, the dark fringes must be perpendicular to $\mathbf{c}$. For $\mathbf{c} = (\delta/2, 0)$, the distance between successive dark fringes is $\lambda D/\delta$ [this follows from Formula (6.90)]. By rotation (see the solution to Exercise 6.19), it follows that the distance between the dark fringes is $\lambda D/\delta$ for every $\mathbf{c}$ that satisfies $|\mathbf{c}| = \delta/2$.

Section 6

6.31 The instrument function for (a) is shown in Figure D.43.

6.33 The instrument function for (a) is shown in Figure D.44.

Section 7

6.35 There is a broadening of the spectral line widths, hence there will be less resolution of distinct wavelengths.

6.37   The error in the central slit position seems to cause a slight decrease in the intensity of the off-center spikes, but no apparent loss of resolution (since the spikes seem to be of the same width as they were for the non-misaligned grating in Example 6.8).

6.39 By shifting and linearity (see Exercise 6.16), we have

$$\hat{A}(\mathbf{u}) = \sum_{m=0}^{M-1} \sum_{n=0}^{N-1} \hat{A}_0(\mathbf{u}) e^{-i2\pi(m\mathbf{a}\cdot\mathbf{u}+n\mathbf{c}\cdot\mathbf{u})}$$

$$= \hat{A}_0(\mathbf{u}) \sum_{m=0}^{M-1} e^{-i2\pi m\mathbf{a}\cdot\mathbf{u}} \sum_{n=0}^{N-1} e^{-i2\pi n\mathbf{c}\cdot\mathbf{u}}$$

Substituting $\mathbf{a} \cdot \mathbf{u}$ (and $\mathbf{c} \cdot \mathbf{u}$) in place of δu in the derivation of Equation (6.97) from Equation (6.94), we obtain

$$I(\mathbf{u}, D) = I_0(\mathbf{u}, D) \frac{\sin^2(\pi M\mathbf{a} \cdot \mathbf{u}/\lambda D)}{\sin^2(\pi \mathbf{a} \cdot \mathbf{u}/\lambda D)} \frac{\sin^2(\pi N\mathbf{c} \cdot \mathbf{u}/\lambda D)}{\sin^2(\pi \mathbf{c} \cdot \mathbf{u}/\lambda D)} .$$

$S_M^{\mathbf{a}}$ and $S_N^{\mathbf{c}}$ have maximum values of M^2 and N^2, whenever $\mathbf{a} \cdot \mathbf{u} = m\lambda D$ and $\mathbf{c} \cdot \mathbf{u} = n\lambda D$, respectively. The equations in (6.212) are just another way of expressing $\mathbf{a} \cdot \mathbf{u} = m\lambda D$ and $\mathbf{c} \cdot \mathbf{u} = n\lambda D$. The two equations in (6.212) describe two collections of parallel lines. The lines in one collection are described by the equation $\mathbf{a} \cdot \mathbf{u} = m\lambda D$, hence are all perpendicular to $\mathbf{a}$ since $\mathbf{a}$ is a normal vector for these equations. Likewise, the lines in the second collection are all perpendicular to $\mathbf{c}$. Because $\mathbf{a}$ and $\mathbf{c}$ are not parallel to each other, the lines in these two collections intersect at a grid of points [the simultaneous solutions of the two equations in (6.212)]. Since, even for very moderate sizes of M and N, the functions $S_M^{\mathbf{a}}$ and $S_N^{\mathbf{c}}$ are sharply peaked at their maxima, the diffraction pattern appears as a collection of high-intensity dots, each dot having intensity $M^2 N^2 I_0(u, v, D)$.

6.41 The instrument function in (6.110) has a sharp peak at $u = N\lambda D$, and its first zero to the right of this peak is at $u = N\lambda D + \lambda D/\beta$. Hence, by Rayleigh's criterion, for $\lambda_1 < \lambda_2$ we must have $N\lambda_1 D + \lambda_1 D/\beta \leq N\lambda_2 D$. Dividing out by D and subtracting $N\lambda_1$ yields $\lambda_1/\beta \leq N(\lambda_2 - \lambda_1)$ from which (6.111) follows.

6.43 First-order spectrum: $1/M$, second-order spectrum: $1/(2M)$, ..., k^{th}-order spectrum: $1/(kM)$. *Note:* these are the same separations as for a vertical slit grating; the relative separations are determined by the structure factor, not the form factor.

6.45 The instrument functions both look similar to the one shown in Figure 6.19. But, for $N = 100$, the spectral lines are *narrower* (by a factor of 2) than those for $N = 50$.

6.47 Using the criterion that the intensity is approximately zero if it is less than 1% of the peak intensity, the width of the first-order spectral line is approximately 0.55.

Section 8

6.49 If we use formula (6.203) for the light entering the lens, since $D = f$ the focal length of the lens, we obtain the following expression for the light exiting the lens:

$$\psi(\mathbf{x}, \epsilon, t) \approx \frac{S(t) e^{i\frac{2\pi}{\lambda}[f^2 + |\mathbf{X}|^2]^{\frac{1}{2}}}}{\lambda f} e^{\frac{-i\pi}{\lambda f}|\mathbf{X}|^2} P(\mathbf{x}).$$

Using the usual Fresnel approximations, we then have

$$\psi(\mathbf{x}, \epsilon, t) \approx \frac{S(t) e^{i\frac{2\pi}{\lambda} f} e^{\frac{i\pi}{\lambda f}|\mathbf{X}|^2}}{\lambda f} e^{\frac{-i\pi}{\lambda f}|\mathbf{X}|^2} P(\mathbf{x}) = \frac{S(t) e^{i\frac{2\pi}{\lambda} f}}{\lambda f} P(\mathbf{x}).$$

Consequently,

$$\gamma(\mathbf{x}, \mathbf{z}) \approx \frac{1}{T} \int_0^T \frac{S(t) e^{i\frac{2\pi}{\lambda} f}}{\lambda f} P(\mathbf{x}) \frac{S^*(t) e^{-i\frac{2\pi}{\lambda} f}}{\lambda f} P^*(\mathbf{z}) \, dt$$

$$= \frac{1}{(\lambda f)^2} P(\mathbf{x}) P^*(\mathbf{z}) \frac{1}{T} \int_0^T |S(t)|^2 \, dt.$$

Letting κ equal $[1/(\lambda f)^2](1/T) \int_0^T |S(t)|^2 \, dt$, this gives $\gamma(\mathbf{x}, \mathbf{z}) \approx \kappa P(\mathbf{x}) P^*(\mathbf{z})$. Hence, using Formula (6.24) we obtain

$$\Gamma(\mathbf{x}, \mathbf{z}) = \frac{\gamma(\mathbf{x}, \mathbf{z})}{\sqrt{\gamma(\mathbf{x}, \mathbf{x})}\sqrt{\gamma(\mathbf{z}, \mathbf{z})}} \approx \left(\frac{P(\mathbf{x})}{|P(\mathbf{x})|} \right) \left(\frac{P(\mathbf{z})}{|P(\mathbf{z})|} \right)^*.$$

If P is the pupil function defined in (6.130), then we obtain $\Gamma(\mathbf{x}, \mathbf{z}) \approx 1$ whenever $\mathbf{x}$ and $\mathbf{z}$ are in the aperture. Hence, the light is coherent across the aperture. Furthermore, since $I(\mathbf{x}, \epsilon) = \gamma(\mathbf{x}, \mathbf{x})$ we have $I(\mathbf{x}, \epsilon) \approx \kappa |P(\mathbf{x})|^2 = \kappa$, when P is the pupil function in (6.130) and $\mathbf{x}$ is in the aperture.

6.51 By the phase transformation due to the lens, we obtain

$$\psi(\mathbf{u}, \epsilon, t) \approx e^{ic} A(\mathbf{u}) \psi(\mathbf{u}, 0, t) e^{\frac{-i\pi}{\lambda f}|\mathbf{u}|^2} P(\mathbf{u}).$$

By Fresnel diffraction, we then have

$$\psi(\mathbf{X}, \Delta + \epsilon, t) \approx \frac{e^{ib}}{\lambda \Delta} \int_{\mathbf{R}^2} \psi(\mathbf{u}, \epsilon, t) e^{\frac{i\pi}{\lambda \Delta}|\mathbf{X} - \mathbf{u}|^2} \, d\mathbf{u}.$$

Combining these two formulas yields (6.214).

6.53 If $\Delta = f$, then

$$e^{\frac{i\pi}{\lambda\Delta}|\mathbf{X}-\mathbf{u}|^2} e^{\frac{-i\pi}{\lambda f}|\mathbf{u}|^2} = e^{\frac{i\pi}{\lambda f}|\mathbf{X}-\mathbf{u}|^2} e^{\frac{-i\pi}{\lambda f}|\mathbf{u}|^2}$$

$$= e^{\frac{i\pi}{\lambda f}|\mathbf{X}|^2} e^{\frac{-i2\pi}{\lambda f}\mathbf{X}\cdot\mathbf{u}} e^{\frac{i\pi}{\lambda f}|\mathbf{u}|^2} e^{\frac{-i\pi}{\lambda f}|\mathbf{u}|^2}$$

$$= e^{\frac{i\pi}{\lambda f}|\mathbf{X}|^2} e^{\frac{-i2\pi}{\lambda f}\mathbf{X}\cdot\mathbf{u}}.$$

Consequently, after factoring $e^{\frac{i\pi}{\lambda f}|\mathbf{X}|^2}$ outside the integral in (6.216) and eliminating this factor using $|\cdot|^2$, we obtain

$$I(\mathbf{X}, f + \epsilon) \approx \left| \frac{1}{\lambda f} \int_{\mathbf{R}^2} A(\mathbf{u}) e^{\frac{-i2\pi}{\lambda f}\mathbf{X}\cdot\mathbf{u}} \, d\mathbf{u} \right|^2.$$

6.55 If the lens has pupil function $\mathcal{P}(\mathbf{X})$ and focal length f, then $\psi(\mathbf{X}, f + \epsilon, t)$ becomes

$$\psi(\mathbf{X}, f + \epsilon, t) e^{\frac{-i\pi}{\lambda f}|\mathbf{X}|^2} \mathcal{P}(\mathbf{X})$$

after passing through the lens. The factor $e^{i\pi|\mathbf{X}|^2/\lambda f}$ will then be canceled in formula (6.218) in Exercise 6.54, yielding formula (6.219).

Section 9

6.57 Including fourth-power approximations will change the approximations in (6.121) to the following:

$$\left[R_1^2 - |\mathbf{u}|^2\right]^{\frac{1}{2}} \approx R_1 - \frac{|\mathbf{u}|^2}{2R_1} + \frac{|\mathbf{u}|^4}{8R_1^3},$$

$$\left[R_2^2 - |\mathbf{u}|^2\right]^{\frac{1}{2}} \approx R_2 - \frac{|\mathbf{u}|^2}{2R_2} + \frac{|\mathbf{u}|^4}{8R_2^3}.$$

With these approximations, (6.122) becomes

$$|AB| \approx \frac{|\mathbf{u}|^2}{2R_1} - \frac{|\mathbf{u}|^4}{8R_1^3}, \qquad |CD| \approx \frac{|\mathbf{u}|^2}{2R_2} - \frac{|\mathbf{u}|^4}{8R_2^3},$$

$$|BC| = \epsilon - |AB| - |CD|.$$

Consequently, (6.125) changes to

$$|AB| + \eta|BC| + |CD| \approx \epsilon\eta - \frac{|\mathbf{u}|^2}{2f} + \kappa|\mathbf{u}|^4$$

where κ is a constant. The term $\kappa|\mathbf{u}|^4$ produces an extra factor in (6.132) of the form $e^{i\pi\sigma|\mathbf{u}|^4}$ where $\sigma = 2\kappa/\lambda$.

Section 10

6.59 The image in each case looks similar to the image shown in Figure 6.23(c). As λ increases, the oscillations in the image decrease in frequency (due to widening of the *PSF*).

6.61 The graph of the image is shown in Figure D.45. There is ringing due to the dust particles and there are interference effects.

6.63 The image from the Gaussian filtered pupil is shown in Figure D.46. This image is a significant improvement over the image obtained in Example 6.9. There is no Gibbs' effect (i.e., there is no extra brightness near the edge; the edge would appear blurred due to a very gradual change in intensity). There is also a complete absence of extraneous oscillation. The image edge (as measured by half of maximum intensity) is closer to the true edge; in fact it appears to be precisely equal to the true edge. This image of an edge is not as good as can be obtained with incoherent illumination [see Figure 6.30(b). Nevertheless, it is a major improvement over the coherent image using an unmodified pupil [see Figure 6.23(c). (In fact, for some of the applications described in [Mi-T], it is a very effective substitute for an incoherent image.)

6.65 $\beta = .04$, $\approx 480\%$ improvement. $\beta = .08$, $\approx 430\%$ improvement. $\beta = 0.16$, $\approx 300\%$ improvement. $\beta = 0.2$, $\approx 250\%$ improvement. $\beta = 0.25$, $\approx 200\%$ improvement. False details appear for $\beta > 0.2$. Raising the value of P_{GF} for $|x| < 3$ causes a decrease in contrast improvement (over the bright field image), but increases the value of β at which false details appear.

Section 11

6.67 Let the coordinates of the back focal plane be specified by $(\mathbf{x}_I, f+F)$, where the subscript I stands for *image*. Then, by repeating the derivation of (6.181) we obtain

$$\psi(\mathbf{x}_I, t) \approx \left[\frac{e^{ib}}{\lambda F} \int \psi(\mathbf{X}, t)e^{\frac{-i2\pi}{\lambda F}\mathbf{X}\cdot\mathbf{x}_I} \, d\mathbf{X} \right] \mathcal{P}(\mathbf{x}_I)$$

where $\mathcal{P}(\mathbf{x}_I)$ is the pupil function of the second lens. Substituting the expression from (6.181) for $\psi(\mathbf{X}, t)$ into this formula, we get

$$\psi(\mathbf{x}_I, t) \approx \frac{e^{ic}}{\lambda^2 Ff} \int \left[\int A(\mathbf{x})\psi(\mathbf{x}, t)e^{\frac{-i2\pi}{\lambda f}\mathbf{x}\cdot\mathbf{X}} \, d\mathbf{x} \right] P(\mathbf{X})e^{\frac{-i2\pi}{\lambda F}\mathbf{X}\cdot\mathbf{x}_I} \, d\mathbf{X} \, \mathcal{P}(\mathbf{x}_I).$$

Computing $(1/T) \int_0^T |\psi(\mathbf{x}_I, t)|^2 dt = I(\mathbf{x}_I)$ in the usual way, using the coherency assumption $\Gamma(\mathbf{x}, \mathbf{z}) = 1$, we obtain

$$I(\mathbf{x}_I) \approx \left| \frac{1}{\lambda^2 F f} \int \left(\int A(\mathbf{x}) e^{\frac{-i2\pi}{\lambda f}\mathbf{x}\cdot\mathbf{X}} d\mathbf{x} \right) P(\mathbf{X}) e^{\frac{-i2\pi}{\lambda F}\mathbf{X}\cdot\mathbf{x}_I} d\mathbf{X}\, P(\mathbf{x}_I) \right|^2.$$

Assuming that $\mathcal{P}$ is a pupil function like the one defined in (6.130), we will have $|\mathcal{P}|^2 = \mathcal{P}$. And, the innermost integral is a Fourier transform of the aperture function A. Hence

$$I(\mathbf{x}_I) \approx \left| \frac{1}{\lambda^2 F f} \int \hat{A}\left(\frac{1}{\lambda f}\mathbf{X}\right) P(\mathbf{X}) e^{\frac{-i2\pi}{\lambda F}\mathbf{X}\cdot\mathbf{x}_I} d\mathbf{X} \right|^2 \mathcal{P}(\mathbf{x}_I).$$

Using the convolution theorem (Theorem 5.7 in Chapter 5), we obtain [*Note:* the transform of $\hat{P}(-\mathbf{x})$ is $P(\mathbf{X})$]:

$$I(\mathbf{x}_I) \approx \left| \frac{1}{\lambda^2 F f} \int (\lambda f)^2 A(\lambda f s) \hat{P}\left(-\left(\frac{-\mathbf{x}_I}{\lambda F} - \mathbf{s}\right)\right) d\mathbf{s} \right|^2 \mathcal{P}(\mathbf{x}_I).$$

Making some changes of variables, we then have

$$I(\mathbf{x}_I) \approx \left| \int \frac{1}{F/f} A\left(\frac{-1}{F/f}\mathbf{u}\right) \hat{P}\left(\frac{\mathbf{x}_I - \mathbf{u}}{\lambda F}\right) \frac{d\mathbf{u}}{\lambda^2 F^2} \right|^2 \mathcal{P}(\mathbf{x}_I).$$

The integral in the formula above is analogous to formula (6.152), hence a magnified, inverted image is formed. The magnification factor is F/f and the *PSF* is

$$\frac{1}{\lambda^2 F^2} \hat{P}\left(\frac{\mathbf{x}_I}{\lambda F}\right).$$

Section 12

6.69 The *PSF* is graphed in Figure D.47. There is much less oscillation in the incoherent *PSF*, it has no negative values, and is more narrowly concentrated at the origin. The incoherent images of tiny point-like objects would look like tiny point-like objects (although slightly wider), while the coherent images would look like ripples spreading out from a central point (like when a pebble is dropped in a pond).

6.71 The image is shown in Figure D.48. It is clearly a better image than the one shown in Figure D.45. This superiority is due to the lack of ringing and the lack of interference effects in the incoherent image.

6.73 When $c = 17$, or 19, the relative contrast is about 1 for coherent illumination. For incoherent illumination it is about 0.577 for $c = 17$, and about 0.524 for $c = 19$.

When $c = 21$, 23, or 25, the relative contrast is 0 for coherent illumination. For incoherent illumination it is about 0.476 for $c = 21$, about 0.426 for $c = 23$, and about 0.376 for $c = 25$. The advantage of incoherent illumination is that it produces some relative contrast for each of the given frequencies. For $c = 17$ and 19, coherent illumination exhibits almost twice as much contrast as incoherent illumination.

6.75 The *PSF* (see Example 6.13) is $[40 \sin c\,(40X)]^2[40 \sin c\,(40Y)]^2$. As we showed in that example, $\int_{-\infty}^{\infty}[40 \sin c\,(40X)]^2\,dX = 40$. Consequently, the normalized *PSF* is

$$40 \sin c^2(40X)\,40 \sin c^2(40Y).$$

Furthermore, because $\int_{-\infty}^{\infty} 40 \sin c^2(40X)\,dX = 1$ and $\int_{-\infty}^{\infty} 40 \sin c^2(40Y)\,dY = 1$, it makes sense to define the normalized *PSF* along the X-direction to be $40 \sin c^2(40X)$. Using this normalized *PSF* the image of the edge function given in Example 6.13 has exactly the same form as the image in Figure 6.30. The difference, however, is that the intensity (Y-scale) is now the same one needed for the original object (e.g., a Y-interval of $[-2, 2]$ works well for both the object and the image).

6.77 The coherent *PSF* has a transform of $P(\lambda\Delta u) = \text{rect}(u/20)$. The object function $A_k(x) = \exp(-(x/3)^{10})(1 + \cos k\pi x)$ has a transform of $\hat{A}_k(u) = \hat{a}(u) + 0.5\hat{a}(u + k/2) + 0.5\hat{a}(u - k/2)$, where $\hat{a}(u)$ is the transform of $a(x) = \exp(-(x/3)^{10})$. The transform $\hat{a}(u)$ essentially consists of a narrow spike at the origin. Hence $\hat{A}_k(u)$ consists of three spikes at $u = 0$ and $u = \pm k/2$. The transform of the coherent image is $\hat{A}_k(u)P(\lambda\Delta u) = \hat{A}_k(u)\,\text{rect}(u/20)$. When $k = 17$, and 19, then $k/2 = 8.5$ and 9.5. Hence, all three spikes are unchanged by the multiplication by $\text{rect}(u/20)$, since $\text{rect}(u/20) = 1$ for $u = 0$, $u = \pm 8.5$, and $u = \pm 9.5$. Thus, the coherent image will be nearly identical to the object, and the relative contrast will be 1. When $k = 21$, or 23, or 25, however, $\text{rect}(u/20) = 0$ for $u = \pm k/2$. Hence $\hat{A}_k(u)\,\text{rect}(u/20) \approx \hat{a}(u)$, so the image is approximately $a(x) = \exp(-(x/3)^{10})$ which has a relative contrast of 0.

If we let $G(u)$ stand for the transform of the coherent *PSF*, then $G(u) = P(\lambda\Delta u) = \text{rect}(u/20)$. It follows that the transform of the *incoherent PSF* is

$$\frac{G * G(u)}{G * G(0)} = \Lambda\left(\frac{u}{20}\right).$$

(It is necessary to divide by $G * G(0)$ so that we obtain a value of 1 at $u = 0$, since the integral of the normalized incoherent PSF is 1.) From this last formula, it follows that the transform of the normalized incoherent image is $\hat{I}_k(u)\Lambda(u/20)$ where $I_k(x) = \exp(-(x/3)^{10})(1 + \cos k\pi x)$. Hence

$$\hat{I}_k(u)\Lambda\left(\frac{u}{20}\right) = \hat{a}(u) + \frac{1}{2}\hat{a}\left(u - \frac{k}{2}\right)\Lambda\left(\frac{u}{20}\right) + \frac{1}{2}\hat{a}\left(u + \frac{k}{2}\right)\Lambda\left(\frac{u}{20}\right).$$

For $k = 17$, 19, 21, 23, and 25, the values of $\Lambda(u/20)$ are 0.575, 0.525, 0.475, 0.425, and 0.375, which accord nicely with the contrasts obtained for Exercise 6.73.

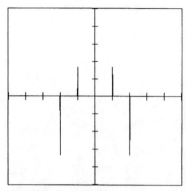

Figure D.1.   Exercise 1.3.
X interval: [-45, 45]   X increment = 9
Y interval: [-30, 30]   Y increment = 6

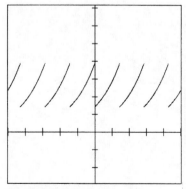

Figure D.2.   Exercise 1.5(a).
X interval: [-25, 25]   X increment = 5
Y interval: [-45, 45]   Y increment = 9

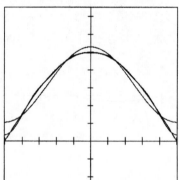

Figure D.3.   Exercise 1.11(a).
X interval: [-1.57, 1.57]   X increment = .314
Y interval: [-.5, 1.5]   Y increment = .2

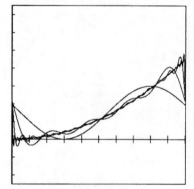

Figure D.4.   Exercise 1.11(b).
X interval: [0, 3]   X increment = .3
Y interval: [-5, 15]   Y increment = 2

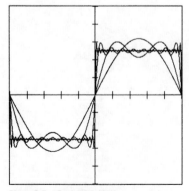

Figure D.5.   Exercise 1.11(d).
X interval: [-.5, .5]   X increment = .1
Y interval: [-2, 2]   Y increment = .4

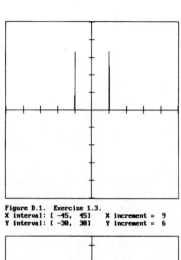

Figure D.6.   Exercise 1.15(a).
X interval: [-3.5, 3.5]   X increment = .7
Y interval: [-2, 5]   Y increment = .7

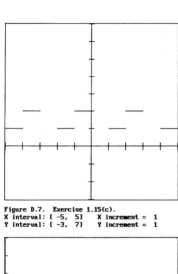

Figure D.7. Exercise 1.15(c).
X interval: [-5, 5]   X increment = 1
Y interval: [-3, 7]   Y increment = 1

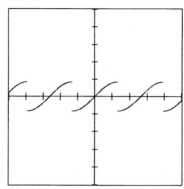

Figure D.8. Exercise 1.15(e).
X interval: [-6, 6]   X increment = 1.2
Y interval: [-6, 6]   Y increment = 1.2

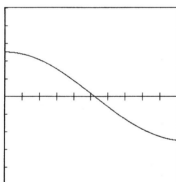
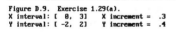

Figure D.9. Exercise 1.29(a).
X interval: [0, 3]   X increment = .3
Y interval: [-2, 2]   Y increment = .4

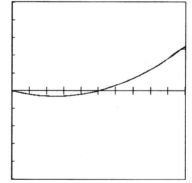

Figure D.10. Exercise 1.29(c).
X interval: [0, 2]   X increment = .2
Y interval: [-4, 4]   Y increment = .8

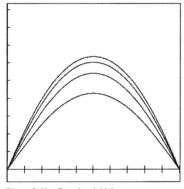

Figure D.11. Exercise 4.1(a).
X interval: [0, 10]   X increment = 1
Y interval: [-100, 1400]   Y increment = 150

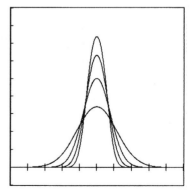

Figure D.12. Exercise 4.1(b).
X interval: [0, 16]   X increment = 1.6
Y interval: [-10, 90]   Y increment = 10

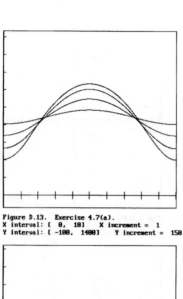

Figure D.13. Exercise 4.7(a).
X interval: [0, 10]   X increment =   1
Y interval: [-100, 1400]   Y increment = 150

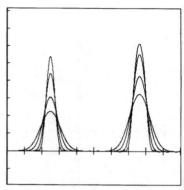

Figure D.14. Exercise 4.7(d).
X interval: [0, 40]   X increment =   4
Y interval: [-30, 130]   Y increment = 16

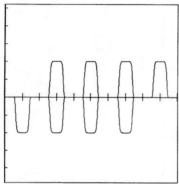

Figure D.15. Exercise 4.9(a).
X interval: [0, 10]   X increment =   1
Y interval: [-.12, .13]   Y increment = .025

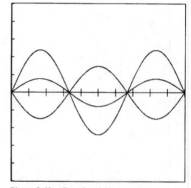

Figure D.16. Exercise 4.11.
X interval: [0, 10]   X increment =   1
Y interval: [-2, 2]   Y increment = .4

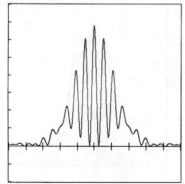

Figure D.17. Exercise 4.14(a).
X interval: [24, 40]   X increment = 1.6
Y interval: [-.1, .4]   Y increment = .05

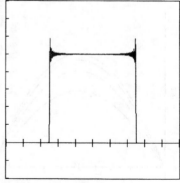

Figure D.18. Exercise 4.17.
X interval: [28, 36]   X increment = .8
Y interval: [-.1, .4]   Y increment = .05

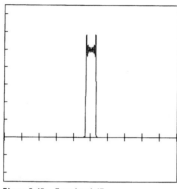

Figure D.19. Exercise 4.17.
X interval: [28, 36]   X increment = .8
Y interval: [-1, 3]   Y increment = .4

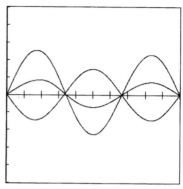

Figure D.20. Exercise 4.23.
X interval: [0, 10]   X increment = 1
Y interval: [-.0003, .0003]   Y increment = .00006

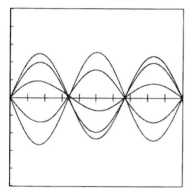

Figure D.21. Exercise 4.27.
X interval: [0, 10]   X increment = 1
Y interval: [-.0003, .0003]   Y increment = .00006

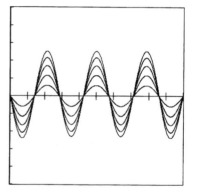

Figure D.22. Exercise 4.29.
X interval: [0, 10]   X increment = 1
Y interval: [-.001, .001]   Y increment = .0002

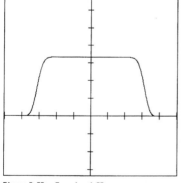

Figure D.23. Exercise 4.33.
X interval: [-3.14, 3.14]   X increment = .628
Y interval: [-1, 2]   Y increment = .3

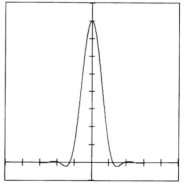

Figure D.24. Exercise 4.39.
X interval: [-3.14, 3.14]   X increment = .628
Y interval: [-1, 9]   Y increment = 1

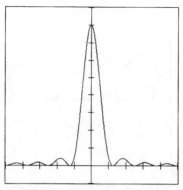

Figure D.25. Exercise 4.41.
X interval: [-3.14, 3.14]   X increment = .628
Y interval: [-1, 9]   Y increment = 1

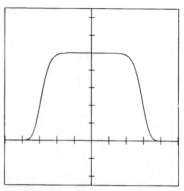

Figure D.26. Exercise 4.45.
X interval: [-.5, .5]   X increment = .1
Y interval: [-.5, 1.5]   Y increment = .2

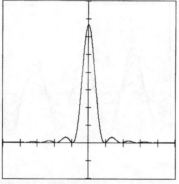

Figure D.27. Exercise 5.3(a).
X interval: [-16, 16]   X increment = 3.2
Y interval: [-.1, .4]   Y increment = .05

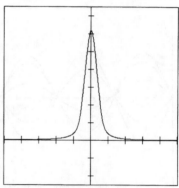

Figure D.28. Exercise 5.21.
X interval: [-8, 8]   X increment = 1.6
Y interval: [-.25, .75]   Y increment = .1

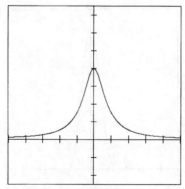

Figure D.29. Exercise 5.21.
X interval: [-8, 8]   X increment = 1.6
Y interval: [-.5, 1.5]   Y increment = .2

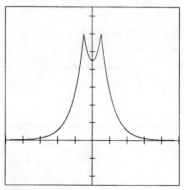

Figure D.30. Exercise 5.23.
X interval: [-16, 16]   X increment = 3.2
Y interval: [-.25, .75]   Y increment = .1

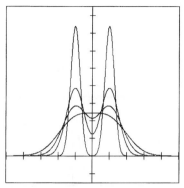

Figure D.31. Exercise 5.25(b).
X interval: [-10, 10]   X increment = 2
Y interval: [-3, 17]   Y increment = 2

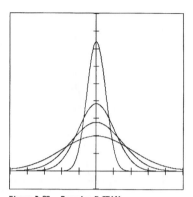

Figure D.32. Exercise 5.25(d).
X interval: [-10, 10]   X increment = 2
Y interval: [-2, 18]   Y increment = 2

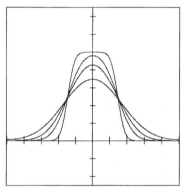

Figure D.33. Exercise 5.25(f).
X interval: [-10, 10]   X increment = 2
Y interval: [-.5, 1.5]   Y increment = .2

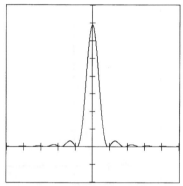

Figure D.34. Exercise 5.31.
X interval: [-16, 16]   X increment = 3.2
Y interval: [-.1, .4]   Y increment = .05

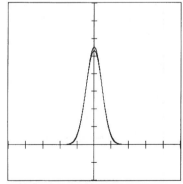

Figure D.35. Exercise 5.33.
X interval: [-16, 16]   X increment = 3.2
Y interval: [-.1, .4]   Y increment = .05

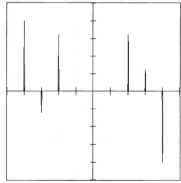

Figure D.36. Exercise 5.37.
X interval: [-10, 10]   X increment = 2
Y interval: [-10, 10]   Y increment = 2

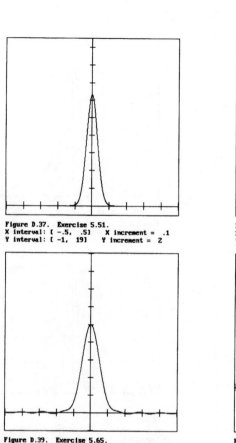

Figure D.37. Exercise 5.51.
X interval: [-.5, .5]   X increment = .1
Y interval: [-1, 19]   Y increment = 2

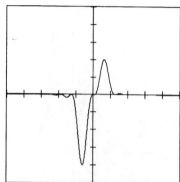

Figure D.38. Exercise 5.63.
X interval: [-4, 4]   X increment = .8
Y interval: [-5, 5]   Y increment = 1

Figure D.39. Exercise 5.65.
X interval: [-8, 8]   X increment = 1.6
Y interval: [-.1, .9]   Y increment = .1

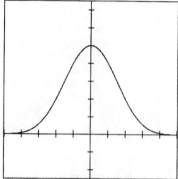

Figure D.40. Exercise 5.67.
X interval: [-2, 2]   X increment = .4
Y interval: [-.25, .75]   Y increment = .1

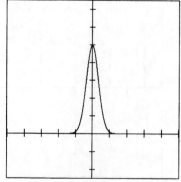

Figure D.41. Exercise 6.11.
X interval: [-4, 4]   X increment = .8
Y interval: [-.5, 1.5]   Y increment = .2

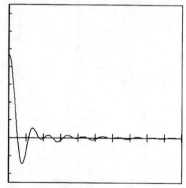

Figure D.42. Exercise 6.25.
X interval: [0, 5]   X increment = .5
Y interval: [-5, 15]   Y increment = 2

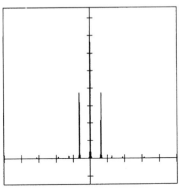

Figure D.43. Exercise 6.31.
X interval: [-4, 4]   X increment = .8
Y interval: [-100, 600]   Y increment = 70

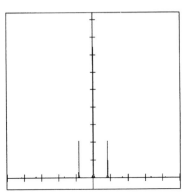

Figure D.44. Exercise 6.33.
X interval: [-4, 4]   X increment = .8
Y interval: [-100, 1500]   Y increment = 160

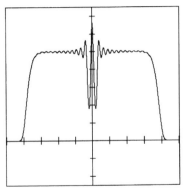

Figure D.45. Exercise 6.61.
X interval: [-1, 1]   X increment = .2
Y interval: [-.5, 1.5]   Y increment = .2

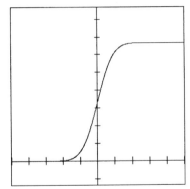

Figure D.46. Exercise 6.63.
X interval: [-.5, .5]   X increment = .1
Y interval: [-.2, 1.3]   Y increment = .15

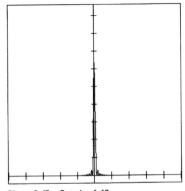

Figure D.47. Exercise 6.69.
X interval: [-1, 1]   X increment = .2
Y interval: [-100, 2400]   Y increment = 250

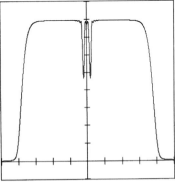

Figure D.48. Exercise 6.71.
X interval: [-1, 1]   X increment = .2
Y interval: [-5, 45]   Y increment = 5

Bibliography

[A-S]   Abramowitz, M. and Stegun, I. A., *Handbook of Mathematical Functions,* Dover, New York, 1965.

[Ba]   Bartle, R. W., *The Elements of Real Analysis,* John Wiley & Sons, 1965.

[Bas]   Baskakov, S. I., *Signals and Circuits,* translated from the Russian by B. V. Kuznetsov, Mir, Moscow, 1986.

[Be]   Bell, R. J., *Introductory Fourier Transform Spectroscopy,* Academic Press, New York, 1972.

[Bo-W]   Born, M. and Wolf, E., *Principles of Optics,* Pergamon Press, Oxford, 1965.

[Br]   Bracewell, R. N., *The Fast Hartley Transform,* Oxford University Press, Oxford, 1986.

[Br,2]   Bracewell, R. N., *The Fourier Transform and Its Applications,* McGraw-Hill, New York, 1978.

[Br-H]   Briggs, W. L. and Henson, V. E., *The DFT,* SIAM, Philadelphia, 1995.

[Bri]   Brigham, E. O., *The Fast Fourier Transform,* Prentice-Hall, Englewood Cliffs, NJ, 1974.

[Ch-B]   Churchill, R. V. and Brown, J. W., *Fourier Series and Boundary Value Problems,* McGraw-Hill, New York, 1978.

[Da]   Davis, H. F., *Fourier Series and Orthogonal Functions,* Allyn and Bacon, Boston, 1963.

[Da-H]   Davis, P. J. and Hersh, R., *The Mathematical Experience,* Houghton-Mifflin, 1982.

[Dau]   Daubechies, I., *Ten Lectures on Wavelets,* SIAM, Philadelphia, 1992.

[El-R]   Elliot, D. F. and Rao, K. R., *Fast Fourier Transforms: Algorithms, Analysis, and Applications,* Academic Press, New York, 1982.

[Fa]   Fante, R. L., *Signal Analysis and Estimation,* John Wiley & Sons, New York, 1988.

[Fe]   Feynman, R. P., *QED, the Strange Theory of Light and Matter,* Princeton University Press, Princeton, NJ, 1985.

[Fo]   Fourier, J., *The Analytical Theory of Heat,* translated by A. Freeman, Dover, New York, 1955.

[Go]   Goodman, J. W., *Introduction to Fourier Optics,* McGraw-Hill, New York, 1968.

[Go,2]   Goodman, J. W., *Statistical Optics,* John Wiley & Sons, New York, 1985.

[Ha]   Hamming, R. W., *Digital Filters,* Prentice-Hall, Englewood Cliffs, NJ, 1977.

[Hi]   Higgins, J. R., *Sampling Series in Fourier Analysis and Signal Theory,* Oxford University Press, Oxford, 1995.

[HTW]   Harburn, G., Taylor, C. A., and Welberry, T. R., *Atlas of Optical Transforms,* Cornell University Press, Ithaca, NY, 1975.

[Ii]   Iizuka, K., *Engineering Optics,* Springer, New York, 1985.

[Je]   Jerri, A., The Shannon sampling theorem—its various extensions and applications: a tutorial review, *Proc. IEEE,* 65(11), 1565, 1977.

[Ka]   Katznelson, Y., *An Introduction to Harmonic Analysis,* John Wiley & Sons, New York, 1968.

[Kr]   Krantz, S. G., *Real Analysis and Foundations,* CRC Press, Boca Raton, FL, 1992.

[Me]   Meyer-Arendt, J. R., Microscopy as a spatial filtering process, *Advances in Optical and Electron Microscopy,* vol. 8, Academic Press, London, 1982.

[Mey]   Meyer, Y., *Wavelets, Algorithms and Applications,* SIAM, Philadelphia, 1993.

[Mi] Misell, D. L., The phase problem in electron microscopy, *Advances in Optical and Electron Microscopy*, vol. 7, Academic Press, London, 1978.

[Mi-T] Mills, J. P. and Thompson, B. J., Effect of aberrations and apodizations on the performance of coherent optical systems (I and II), *J. Opt. Soc. Am., A,* 3, 694, 1986.

[Mo] Monforte, J., The digital reproduction of sound, *Sci. Am.,* 251(6), 78, 1986.

[Mo-W] Moss, F. and Wiesenfeld, K., The benefits of background noise, *Sci. Am.,* 273(2), 66, 1995.

[Nu] Nussbaumer, H. J., *Fast Fourier Transform and Convolution Algorithms,* Springer, New York, 1982.

[Op-S] Oppenheim, A. V. and Schaffer, R. W., *Digital Signal Processing,* Prentice-Hall, Englewood Cliffs, NJ, 1975.

[Pi] Pincus, H. J., Optical diffraction analysis in microscopy, in *Advances in Optical and Electron Microscopy*, vol. 7, Academic Press, London, 1978.

[PFTV] Press, W. H., Flannery, B. P., Teukolsky, S. A., and Vetterling, W. T., *Numerical Recipes, 2nd ed,*. Cambridge University Press, Cambridge, 1992. [*Note:* there are editions of this book with computer programs in Fortran, Pascal, or C.]

[Ra] Rao, K. R., Ed., *Discrete Fourier Transforms and their Applications,* Van Nostrand Reinhold, New York, 1985.

[Ra-G] Rabiner, L. R. and Gold, B., *Digital Signal Processing,* Prentice-Hall, Englewood Cliffs, NJ, 1975.

[Ra-R] Rabiner, L. R. and Radar, C. M., Eds., *Digital Signal Processing,* IEEE Press, New York, 1972.

[RDPT] Reynolds, G. O., DeVelis, J. B., Parrent, G. B. Jr., and Thompson, B. J., *Physical Optics Notebook: Tutorials in Fourier Optics,* SPIE Optical Engineering Press, Washington, D.C., 1989.

[Ru] Rudin, W., *Real and Complex Analysis,* McGraw-Hill, New York, 1974.

[Ru,2] Rudin, W., *Principles of Mathematical Analysis,* McGraw-Hill, New York, 1964.

[Sn]   Sneddon, I. N., *Fourier Transforms,* McGraw-Hill, New York, 1951.

[St]   Stearns, S. D., *Digital Signal Processing,* Hayden Book Co., New York, 1975.

[Str]   Strang, G., *Introduction to Applied Mathematics,* Wellesley-Cambridge Press, Wellesley, MA, 1986.

[St-W] Stein, E. and Weiss, G., *Fourier Analysis on Euclidean Spaces,* Princeton University Press, Princeton, NJ, 1971.

[To]   Tolstov, G. P., *Fourier Series,* Prentice-Hall, Englewood Cliffs, NJ, 1962.

[Wa]   Walker, J. S., *Fourier Analysis,* Oxford University Press, Oxford, 1988.

[Wa,2] Walker, J. S., A new bit reversal algorithm, *IEEE Trans. Acoust. Speech Signal Process.,* 38(8), 1472, 1990.

[Walt] Walter, G. G., *Wavelets and Other Orthogonal Systems with Applications,* CRC Press, Boca Raton, FL, 1994.

[We]   Weinberger, H. F., *A First Course in Partial Differential Equations,* John Wiley & Sons, New York, 1965.

[Zy]   Zygmund, A., *Trigonometric Series,* Cambridge University Press, Cambridge, 1968.

Index